AF709144

WITHDRAWN

WAVELET, SUBBAND AND BLOCK TRANSFORMS IN COMMUNICATIONS AND MULTIMEDIA

THE KLUWER INTERNATIONAL SERIES IN ENGINEERING AND COMPUTER SCIENCE

WAVELET, SUBBAND AND BLOCK TRANSFORMS IN COMMUNICATIONS AND MULTIMEDIA

edited by

Ali N. Akansu
New Jersey Institute of Technology
Newark, New Jersey

Michael J. Medley
United States Air Force Research Laboratory
Rome, New York

KLUWER ACADEMIC PUBLISHERS
Boston / Dordrecht / London

Distributors for North, Central and South America:
Kluwer Academic Publishers
101 Philip Drive
Assinippi Park
Norwell, Massachusetts 02061 USA
Telephone (781) 871-6600
Fax (781) 871-6528
E-Mail <kluwer@wkap.com>

Distributors for all other countries:
Kluwer Academic Publishers Group
Distribution Centre
Post Office Box 322
3300 AH Dordrecht, THE NETHERLANDS
Telephone 31 78 6392 392
Fax 31 78 6546 474
E-Mail <services@wkap.nl>

Electronic Services <http://www.wkap.nl>

Library of Congress Cataloging-in-Publication Data

A C.I.P. Catalogue record for this book is available
from the Library of Congress.

TK
5102.92
.W38
1999

Copyright © 1999 by Kluwer Academic Publishers

All rights reserved. No part of this publication may be reproduced, stored in a retrieval system or transmitted in any form or by any means, mechanical, photocopying, recording, or otherwise, without the prior written permission of the publisher, Kluwer Academic Publishers, 101 Philip Drive, Assinippi Park, Norwell, Massachusetts 02061

Printed on acid-free paper.

Printed in the United States of America

The figure used in the cover design was generated by using matlab and gammatone cochlear filters. This design is a courtesy of Shyh-shiaw Kuo and appears in the chapter authored by James D. Johnston, et al.

Contents

List of Figures

List of Tables

Contributing Authors

Kenneth J. Hetling
MIT/Lincoln Laboratory — Lexington,MA.

Gary J. Saulnier
Rensselaer Polytechnic Institute — Troy, NY.

Ali N. Akansu
New Jersey Institute of Technology — Newark, NJ.

Xueming Lin
GlobeSpan Semiconductors Incorporated — Middletown, NJ.

Marc de Courville
Motorola CRM — Paris, France.

Pierre Duhamel
ENST — Paris, France.

Michael J. Medley
Air Force Research Laboratory — Rome, NY.

Mehmet V. Tazebay
Sarnoff Corporation — Princeton, NJ.

Richard S. Orr
Stanford Telecom — Reston, VA.

Glenn E. Prescott
University of Kansas — Lawrence, KA.

Thomas C. Farrell
TRW — Redondo Beach, CA.

Massimo Sorbara
GlobeSpan Semiconductors Incorporated — Middletown, NJ.

Nurgün Erdöl
Florida Atlantic University — Boca Raton, FL.

James D. Johnston
AT&T Laboratories - Research — Florham Park, NJ.

Schuyler R. Quackenbush
AT&T Laboratories - Research — Florham Park, NJ.

Grant A. Davidson
AT&T Laboratories - Research — Florham Park, NJ.

Karlheinz Brandenburg
AT&T Laboratories - Research — Florham Park, NJ.

Jurgen Herre
AT&T Laboratories - Research — Florham Park, NJ.

Aria Nosratinia
Rice University — Houston, TX.

Geoffrey Davis
Dartmouth College — Hanover, NH.

Zixiang Xiong
University of Hawaii — Honolulu, HI.

Rajesh Rajagopalan
Lucent Technologies — Murray Hill, NJ.

Iraj Sodagar
Sarnoff Corporation — Princeton, NJ.

Ya-Qin Zhang
Sarnoff Corporation — Princeton, NJ.

Mitchell D. Swanson
Cognicity Incorporated — Minneapolis, MN.

Bin Zhu
Cognicity Incorporated — Minneapolis, MN.

Ahmed H. Tewfik
University of Minnesota — Minneapolis, MN.

Mark J. T. Smith
Georgia Institute of Technology — Atlanta, GA.

Alen Docef
University of British Columbia — Vancouver, BC.

Preface

Wavelet and subband transforms have been of great interest in the fields of engineering and applied mathematics. The theories of these powerful signal processing tools have matured and many applications utilizing them are emerging in different disciplines. This book, comprised of eleven chapter contributions from prominent researchers in the field, focuses on communications and multimedia applications of wavelet and subband transforms.

The first six chapters of this book deal with a variety of communications applications that significantly benefit from wavelet and subband theories. Similarly, the remaining five chapters present recent advances in multimedia applications of wavelet and subband transforms. These chapters interconnect the requirements of applications with the underlying theory and their engineering solutions. Hence, the reader can easily trace the entire path from fundamentals to the purpose and merit of application in hand.

A combined list of references for the entire volume is given at the end of the text that should be helpful to the interested reader for a further study.

This book is anticipated to be of particular interest to engineers and scientists who want to learn about state-of-the-art subband and wavelet transform applications as well as their theoretical underpinnings. It can also serve as a supplementary book for graduate level engineering and applied mathematics courses on wavelet and subband transforms.

We are grateful to all chapter authors for their very valuable contributions to this volume. We have been fortunate to have their associations and friendships which have made this book possible. We are sure that the readers of this book will greatly benefit from their expertise as we always do. The editors would also like to express their sincerest gratitude to Chester A. Wright for his LaTeX expertise and assistance in integrating each of the authors' contributions into the final version of this book.

Ali N. Akansu Michael J. Medley

1 TRANSMULTIPLEXERS: A UNIFYING TIME-FREQUENCY TOOL FOR TDMA, FDMA, AND CDMA COMMUNICATIONS

Kenneth J. Hetling[1], Gary J. Saulnier[2], Ali N. Akansu[3] and Xueming Lin[4]

[1]MIT Lincoln Laboratory
Lexington, MA*
hetling@ll.mit.edu

[2]Rensselaer Polytechnic Institute
Troy, NY
saulng@rpi.edu

[3]New Jersey Institute of Technology
Newark, NJ
ali@hertz.njit.edu

[4]GlobeSpan Semiconductors Incorporated
Middletown, NJ
xlin@globespan.com

1.1 INTRODUCTION

Prior to the mid 1970's, the telephone system in North America was primarily an analog system, utilizing frequency-division multiplexing (FDM) with single-sideband (SSB) modulation and 4 KHz frequency slots for each voice channel. A hierarchical structure was used in assembling multiple voice channels where

*Opinions, interpretations, conclusions, and recommendations are those of the author and not necessarily endorsed by the United States Air Force.

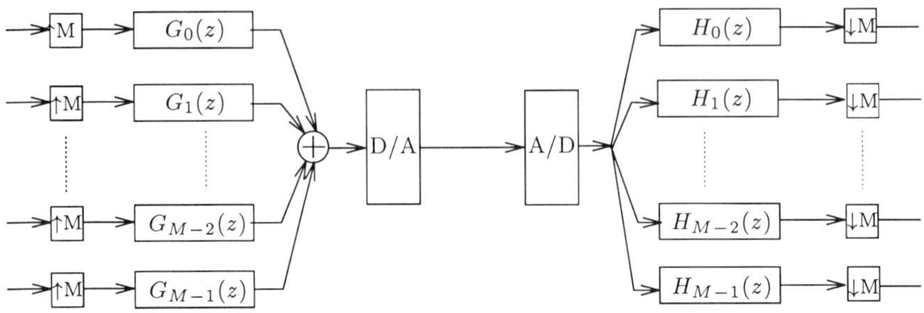

Figure 1.1 Block diagram of a transmultiplexer.

12 channels formed a *group*, 5 groups formed a *supergroup*, 10 supergroups formed a *mastergroup* and 6 mastergroups formed a *jumbogroup*. Today, the system is primarily a digital system utilizing time-division multiplexing (TDM) of pulse-code modulated (PCM) voice signals. Now, the basic unit is the *DS-1* signal which carries 24 voice signals and has a rate of 1.544 Mbps. In this case, 4 DS-1 signals are time-domain multiplexed to form a *DS-2*. Likewise, 7 DS-2 signals are combined to form a *DS-3*, 6 DS-3 signals form a *DS-4*, and 2 DS-4 signals form a *DS-5*.

During the transition from the analog FDM system to the digital TDM system the need developed for systems to efficiently perform the conversion between the two formats. A device, called a *transmultiplexer*, was developed as an efficient, bi-directional translator between the two signaling formats with the most basic transmultiplexer working between two groups and a single DS-1 signal [229, 291]. More complex versions were developed to operate on larger numbers of voice signals. The drive for better transmultiplexers is linked to many developments in the digital signal processing and, more specifically, the filter bank areas. Although no longer in use in the North American phone system, transmultiplexers are still of interest and have found applications in other areas, most notably in orthogonal frequency division multiplexing (OFDM).

Figure 1.1 is a block diagram which shows the structure of a digitally-implemented transmultiplexer. Each signal at the input and output consists of baseband samples of one voice signal. These signals are interleaved, i.e. time multiplexed, to form the TDM signal. The FDM signal is at the center of the structure. The analog version of the FDM signal, between the D/A and A/D converters, is transmitted on the analog lines. For example, to translate a single analog group into a DS-1 signal, the right half of the structure would be used with M=12 (actually M=14 was often used to allow for transition bands on filters which selected the groups). In brief, the TDM-to-FDM translation uses up-sampling to create spectral images at each of the FDM carrier frequencies followed by bandpass filters which select one-half of the spectral image for each voice signal, thereby isolating a SSB waveform on the desired carrier frequency for that signal. The FDM-to-TDM translation uses bandpass filters to isolate the individual carriers and down-sampling to frequency translate each carrier to baseband.

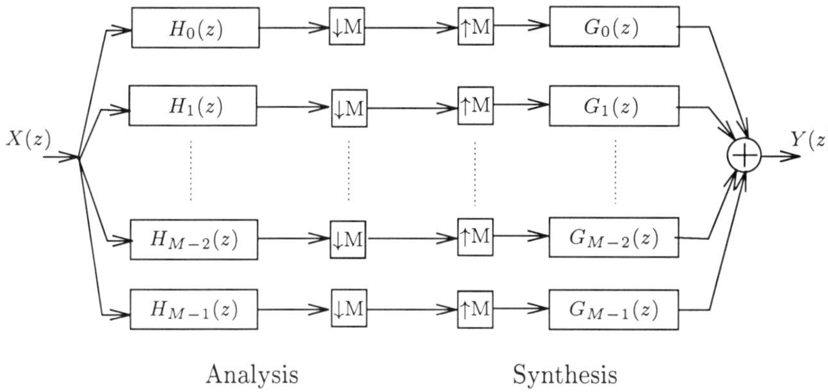

Figure 1.2 Block diagram of a subband filter bank.

The primary concerns in transmultiplexer design are *distortion* and *cross-talk*. Distortion describes the introduction of any change in a signal other than gain or delay. Cross-talk refers to the leakage of one voice signal into another. Both distortion and cross-talk can be minimized by making the bandpass filters $\{G_i(z)\}$ and $\{H_i(z)\}$ nearly ideal. Finite impulse response (FIR) or infinite impulse response (IIR) filters with orders exceeding 2000 or 7, respectively, can be required for acceptable cross-talk performance [17]. It was the development of the polyphase network and, in particular, its use in implementing uniform filter banks based either on the DCT or DFT, that made the implementation of such high-order FIR filters practical [19, 18, 341].

As is evident from Figure 1.1, the transmultiplexer is constructed using M-band multirate filter banks, with TDM-to-FDM translation effected using a synthesis bank and FDM-to-TDM translation accomplished via an analysis bank. In [345], Vetterli showed that the transmultiplexer is the *dual* of the maximally-decimated subband filter bank and that cross-talk is the dual of *aliasing*. Figure 1.2 shows a block diagram of a subband filter bank. Comparison of this diagram with that of the transmultiplexer in Figure 1.1 shows that the transmultiplexer is essentially the same structure with the position of the analysis and synthesis filter banks reversed.

This relationship between the transmultiplexer and the subband filter bank makes it possible to apply results obtained for subband filter banks to the transmultiplexer problem. Most importantly, the duality between the subband filter bank and the transmultiplexer allows the results of perfect reconstruction, in which aliasing in the subbands is canceled during reconstruction, to be applied to transmultiplexers in the form of *cross-talk cancellation*. Now, simpler, less-costly filter banks incorporating perfect reconstruction quadrature-mirror filters (PR-QMFs) and pseudo-QMF's [337], which provide near-perfect reconstruction, can be used in transmultiplexers with little or no loss in performance. However, a new constraint imposed by the transmultiplexer structure is that the sampling in the FDM-to-TDM conversion must be synchronized with the sampling in the TDM-to-FDM conversion and, as a result, a timing signal must be sent along with the voice signals. Additionally, it is important that the channel not introduce an excessive amount of frequency-selective amplitude or group delay distortion into the signal which would hinder the cross-talk cancellation [345]. In essence, timing alignment must not be lost nor disparate gains introduced *between* the various FDM components by the channel.

As its name indicates, the transmultiplexer is a solution to a *multiplexing* problem whereby all the signals originate at the same transmitter and are sent to the same receiver. Of greater interest in the area of wireless communications is the *multiple-access* problem whereby multiple signals originating at different transmitters share the channel. Conventional approaches for sharing the channel under these conditions include time-division multiple-access (TDMA), frequency-division multiple-access (FDMA), and code-division multiple-access (CDMA). In TDMA, the channel is shared in time with each user occupying the channel in an assigned time slot. Clearly, this approach requires that time synchronization be maintained between the various users in order to avoid overlap (in time) among the users. Provided that time orthogonality is maintained, this technique does not specifically restrict the frequency content of the signal. Conversely, in FDMA the channel is shared in frequency with each user occupying an assigned frequency slot. As long as frequency orthogonality is maintained, timing synchronization is not needed between the users. In CDMA, multiple users overlap in both time and frequency but are able to minimize cross-talk, generally called *multi-user interference* in this context, by using orthogonal or nearly-orthogonal *spreading codes*. CDMA systems can be classified as *synchronous*, *quasi-synchronous* or *asynchronous* depending on the level of timing that is maintained between the users. Although in all cases the users utilize the channel at the same time, timing is still important because it affects the difficultly in achieving orthogonality or near-orthogonality among the users. In other words, it is much easier to derive a set of codes that are orthogonal when time aligned than it is to derive a set of codes which are orthogonal over all relative time shifts. Generally, perfect orthogonality cannot be obtained in this latter case and some interference between the users is unavoidable. In a synchronous CDMA system, precise timing is maintained, thereby making it easier to obtain orthogonality between the users and lessening the problem of multi-user interference. In a quasi-asynchronous system, some timing misalignment is allowed among the users thus making code design somewhat more difficult than in the synchronous case, generally raising the level of multi-user interference. Finally, in an asynchronous system orthogonality is extremely difficult to obtain since no attempt is made to synchronize users.

The common theme in the design of transmultiplexers, as well as other multiplexing and multiple-access techniques, is producing orthogonality between the signals in order to minimize cross-talk or multi-user interference. The conventional transmultiplexer uses frequency domain orthogonality, i.e. FDM. The results in [345] lead to a situation where the multiplexed signals overlap in both time and frequency while still providing low levels of distortion or cross-talk.

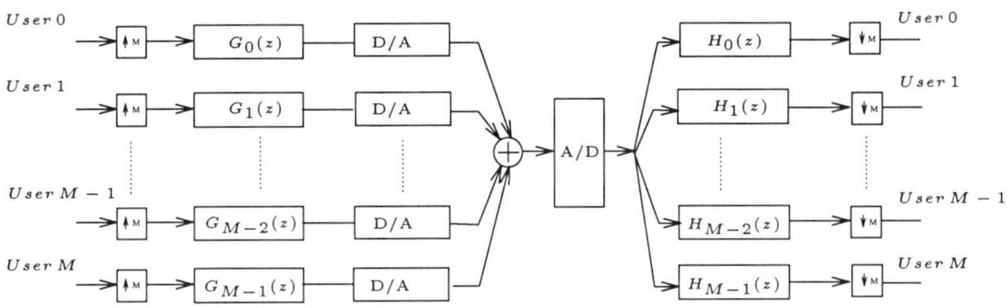

Figure 1.3 A CDMA system put into a filter bank framework.

In actuality, the filters need not approximate bandpass responses at all for the signals to be recovered with the requisite low levels of distortion and cross-talk when cross-talk cancellation is used. In this case, the signaling is more aptly called code-division multiplexing (CDM) as opposed to FDM. Additionally, systems can be constructed where the filter bank is non-uniform and, for example, implements the wavelet transform [45, 44].

This chapter considers the design of filter banks that can be used in the configuration of Figure 1.1 to produce signals that are suitable for CDMA. In this case, depicted in Figure 1.3, a particular user would use one of the filters in the synthesis bank to produce its particular spreading code and the summing of the signals would be performed in the channel.

A full analysis bank could be used to receive all the signals or a single filter (correlator) could be used to receive a single signal. Even though each user would use only part of the synthesis bank to generate its signal, the design of the bank as a whole is important since it determines the relationship between the various spreading codes or, more precisely, non-binary spreading waveforms. These spreading waveforms are the impulse responses of the filters in the bank. For the case of a synchronous CDMA system with a well-behaved channel, that is, a channel with little frequency-selective amplitude distortion or group-delay distortion, the problem is very much like the multiplexing problem and any perfect reconstruction filter bank will work. This chapter will focus instead on the more difficult problem of asynchronous and quasi-synchronous CDMA and multipath interference channels.

As part of the design of CDMA signals, it will be shown that CDMA, TDMA, and FDMA systems, as well as transmultiplexers, can all be derived using a common filter bank framework, i.e. all these techniques can be produced by designing filter banks under the appropriate constraints. Recent advances in the theory of subband filter banks (transforms) provide the mathematical tools for the design of optimal filter banks or, equivalently, orthogonal bases [337]. It is commonly understood in the signal processing community that there are infinitely many bases available in the solutions space of filter banks. The engineering art is to define the objective function for a given application and to identify the best possible solutions in that space. These objective functions must address appropriate time-frequency properties as well as requirements for orthogonality. These fundamental concepts and their mathematical descriptors along with their significance in applications under consideration are discussed in the following section and then applied to the CDMA problem in subsequent sections.

1.2 MATHEMATICAL PRELIMINARIES AND EXAMPLES

The time-frequency and orthogonality properties of function sets, or filter banks, are the unifying theme of the topics presented in this chapter. Here it is shown from a signal processing perspective how the communications systems discussed earlier are merely variations of the same theoretical concept. The subband transform theory and its extensions provide the theoretical framework which serves all these variations.

1.2.1 Time-Frequency Measures for a Discrete-Time Function

The time and frequency domain energy concentration or selectivity of a function is a classic signal processing problem. The *uncertainty principle* states that no function can simultaneously concentrate its energy in both time and frequency [114]. The time-spread of a discrete-time function, $h_0(n)$, is defined by

$$\sigma_n^2 = \frac{1}{E}\sum_n (n-\bar{n})^2 |h_0(n)|^2.$$

The energy, E, and time center, $\bar{n}$, of the function $h_0(n)$ are given as

$$\begin{aligned} E &= \sum_n |h_0(n)|^2 = \frac{1}{2\pi}\int_{-\pi}^{\pi} |H_0(e^{j\omega})|^2\, d\omega \\ \bar{n} &= \frac{1}{E}\sum_n n|h_0(n)|^2, \end{aligned}$$

where $H_0(e^{j\omega})$ is the Fourier transform of $h_0(n)$ expressed as

$$H_0(e^{j\omega}) = \sum_n h_0(n)e^{-j\omega n}.$$

Similarly, the frequency domain spread of a discrete-time function is defined as

$$\sigma_\omega^2 = \frac{1}{2\pi E}\int_{-\pi}^{\pi} (\omega-\bar{\omega})^2 |H_0(e^{j\omega})|^2\, d\omega,$$

where its frequency center is written as

$$\bar{\omega} = \frac{1}{2\pi E}\int_{-\pi}^{\pi} \omega |H_0(e^{j\omega})|^2\, d\omega.$$

Figure 1.4 displays a time-frequency tile of a typical discrete-time function. The shape and location of the tile can be adjusted by properly designing the

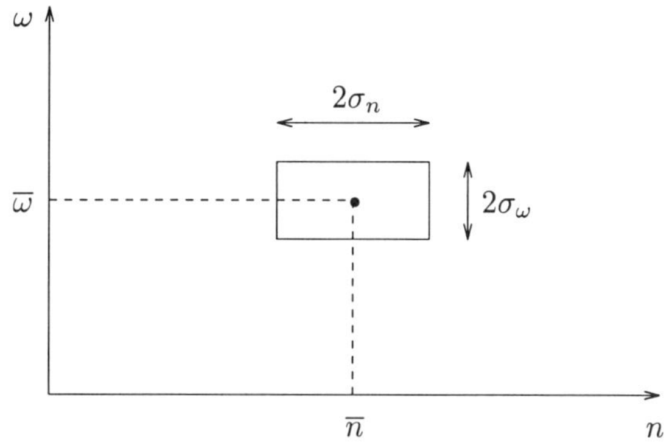

Figure 1.4 Time-frequency plane showing resolution cell tile of a typical discrete-time function.

time and frequency centers and spreads of the function under construction. Note that one can design a discrete-time function, $h_0(n)$, which approaches the desired set of time-frequency measures $\{\overline{n}_d, \overline{\omega}_d, \sigma^2_{dn}, \sigma^2_{d\omega}\}$. This methodology can be further extended in the case of basis design. In addition to shaping time-frequency tiles, the completeness requirements (orthogonality properties) of a basis can be imposed on the design. The orthogonality requirements of a synthesis/analysis filter bank are discussed in the following section.

1.2.2 Orthogonal Synthesis/Analysis Filter Banks

Consistent with the M-band, maximally decimated, FIR, PR-QMF analysis/synthesis structure displayed in Figure 1.2, the PR filter bank output is a delayed version of the input, namely

$$y(n) = x(n - n_0) \tag{1.1}$$

where n_0 is a delay constant related to the filter duration. In a paraunitary filter bank solution, the synthesis and analysis filters are related as (similar to a matched filter pair)

$$g_r(n) = h_r(p - n) \tag{1.2}$$

where p is a time delay. Therefore, it is easily shown that the PR-QMF filter bank conditions can be written based on the analysis filters in the time domain as [2]

$$\sum_k h_r(k)h_r(k + Mn) = \delta(n), \quad r = 0, 1, ..., M - 1, \quad \forall n \tag{1.3}$$

$$\sum_k h_r(k)h_s(k + Mn) = 0, \quad \forall n. \tag{1.4}$$

References [2, 337] provide a detailed treatment of perfect reconstruction filter banks and their extensions.

In contrast to the analysis/synthesis configuration in Figure 1.2, Figure 1.1 displays a synthesis/analysis filter bank where there are M inputs and M outputs of the system. It is shown for the critically sampled case that if the synthesis, $\{g_i(n)\}$, and analysis filters, $\{h_i(n)\}$, satisfy the PR-QMF conditions of (1.2), (1.3) and (1.4), the synthesis/analysis filter bank yields an equal input and output for all the branches of the structure as

$$\hat{x}_i(n) = x_i(n - n_0)$$

where n_0 is a time delay.

1.2.3 *Autocorrelation and Cross-Correlation Functions of Basis Functions*

For the purposes of the filter designs considered here, the two most important properties of the basis functions, or spreading codes, are their autocorrelation and the cross-correlation properties. The autocorrelation response of a basis function, $h_i(n)$, is defined as

$$R_{ii}(k) = \sum_n h_i(n)h_i(n+k), \quad k > 0, k \in Z.$$

A more convenient form (which will be used in the design procedure) is achieved using vector notation. Let $\mathbf{h}_i$ be a column vector of the basis function elements $h_i(n)$. The autocorrelation vector, $\mathbf{R}_{ij}$, can now be compactly represented as

$$\mathbf{R}_{ii} = \begin{bmatrix} (\mathbf{h}_i)_0^{\mathrm{T}}\mathbf{h}_i \\ (\mathbf{h}_i)_1^{\mathrm{T}}\mathbf{h}_i \\ \vdots \\ (\mathbf{h}_i)_K^{\mathrm{T}}\mathbf{h}_i \end{bmatrix}$$

where $(\mathbf{h}_i)_k$ is a downward circular shift of the vector elements.

Similarly, the cross-correlation response of basis functions $\{h_i(n)\}$ and $\{h_j(n)\}$ is defined as

$$R_{ij}(k) = \sum_n h_i(n)h_j(n+k) \quad \forall i, j, k \in Z.$$

Similar to the autocorrelation function, the cross-correlations can be represented in vector notation by

$$\mathbf{R}_{ij} = \begin{bmatrix} (\mathbf{h}_i)_0^{\mathrm{T}}\mathbf{h}_j \\ (\mathbf{h}_i)_1^{\mathrm{T}}\mathbf{h}_j \\ \vdots \\ (\mathbf{h}_i)_K^{\mathrm{T}}\mathbf{h}_j \end{bmatrix}.$$

The orthogonality conditions expressed in (1.3) and (1.4) null every M^{th} coefficient of the autocorrelation and cross-correlation functions, respectively, associated with the basis functions. Note that only for cases of the Kronecker delta function set and the ideal brick wall filter bank (set of *sinc* functions) autocorrelation and cross-correlation functions are ideal. For example,

$$R_{ii}(k) = \delta(k)$$

for the Kronecker delta function set and

$$R_{ij}(k) = \delta(i-j)\delta(k)$$

for the ideal brick wall filter bank. The Kronecker delta function set (functions of duration one with unique delays) is identical to an ideal orthogonal TDMA scenario. On the other hand, an ideal brick-wall filter bank (functions of infinite durations with non-overlapping frequency bands) is an ideal orthogonal FDMA structure.

1.2.4 Some Orthogonal Sets and Their Time-Frequency Properties

In this section three well-known orthogonal function sets are presented in conjunction with their time and frequency domain features.

I. Ideal Filter Bank. An ideal filter bank consists of brick-wall shaped frequency responses displayed in Figure 1.5(a). These functions do not overlap in frequency and are orthogonal. The time-domain counterpart of a brick-wall frequency function, $g_i(n)$, is displayed in Figure 1.5(b). Note that these functions have an infinite time duration and can not be implemented. In practice, this type of structure has resulted in the long transmultiplexer FIR filter lengths mentioned earlier. The autocorrelation and cross-correlation functions of this set are displayed in Figures 1.5(c) and 1.5(d), respectively. It is observed from this figure that the ideal brick-wall function set in the frequency domain perfectly fits the requirements of an FDMA system. In this scenario, orthogonal subcarrier functions are perfectly localized in the frequency domain although they have poor localization in time. The time-frequency cells (tiles) of this function family are illustrated in Figure 1.5(e).

II. Kronecker Delta Function Set. This set consists of functions with a single non-zero sample value and unique time shifts. Therefore, each one of these functions occupies the full spectrum. This is an ideal subcarrier set for TDMA communications scenarios. These functions, in frequency and time, along with their autocorrelation and cross-correlation functions, are displayed in Figure 1.6. In contrast to FDMA, each subcarrier is perfectly time localized without any restrictions in frequency, (cf. Figure 1.6). In this case, perfect orthogonality is imposed in the time domain.

III. Orthogonal Walsh-Hadamard Set (M=4). Walsh-Hadamard function sets are used in practice for multiuser communications. They are finite duration and very efficient to implement (each function consists of ± 1 valued samples). Similar to the previous two examples, Figure 1.7 displays time and frequency functions associated with the Walsh-Hadamard set for size, M=4. Their

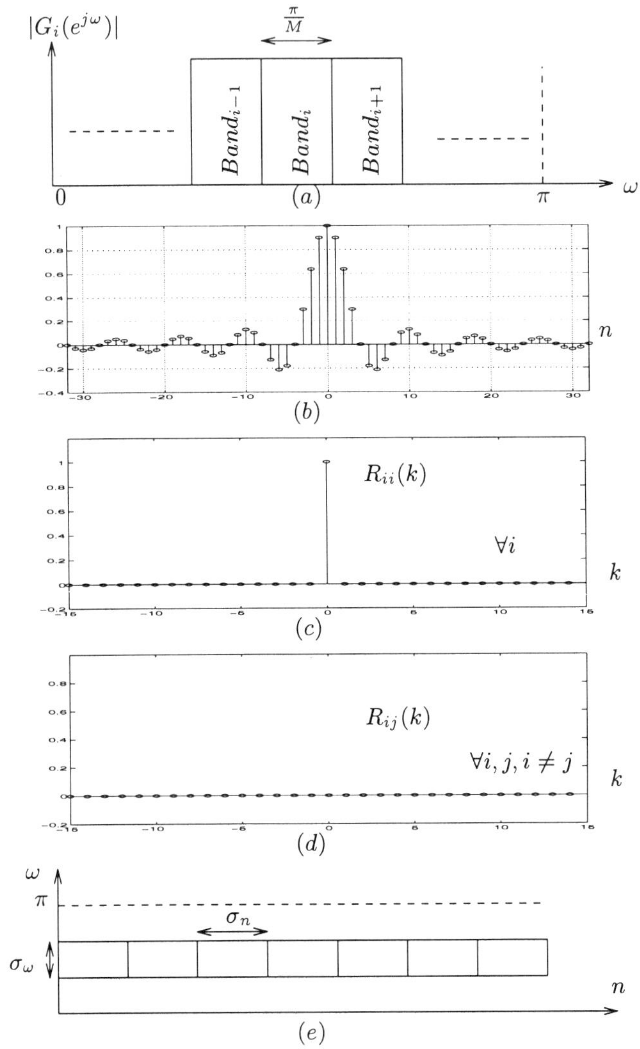

Figure 1.5 The ideal FDMA subscriber set in (a) frequency and (b) time. Their (c) autocorrelation, (d) cross-correlation functions and (e) time-frequency tiles.

autocorrelation and cross-correlation functions are displayed in Figure 1.8. It is observed from this figure that these functions are not well localized in time or frequency. Therefore, they have found use in CDMA standards for base station to mobile station (down-link, synchronous) communications.

Note that although the Walsh-Hadamard basis set functions are orthogonal and easy to implement, their autocorrelation and cross-correlation properties are unacceptable for the asynchronous communications scenario (up-link) since they dominate the performance of the system. The orthogonality of the Walsh-

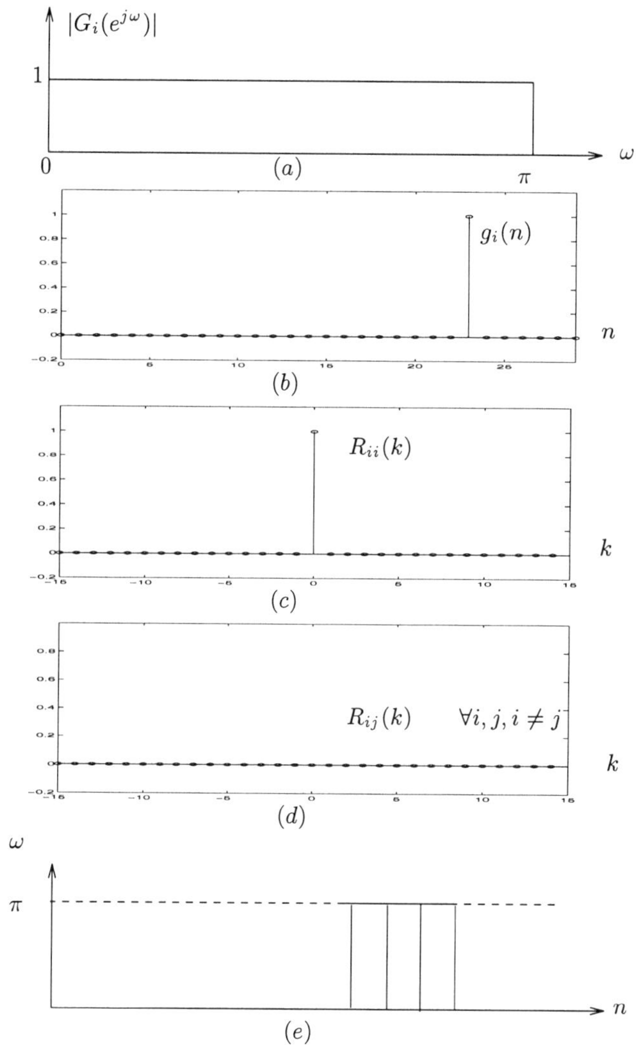

Figure 1.6 The ideal TDMA subcarrier set in (a) frequency and (b) time. Their (c) autocorrelation, (d) cross-correlation functions and (e) time-frequency tiles.

Hadamard basis set is distributed in both time and frequency. This is the case for any finite duration basis set (with more than one tap).

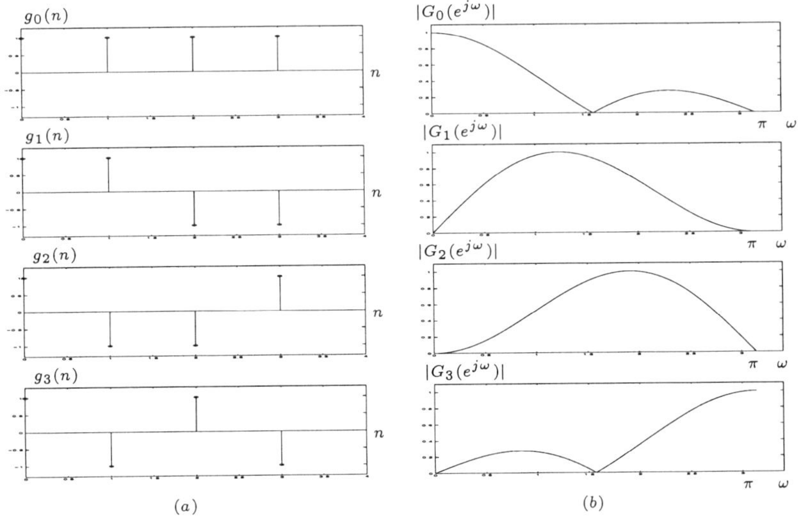

Figure 1.7 $M = 4$ Walsh-Hadamard basis set representations in (a) time and (b) frequency.

1.3 ASYNCHRONOUS WAVEFORM DESIGN

1.3.1 The Asynchronous Multiuser Channel

Attention can now be turned towards the actual design of the basis functions. Clearly if the users are synchronized to a common clock, then the PR-QMF bank conditions will result in perfect performance [345]. A more interesting situation, however, occurs in the *asynchronous multiuser channel* [118] where the user transmissions are not synchronized meaning that the symbol transitions are not aligned among all the users. This is illustrated for the simple two user case in Figure 1.9. In this situation, the signal of interest is interfered with by a second user transmission whose bit transitions are delayed by a time period, δ. Two assumptions are made to simplify the problem. First, the codes generated only span the bit duration meaning that a full length of a code is used to spread each data bit. Second, the delay time, δ, is constrained to be an integer multiple of the "chip" interval. Thus, for a length N code, the only delays considered are $\delta = \frac{nT_b}{N}$ where T_b is the bit interval and $n = \{1, 2, \cdots, N-1\}$.

There are two different polarity cases which must be considered for the situation in Figure 1.9. In the first case, the polarity of the interfering bits do not change during the bit interval of the signal of interest. This arrangement is called polarity case one (PC1). In polarity case two (PC2), the interfering bit polarity changes during the bit interval. Only the presence or absence of a data bit transition in the interfering bits is important, since the actual polarities of the bits do not affect the *magnitude* of cross-correlation and autocorrelation values used in the spreading code design.

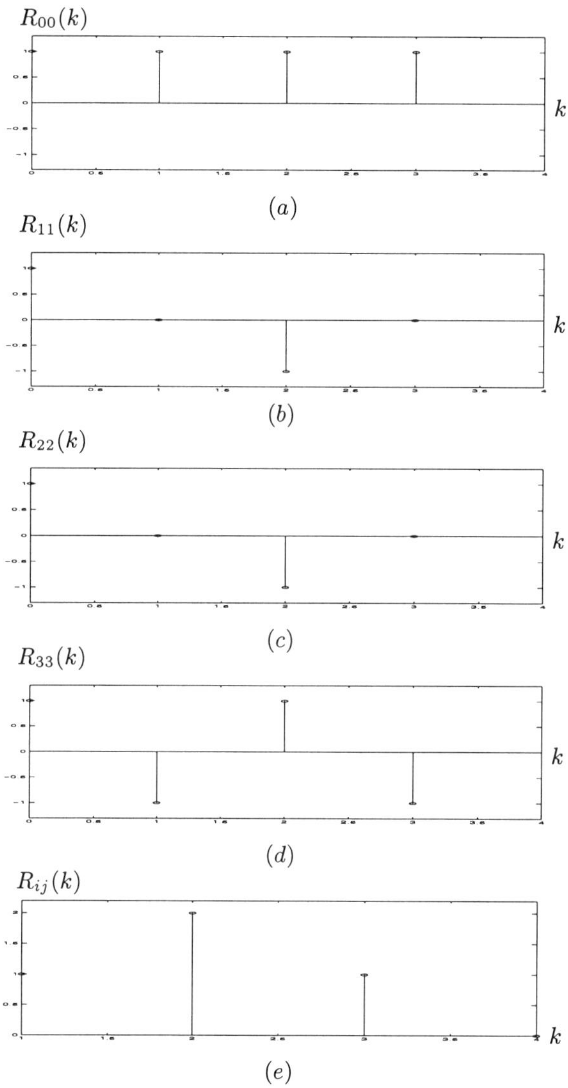

Figure 1.8 $M = 4$ Walsh-Hadamard basis set (a) – (d) autocorrelation and (e) cross-correlation functions for the first and second basis functions.

In general, spreading codes can be represented as transforms of a "unit" vector [133], $\mathbf{x}_n$, a column vector composed of all zeros except for the nth element which is set equal to unity. The spreading code, $\mathbf{c}_n$, is generated by taking a linear transformation, represented by the matrix $\mathbf{T}$, of the vector $\mathbf{x}_n$. This matrix $\mathbf{T}$ contains the spreading codes in its columns, that is,

$$\mathbf{T} = [\mathbf{t}_0 \quad \mathbf{t}_1 \quad ... \quad \mathbf{t}_{N-1}], \tag{1.5}$$

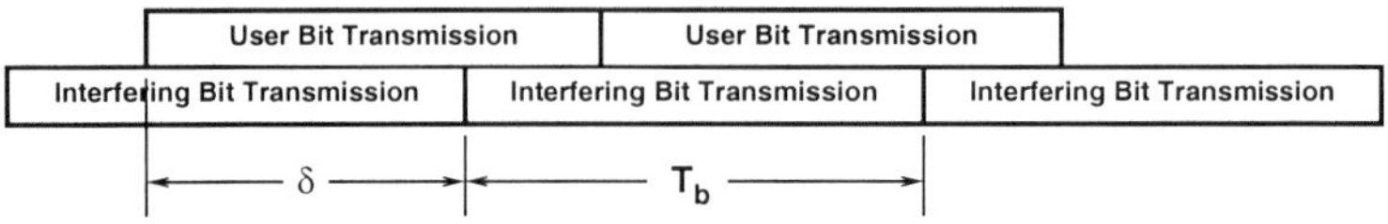

Figure 1.9 Asynchronous multiuser communications.

where $\mathbf{t}_0, \mathbf{t}_1, \ldots, \mathbf{t}_{N-1}$ are column vectors, each of which is a different spreading code. A particular spreading code can now be represented as

$$\mathbf{c}_n = \mathbf{T}\mathbf{x}_n. \tag{1.6}$$

There are no restrictions on $\mathbf{T}$ except that it must be a linear transformation. The number of users (or number of spreading codes) determines the number of columns of $\mathbf{T}$ while the length of the codes determines the number of rows.

Several common multiuser systems can be fit into this framework. For TDMA, the transformation uses the standard basis and $\mathbf{T}$ is the identity matrix. For FDMA, $\mathbf{T}$ is an unrealizable matrix in which the columns are of infinite length and consist of sampled sinusoids appropriately spaced apart in frequency. In each of these two cases, the transformation is a unitary transformation where $\mathbf{T}^{\mathrm{T}}\mathbf{T} = \mathbf{I}$ and $\mathbf{I}$ is the identity matrix. An example of a non-unitary transformation is a CDMA system using Gold codes [110]. In this case, the transformation matrix, $\mathbf{T}_g$, contains the Gold codes in its columns.

The interference caused by the situation shown in Figure 1.9 is directly related to the cross-correlation between the two signals. If $\mathbf{c}_n$ and $\mathbf{c}_m$ represent two different spreading codes of length N, the cross-correlation between them, for a given delay δ and PC1, can be expressed as

$$\rho_\delta = (\mathbf{c}_n)_\delta^{\mathrm{T}} \mathbf{c}_m \tag{1.7}$$

where δ is an integer between 0 and $N-1$ and $(\mathbf{c}_n)_\delta$ is a vector formed by performing a downward circular shift of the vector $\mathbf{c}_n$ by δ positions. In a similar fashion, the cross-correlations for PC2 can be represented by

$$\rho_{\bar{\delta}} = (\mathbf{c}_n)_\delta^{\mathrm{T}} \mathbf{K}_\delta \mathbf{c}_m \tag{1.8}$$

where $\mathbf{K}_\delta$ is a diagonal matrix whose first δ diagonal elements are negative one and the remaining diagonal elements are positive one. Expanding (1.7) using (1.6) gives

$$\rho_\delta = \mathbf{x}_n^{\mathrm{T}} (\mathbf{T})_\delta^{\mathrm{T}} \mathbf{T}\mathbf{x}_m. \tag{1.9}$$

(1.8) can be expanded in a similar fashion. If δ is zero (a synchronous system) and $\mathbf{T}$ is a unitary transformation, then the cross-correlations between any two codes is identically zero. For the asynchronous case, however, it is impossible (with finite length codes) for ρ_δ to be zero for all delays. The objective, therefore, is to design a transform, $\mathbf{T}$, such that the elements of $(\mathbf{T})_\delta^{\mathrm{T}}\mathbf{T}$ and $(\mathbf{T})_\delta^{\mathrm{T}}\mathbf{K}_\delta\mathbf{T}$ are as small as possible for all δ.

One method of implementing a linear transformation is via a multirate filter bank. It is within the framework of a filter bank design problem that the **T** matrix and, consequently, the spreading codes, are designed. Specifically, an M-band filter bank is constructed through the concatenation of many two-band PR-QMF filter banks to form a full binary tree. Although the results obtained using the full binary tree are suboptimal as compared to a direct M-band design, the design problem is greatly simplified. Additionally, the results obtained show promising performance gains over conventional codes and help illustrate the time-frequency duality as it pertains to the communications problem. The design methodology, therefore, is as follows. First, design two band PR-QMF filter banks in accordance with the particular communications channel under consideration, such as a multiuser channel or a multipath channel. Second, construct the full binary tree from the two-band structures. Third, determine the linear transform implemented by the M-band tree. Finally, extract the spreading codes in accordance with (1.5).

1.3.2 Matrix Representation of the Spreading Codes

In order to develop a convenient expression for the filter bank based transform, the decimation and interpolation operations of the filter bank are represented by finite matrix operators. Let $H_0(z)$ be the transfer function of an FIR filter from one branch of a two channel PR-QMF filter bank. If a finite vector $\mathbf{x}$ is input to the analysis filter bank, the output, $\mathbf{y}$, of the decimator which implements $H_0(z)$ can be represented by

$$\mathbf{y} = \mathbf{H}_0\mathbf{x}$$

where

$$\mathbf{H}_0 = \begin{bmatrix} h_{0,N-1} & h_{0,N-2} & \cdots & h_{0,0} & & & & \\ & & h_{0,N-1} & h_{0,N-2} & \cdots & h_{0,0} & & \\ & & & \vdots & & & & \\ h_{0,N-3} & \cdots & h_{0,0} & & & & h_{0,N-1} & h_{0,N-2} \end{bmatrix}$$

and the $\{h_{0,n}\}$ are the elements of the impulse response of $H_0(z)$. Blank entries are assumed to be zero. The actual size of the matrix is clearly dependent upon the input vector length.

For the synthesis bank, assume the $G_0(z)$ is the transfer function of one of the filters in a PR-QMF filter bank. The operation of the filter can be expressed as

$$\mathbf{x} = \mathbf{G}_0\mathbf{y}$$

where $\mathbf{x}$ and $\mathbf{y}$ are the output and input respectively of the interpolator and

$$\mathbf{G}_0^{\mathrm{T}} = \begin{bmatrix} g_{0,N-1} & g_{0,N-2} & \cdots & g_{0,0} & & & & \\ & & g_{0,N-1} & g_{0,N-2} & \cdots & g_{0,0} & & \\ & & & \vdots & & & & \\ g_{0,N-3} & \cdots & g_{0,0} & & & & g_{0,N-1} & g_{0,N-2} \end{bmatrix}$$

is the matrix operator for the interpolator. The impulse response of the filter $G_0(z)$ is represented by the elements $\{g_{0,n}\}$.

In a PR-QMF based transmultiplexer, the spreading codes are obtained by taking the impulse response of the synthesis bank. Using the time domain operator matrices defined above, the spreading codes can be generated directly without using product filters or the Z-transform. In general, for a filter bank tree of depth L, the spreading code along a branch with filters $A_1(z), A_2(z), \ldots, A_L(z)$, and $A_1(z)$ as the first stage filter is [130]

$$\mathbf{c} = \left[\prod_{l=1}^{L} (\mathbf{A}_l)^{\mathrm{T}}\right] \mathbf{x_1} \tag{1.10}$$

where $\mathbf{x_1}$ is a unit vector with the first element equal to unity, the matrices $\mathbf{A}_1$ to $\mathbf{A}_L$ are the decimation matrices for the filters, each of appropriate dimension, and the product is carried out in ascending order as

$$\prod_{l=1}^{L} (\mathbf{A}_l)^{\mathrm{T}} = (\mathbf{A}_1)^{\mathrm{T}}(\mathbf{A}_2)^{\mathrm{T}} \ldots (\mathbf{A}_L)^{\mathrm{T}}.$$

Using (1.10) to represent a particular spreading code, the properties of the codes can be directly formulated in terms of the operator matrices. This allows a direct connection between the spreading code properties and the filter coefficients used in the subband tree.

The remaining codes are constructed by substituting the decimation matrices which correspond to each path through the synthesis tree in (1.10). Each of the resulting column vectors, that is, the resulting spreading codes, can then be used to form the matrix in (1.5) which represents the linear transform.

1.3.3 The Filter Design Problem

The aforementioned representation of the spreading codes can now be used in the design problem for the filters. In particular, for the asynchronous multiuser communications channel, the filter coefficients are designed such that the cross-correlations between the codes are minimized for all possible time delays between them. In other words, the matrix $\mathbf{R} = (\mathbf{T})_\delta^T \mathbf{T}$, formed by the product between the inner two matrices in (1.9), should have elements which are close to zero for all possible values of δ. Since generally it is impossible to drive all cross-correlation values to zero, it must be determined how small the individual values must be and/or how the values should be distributed for good performance, namely low bit-error-rate (BER). Finally, and most importantly, a suitable objective function must be developed which achieves these goals.

In order to derive an appropriate objective function, some assumptions about the multiuser communications system are made. First, the relative delays between the users are assumed to be uniformly distributed. Additionally, for two given users, the relative delay between them is independent of the relative delay between any other set of users. Finally, the received power from each of the users is normalized to one. With these assumptions in mind, it can be

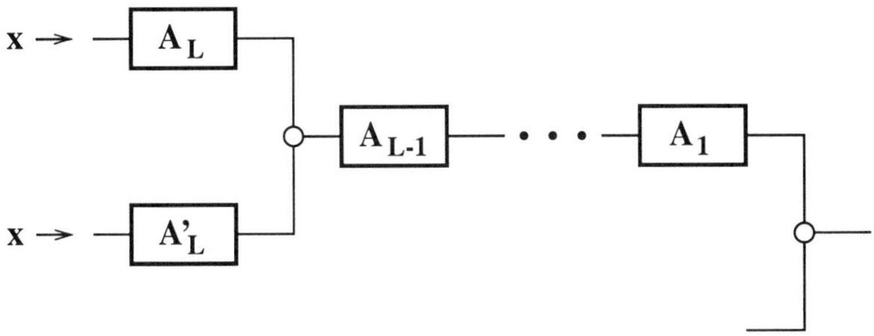

Figure 1.10 Spreading code branches.

shown that for a large number of users and sufficiently long spreading codes the optimal objective function is [133]

$$F_{MU} = \min \left\{ \sum_{\delta} \left[(\rho_\delta)^2 + (\rho_{\bar{\delta}}^-)^2 \right] \right\} \tag{1.11}$$

where ρ_δ and $\rho_{\bar{\delta}}^-$ are given in (1.7) and (1.8), respectively. For the pure multiuser communications problem, no other terms are necessary. Objective functions have also been developed for a small number of users with shorter length spreading codes [133], however, they are more computationally complex and are not discussed here.

Traditionally, filter bank structures reuse the same set of filters at each stage of the tree when constructing an M-band tree from simpler two-band filter banks. It can be shown, however, that this is not necessarily optimal and that a technique called *progressive optimality* can produce better results [321]. Therefore, when constructing the full binary tree, the initial split in the analysis bank is designed first with each subsequent stage designed in the order in which it occurs. Additionally, only a single set of codes, those which are generated by the two branches of a particular QMF pair, are considered at any one time. In this way, the design procedure becomes much less computationally demanding and yet, for the multiuser communications problem, is still mathematically justified [133].

The procedure is best seen by example. Consider two codes, $\mathbf{c}_1$ and $\mathbf{c}_2$, generated from a tree of depth L. The two branches which generate these codes have $L-1$ filters in common and each has one filter from a QMF pair at level L as shown in Figure 1.10. Let the operator matrices for the first $L-1$ filters be represented by $\mathbf{A}_1 \ldots \mathbf{A}_{L-1}$ while the QMF pair at level L is denoted by matrix operators $\mathbf{A}_L$ and $\mathbf{A'}_L$.

Recall that there are two different polarity cases which must be considered, PC1 and PC2. For a given delay time of δ chips, the cross-correlation between the two codes is given in (1.7) and (1.8). Substituting (1.10) into these two equations gives the cross-correlation values [130]

$$\rho_\delta = \mathbf{a}_L^{\mathrm{T}} \mathbf{P}^{\mathrm{T}} \mathbf{Q}_\delta \mathbf{P} \mathbf{a}'_L$$
$$\rho_{\bar{\delta}}^- = \mathbf{a}_L^{\mathrm{T}} \mathbf{P}^{\mathrm{T}} \mathbf{Q}_{\bar{\delta}}^- \mathbf{P} \mathbf{a}'_L$$

for PC1 and PC2, respectively, where

$$\mathbf{P} = \prod_{l=2}^{L-1} (\mathbf{A}_l)^{\mathrm{T}},$$
$$\mathbf{Q}_\delta = (\mathbf{A}_1)_\delta \mathbf{A}_1^{\mathrm{T}},$$

and

$$\mathbf{Q}_{\bar{\delta}}^- = (\mathbf{A}_1)_\delta \mathbf{K}_\delta \mathbf{A}_1^{\mathrm{T}}.$$

The column vectors $\mathbf{a}_L$ and $\mathbf{a}'_L$ contain the target filter coefficients of $A_L(z)$ and $A'_L(z)$, respectively.

These cross-correlation values can now be used in the objective function given in (1.11). Of course, the perfect reconstruction conditions for a PR-QMF two-channel filter bank must also be used in the design process. The result is a multivariable non-linear constrained optimization problem with the filter coefficients as the variables of optimization. For example, the optimization problem for the multiuser channel is to minimize [130]

$$F_{MU} = \left\{ \sum_\delta [(\rho_\delta)^2 + (\rho_{\bar{\delta}}^-)^2] \right\}$$

subject to the perfect reconstruction conditions of

$$\sum_{k=0}^{N-1} a_k a_{k+2n} = \delta_n \quad \forall n \geq 0$$

where the $\{a_k\}$ are the filter coefficients of the QMF pair being designed. This process is repeated for each QMF pair as the tree is constructed. When completed, the transform implemented by the tree can be derived and the codes extracted accordingly.

1.3.4 Multipath/Multiuser Interference

A second type of interference source considered here is the result of multipath propagation. In this case, multiple delayed copies of the signal arrive at the receiver and interfere with each other. In this case, it is the autocorrelation function of the signal which is of interest. Specifically, the autocorrelation function of the signal should be an impulse function such that any delayed copy of the signal will be uncorrelated with the "primary signal", which is generally the received signal copy with the most power.

For the signals under consideration, the autocorrelation function is formulated using the operator matrix notation described earlier. As with the cross-correlations, both polarity cases must be addressed. For a code generated by one branch of a PR-QMF tree of depth L, the autocorrelation functions are [131]

$$\xi_\delta = \mathbf{a}_L^{\mathrm{T}} \mathbf{P}^{\mathrm{T}} \mathbf{Q}_\delta \mathbf{P} \mathbf{a}_L$$
$$\xi_\delta^- = \mathbf{a}_L^{\mathrm{T}} \mathbf{P}^{\mathrm{T}} \mathbf{Q}_\delta \mathbf{P} \mathbf{a}_L$$

for PC1 and PC2, respectively. The matrices $\mathbf{P}$, $\mathbf{Q}$, and $\mathbf{Q}^-$ are defined in the same manner as that used in the multiuser channel scenario.

The objective function, which should drive autocorrelation values to zero (for $\delta \neq 0$), is similar to that used in the multiuser channel scenario and is given by [131]

$$F_{MP} = \left\{ \sum_{\delta=1}^{N-1} [(\xi_\delta)^2 + (\xi_\delta^-)^2] \right\}.$$

This objective function is a simplified objective in that each delay is assumed to be equally likely with identically distributed attenuation characteristics. Once again, the perfect reconstruction conditions are used as constraints. Notice that only one code, and therefore, only one filter from the QMF pair is considered in the objective function. The second filter in the pair, therefore, should be derived from the QMF conditions. It can be shown that this second filter also has the desired properties as dictated by the objective function [133].

In order to generate codes for a multiuser channel with multipath interference, two individual objective functions are weighted and added together to form the composite objective function [131]

$$F_C = \alpha F_{MU} + (1 - \alpha) F_{MP}. \tag{1.12}$$

No other terms are needed in the objective function. The weighting coefficient, $0 < \alpha < 1$, is chosen to match the particular channel conditions, that is, it is chosen to weight the two objectives according to the relative severity of the expected multiuser or multipath interference.

1.3.5 Other Objective Function Parameters

As discussed above, for the multipath and multiuser channel, only the autocorrelations and cross-correlations need to be considered. It may be desirable, however, to exercise further control over the time-spread, σ_n^2, and frequency-spread, σ_ω^2, of the codes where σ_n^2 and σ_ω^2 are defined in Section 1.2.1. To this end, two other objectives have been suggested for use in the optimization function. Specifically, σ_n^2 and σ_ω^2 are incorporated to form the objective function [3].

$$F_C = \alpha F_U + \beta F_{MP} - \gamma \sigma_n^2 - \eta \sigma_\omega^2.$$

The parameters α, β, γ and η are design parameters which are chosen to emphasize the corresponding metrics of the objective function under the constraint

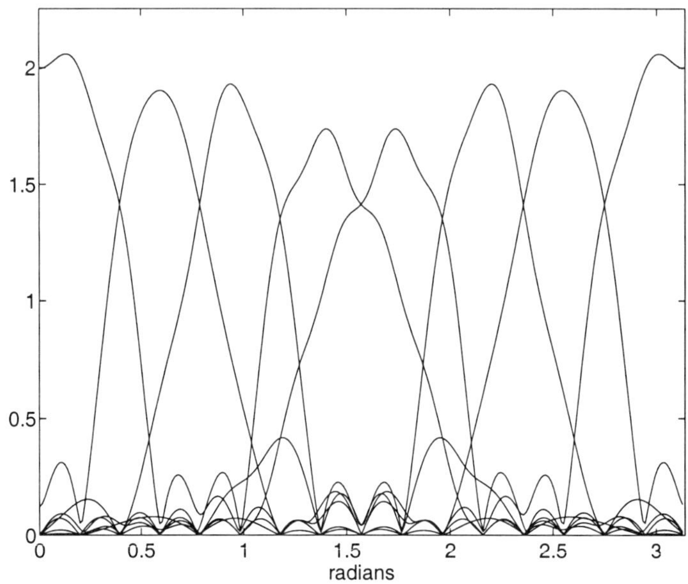

Figure 1.11 Frequency response of multiuser codes.

that $\alpha + \beta + \gamma + \eta = 1$. Additional metrics have also been developed for a narrowband interference channel [129]. The use of these metrics under various channel conditions remains an active area of research.

1.3.6 Results

Several filter banks have been designed using the composite objective function in (1.12) with different values of α. The two limiting cases deserve special attention. When $\alpha = 1$, the objective function becomes the asynchronous multiuser interference problem. A three stage full binary tree using 8-tap FIR filters has been designed using this value for α. The resulting eight codes, each of length 32, have the frequency responses shown in Figure 1.11. In the figure, the individual frequency responses are overlayed and, as can be seen, each of the codes displays narrowband characteristics. When these codes are used, the resulting system operates more like an FDMA system than a CDMA system. This result is not unexpected, however, since the objective function attempted to produce orthogonal codes. Since it also precludes orthogonality from occurring in the time domain (all time shifts were considered), codes which are separated in frequency are produced. Had a direct M-band design been implemented, a pure FDMA system as described earlier would have been generated.

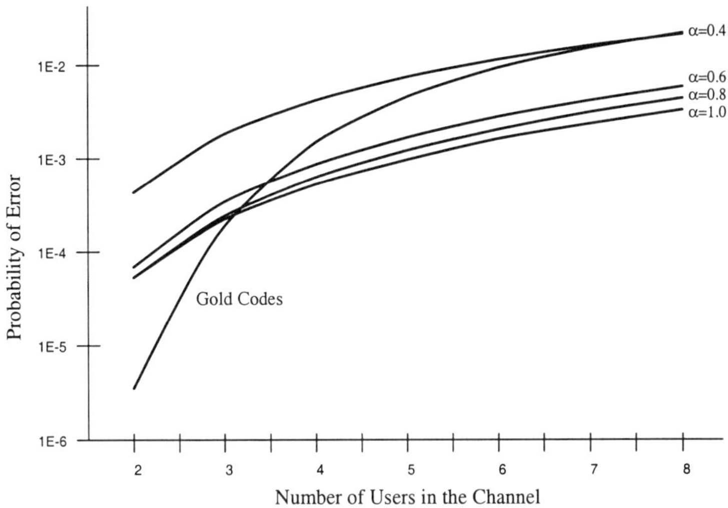

Figure 1.12 BER performance versus the number of users for various weighting factors. $E_b/N_0 = 13$ dB.

The second special case occurs when $\alpha = 0$, meaning that only the multipath interference is considered. In this case, the optimization problem can be solved by inspection with the solution being sets of filter coefficients with only one non-zero value equal to unity. Each code has the non-zero element in a different location. Clearly, these codes have an autocorrelation function that is an impulse and still satisfy the perfect reconstruction conditions. Extending this to an M-band full binary tree results in a TDMA communications system. Notice that orthogonality between the users is now attained by time division (as opposed to frequency division in the multiuser system). As such, the system is unsuitable to asynchronous multiuser communications. Nevertheless, this result is expected since the asynchronous multiuser objective has been eliminated from the optimization.

To examine the effects of composite multiuser/multipath interference objectives, several additional trees have been designed which are similar in structure to the multiuser tree discussed above [132]. The BER performance as a function of the number of users in the channel has been analytically calculated producing the results shown in Figure 1.12. For comparison, the performance of a system using length 31 Gold codes is also provided. All users in the system are assumed to be operating with perfect power control, meaning that the received power for each user is identical. Performance results also have been generated under imperfect power control conditions [133]. As expected, the multiuser performance degrades as α is decreased, however, a frequency analysis of the codes shows that they are becoming more wideband and, thus, are better suited to the multipath channel.

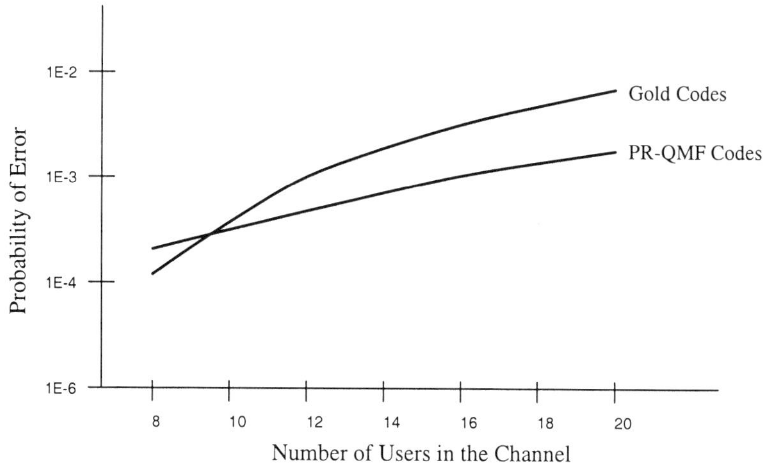

Figure 1.13 BER performance versus the number of users for a quasi-synchronous channel. $E_b/N_0 = 13$ dB.

1.3.7 Designs for a Quasi-Synchronous Channel

Since the multiuser and the multipath channels have different objectives, it is necessary to compromise between the two objectives when considering a channel with both types of interference. As demonstrated above, improved multiuser performance usually results in reduced multipath performance. As a result, it is advantageous to explore the design of codes for use in a multiuser system in which the users attain partial synchronization. In this quasi-synchronous channel, minimization of the cross-correlations between the codes is performed over a smaller range of relative time shifts, leaving more degrees of freedom for other objectives. An example of such a system is a micro-cell cellular system in which the users are synchronized to a common clock. Despite the presence of a common clock, there is still some delay uncertainty since various path lengths from the transmitters to the base station introduce different propagation delays. The resulting timing uncertainty can range from a fraction of a chip to many chips, depending on the size of the cell and the chip rate. The case of multiple-chip uncertainty is considered here.

The development of the objective function for the quasi-synchronous channel is somewhat different than that for the asynchronous channel. Although its complexity precludes it from being presented here, a detailed presentation can be found in [133, 134]. To demonstrate the benefits of considering the quasi-synchronous case, a five stage full binary tree using 8-tap FIR filters is designed and tested. As a result, 32 spreading codes are produced, each of length 128. In this case, the multiuser/multipath weighting factor is taken to be $\alpha = 0.7$ and it is assumed that the timing uncertainty is ± 25 chips. Once again, a comparison to Gold codes (length 127) is performed and the results, shown in Figure 1.13, exhibit an improvement in the number of users supported by the channel for

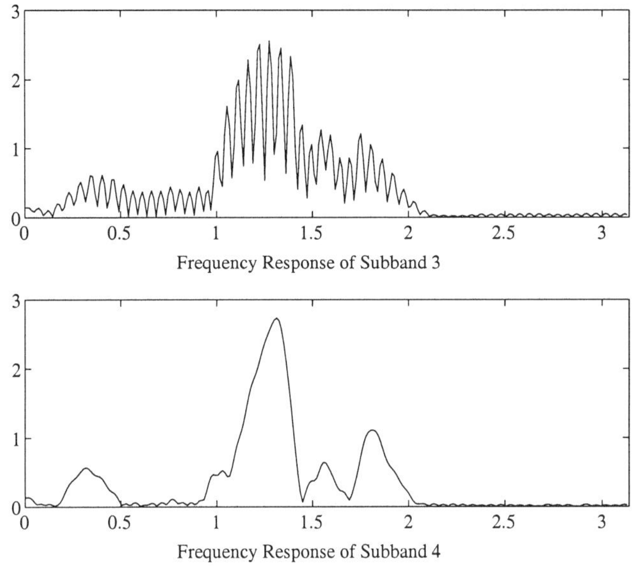

Figure 1.14 Frequency response of codes with partial synchronization.

a particular bit error rate. It is also interesting to view the cross-correlations and frequency responses of the codes. The frequency responses of two sample codes generated by adjacent subbands (one from each decimator of a QMF pair) are plotted in Figure 1.14. Notice that the frequency response of the two codes almost completely overlap. The cross-correlations of interest, however, are still small in magnitude. These cross-correlations can be seen in Figure 1.15 where the cross-correlation values are plotted versus the delay between the two codes. Although some of the cross-correlation values are large, the values over the range of interest, the first 25 and the last 25, were successfully minimized. This result demonstrates the ability of two or more codes to obtain low cross-correlation properties using both time and frequency domain orthogonality.

1.4 CONCLUSION

The transmultiplexer was originally developed as a tool for easing the transition from an analog phone system to today's digital system. Work on the transmultiplexer, which is fundamentally a filter bank, was linked to the work on subband filter banks by Vetterli, resulting in a fundamental change in the way transmultiplexers were designed by eliminating the need for frequency domain orthogonality.

This chapter has extended Vetterli's work by showing that TDMA, FDMA and CDMA can all be derived by performing filter bank designs with the appropriate constraints. In the cases of TDMA and FDMA, this result is useful

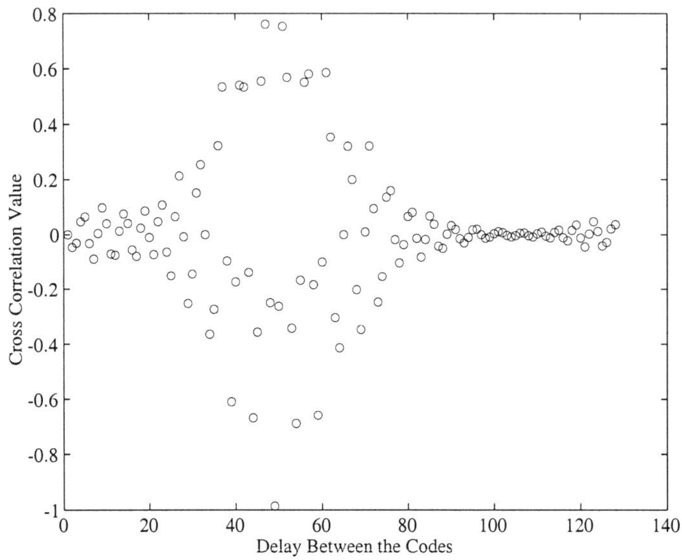

Figure 1.15 Cross-correlations of two codes designed for partial synchronization.

only in the sense that it allows us to unify our thinking about multiple-access techniques since one would not design waveforms for either of those systems in this way. In the case of CDMA, however, the result allows us to derive new spreading waveforms which are optimized to a particular channel, emphasizing properties such as multiple-access interference immunity, multipath resistance, or jamming tolerance. In the examples shown, progressive optimality was used to design several sets of CDMA codes which provided a performance advantage over Gold codes. While progressive optimality greatly simplifies the synthesis of a M-band filter bank, essentially making the problem reasonable from a computational point of view, it does not necessarily produce an optimal filter bank. As a result, the CDMA codes that were designed are not necessarily the "best" codes that are attainable. Future advances in M-band filter bank design techniques and increased computing capability will ultimately enable the design of better CDMA spreading waveforms using the filter bank design framework.

More recently, filter bank and time-frequency theories were jointly utilized to optimally design spread spectrum PR-QMF banks which are promising code solutions for future spread spectrum CDMA communication systems.

2 ORTHOGONAL FREQUENCY DIVISION MULTIPLEXING FOR TERRESTRIAL DIGITAL BROADCASTING

Marc de Courville [1] and Pierre Duhamel[2]

[1]Motorola CRM
Paris, France
Marc.de.Courvill@crm.mot.com

[2]ENST
Paris, France
duhamel@sig.enst.fr

2.1 DIGITAL AUDIO BROADCASTING SYSTEM

This section intends to present how multicarrier modulations are used for broadcasting audio and data programs in Europe. The research and development project known as Eureka 147 launched in 1986 has produced a transmission standard [49] for digital radio broadcasting. The motivation was to provide CD-quality terrestrial audio broadcasting, which is planned to be the successor of today's 40 year old analog radio broadcasting system. The Digital Audio Broadcasting (DAB) technical specification is described briefly below. Besides its high quality, other requirements of the European DAB system are summarized as

- unrestricted mobile, portable and stationary receiver (of course, the system has a limit with respect to the speed of mobile receiver),
- both regional and local service areas with at least six stereo audio programs,
- sufficient capacity for additional data channels (traffic information, program identification, radio text information, dynamic range control).

DAB is a digital radio broadcasting system that relies on a multi-carrier technique known as coded orthogonal frequency division multiplexing (COFDM). In OFDM data is distributed across a large number of narrow-band orthogonal waveforms, each of which is modulated at a lower bit rate and which together occupy a wide bandwidth. Unlike traditional frequency division multiplexing (FDM) systems, these narrow-band carrier signals are transmitted with a maximum spectral efficiency (no *spectral holes* and even with overlapping spectra between two adjacent carriers) due to their orthogonality property.

The proposed DAB system makes use in the digital domain of a discrete Fourier transform (DFT) to perform the modulation and introduces some redundancy in the emitted signal in order to allow a simple equalization scheme (as explained in Section 2.4). Such a system does not perform better than single carrier modulation if used without any coding [280] or frequency interleaving. The real difference comes from the fact that a given symbol is mainly carried by a subcarrier with a precise location (tile or cell) on the time-frequency plane. Furthermore, either a given frequency or the whole spectrum are not likely to be strongly attenuated by a channel fading during a long period of time. Hence, the symbols that are linked by the channel coder are emitted at various time and frequencies so that a small number of them can "simultaneously" be degraded by a fading (like a deterministic frequency hopping procedure). This use of "diversity" (introduced jointly by a time and frequency interleaving procedure) is further increased by multiplexing several broadcast programs into the same signal. As a result, each program has access to the diversity allowed by the full band rather than the one which is strictly necessary to its own transmission. The channel coder is a simple convolutional code.

2.1.1 Frequency Planning

The DAB system is designed in such a way that it is able to operate in any spare frequency slot in the range 30MHz to 3GHz. Furthermore it allows the use of single frequency network (SFN) schemes wherein adjacent transmitters, synchronized in time, can use an identical frequency to broadcast the same programs. These multiple transmissions are considered as artificial echoes provided that the delay does not exceed the maximum that can be handled by the system. Since OFDM allows easy control of this parameter, designing such SFNs is easy and can result in a network gain.

A large scale pilot test involving the consumers is already in place and programs are currently being broadcasted in Europe. Frequency bands allocated by the European Conference of Postal and Telecommunications Administrations (CEPT) regulatory body following Comité Européan de Normalisation Electrotechnique (CENELEC) standard EN 50248 are

- UHF Band L: $[1,452.960\text{GHz (block L1)}, 1,490.624\text{GHz (block 23)}]$
- VHF Band III: $[174.928\text{MHz (block 5A)}, 239.200\text{MHz (block 13F)}]$

Currently VHF Band III and L-Band are used in Europe. In Canada only L-Band is transmitted.

Receivers and integrated circuits (IC) have been manufactured and are currently available on the market (Philips, Grundig, et cetera). The year 2010 is the target date for DAB systems to replace existing FM services (in band 87.5-108 MHz). The new system will have a total bandwidth of 20.5 MHz for audio broadcasting and new services.

2.1.2 Transmission Modes for DAB

In the European Telecommunication Standard (ETS) standard, four transmission modes have been specified depending on the frequency band used for transmission (band L or band III). These bands are required to cope with the effects of Doppler shift of the carriers in fast moving receivers and result in a different tuning of the system parameters as detailed in Table 2.1. As an example, in the case of DAB Mode II, the DAB multiplex is composed of 384 carriers spaced at 4kHz intervals occupying a total bandwidth of 1.536MHz. Whatever the mode used, the system is able to convey a raw data rate of around 2.4 Mbit/s (without taking into account the redundancy introduced by the convolutional coder). For each mode a different number of carriers is employed and the active and guard interval periods change accordingly.

2.1.3 Data-Link Layer Overview

A simplified global DAB system (including the emitter and receiver) is depicted in Figure 2.1. The modulation of each carrier is performed differentially based on a QPSK constellation. Within the multiplex it is possible to transmit up to 6 high quality stereophonic radio programs together with auxiliary data channels. An interleaving operation is performed and scatters in time the symbols linked by the channel coder on 16 common interleaved frames (CIF) representing, for Mode II 1216 OFDM symbols and, in frequency, over 384 subcarriers. A cyclic prefix guard interval (GI) is appended between every block of time domain samples to be sent through the channel. Though this adjunction represents a loss of around 25% in useful bit-rate, it allows the use of a simple equalization scheme. In fact, the purpose of the GI is to “absorb” the multipaths caused by the channel. Its duration is determined by the anticipated reflection periods at the frequency used. The specific design of the GI enables linear convolution in the channel to be performed cyclically. Due to the mathematical property of the DAB demodulator, namely the inverse discrete Fourier transform (IDFT), the suppression of the channel effect is performed at a very low arithmetical complexity cost. This is explained in more detail in the remainder of this chapter.

2.1.4 Error Correcting Codes

Robustness of the transmission is granted by means of time interleaving combined with a powerful forward error protection scheme. This scheme consists of a rate 1/4 convolutional coder which, in order to reduce the large overhead incurred, is punctured according to the nature of the transmitted service. De-

Table 2.1 Transmission mode parameters of DAB

Parameter	Mode I	Mode II	Mode III	Mode IV
Usage	terrestrial, SFN	terrestrial & satellite	satellite & cable	terrestrial
Operational range	< 375 MHz	< 1.5 GHz	< 3 GHz	< 1.5 GHz
Band	VHF Band III	UHF Band L in Europe	S-Band	UHF Band L in Canada
Bandwidth	1.536 MHz			
K useful carriers	1536	384	192	768
Carrier spacing	1 kHz	4 kHz	8 kHz	2 kHz
Gross bit-rate	2.4 Mbit/s			
Sampling rate	$1/2048000 \approx 0.488\mu s$			
L OFDM symbols/frame	76	76	153	76
N FFT size	2048 (1 ms)	512 (250 μs)	256 (125 μs)	1024 (500 μs)
$P = N + D$ TD symbol size	2552 (1.246 ms)	638 (312 μs)	319 (156 μs)	1276 (623 μs)
D GI size	504 (246 ms)	126 (62 ms)	63 (31 ms)	252 (123 ms)
$N_{\text{null}} > N_s = N + D$ null symbol	2656 (1.297 ms)	664 (325 μs)	345 (168 μs)	1328 (648 μs)
$N_f = N_s * L + N_{\text{null}}$ frame duration	196608 (96 ms)	49152 (24 ms)	49152 (24 ms)	98304 (48 ms)
CIFs/frame	4	1	1	2
CIF size	55296 bits			
CU size	64 bits			
OFDM symbols /FIC	3	3	8	3

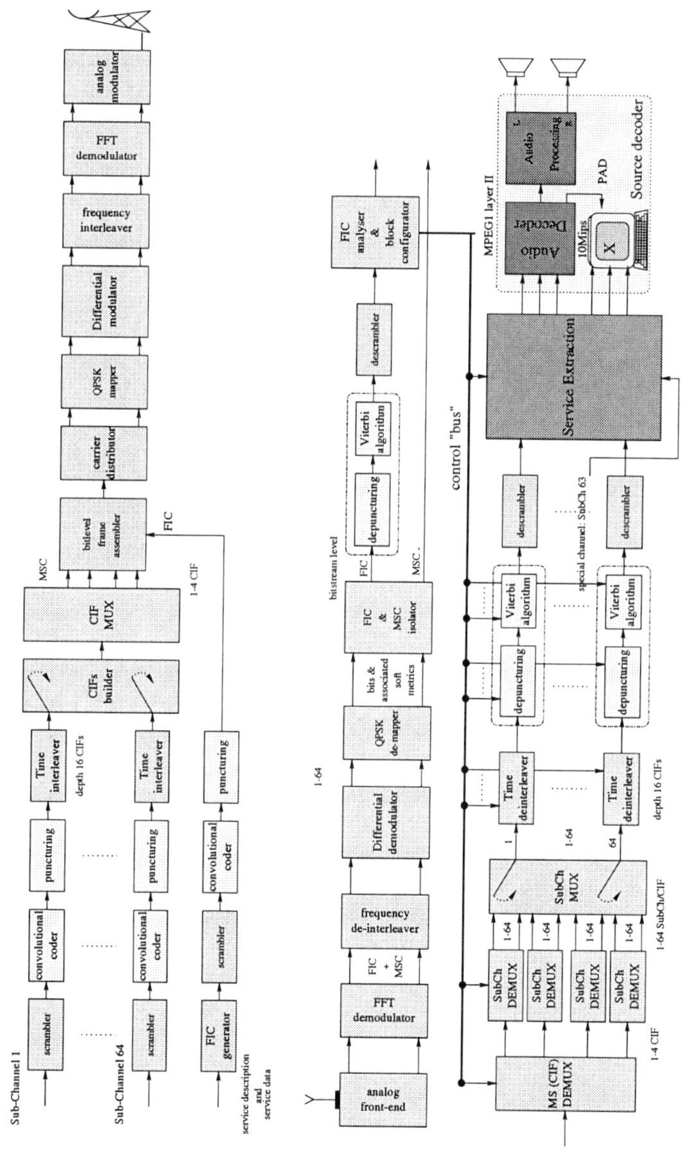

Figure 2.1 Simplified DAB emitter and receiver scheme

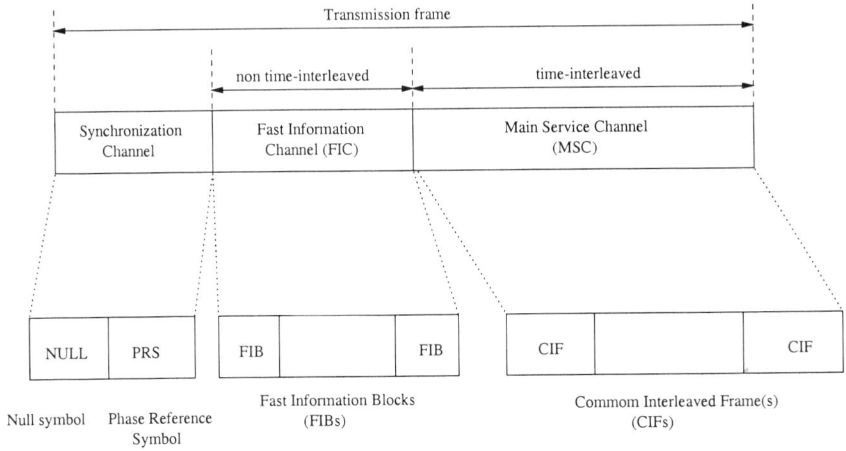

Figure 2.2 Frame structure of the DAB system

coding is performed by the Viterbi algorithm (maximum likelihood decoder). Note that the protection level may not be constant along the frame and can be applied in an extremely versatile manner. Indeed, different parts of the multiplex can be assigned different code rates. Moreover a technique called unequal error protection (UEP) is combined with digital audio coding. Using UEP, header information is more highly protected than the subband samples. Due to the de-interleaving operations, bursts of errors are avoided at the receiver.

2.1.5 *Frame Structure: Transport Layer*

Both multiplexer reconfigurability and the broadcast scenario require the information to be arranged in frames, each of which carry a pilot symbol and a description of the program organization repeated periodically. Indeed, a receiver is supposed to be able to synchronize and lock on an ensemble when switched on at any time in order to receive the broadcast services.

As depicted in Figure 2.2, the typical frame structure of the Mode II DAB system consists of the succession of the following 76 OFDM symbols (each one built from 384 QPSK symbols):

- a null symbol (mainly for performing the timing synchronization)
- a reference symbol for initializing the differential modulation and achieving frequency synchronization
- 3 service symbols called the fast information channel (FIC) provide the DAB decoder multiplex configuration and signal any forthcoming reconfiguration. Protection level, puncturing masks used for each service and many other parameters are detailed within this period. The FIC is not time interleaved since it has to be decoded and analyzed for parameter-

izing the channel decoders before any of the useful information is to be processed.

- 71 "sound and data symbols" form the main service channel (MSC) built from CIFs, each of them consisting (at the bit frame level) of blocks of 55296 bits. Each service bit stream (either data or plain audio) is treated as a separate subchannel embedded in the CIF.

The DAB system is very flexible in the sense that program allocation within the multiplexer is performed dynamically and parameterized within the FIC which enables multiplexer reconfiguration "on the fly". Note that a change in the configuration is signaled in advance with the aid of the main data stream.

2.1.6 Source Coders and Service Types: Application Layer

Different kinds of programs are transmitted (program services, audio data, program associated data (PAD) or other system information (SI)) and the system is able to carry up to 6 CD high quality audio subchannels, each one representing a 192 kbit/s bitstream at the output of the MUSICAM coder. Auxiliary data can be transported in the audio frames as PAD, as a separate subchannel stream, or in small packets (packet mode transmission).

Audio source coding in DAB has been standardized as ISO-MPEG 11172-3 Layer II with the additional facility of half rate sampling, which offers better performance with voice only audio. DAB is effectively a point to multi-point data pipe and, as such, can carry any data type in one of several transport mechanisms. Work currently underway to define transport protocols for future data services is known as multimedia object transfer (MOT), a software protocol that may be used to convey data including HTML, text, JPEG images and, eventually, Java files.

2.1.7 Chapter Outline

As can be understood from the previous discussion, the great flexibility found in the DAB is allowed by characteristics of the chosen modulation format. What follows is the mathematical foundation of an OFDM system using the conventional continuous-time filter bank structure. Since digital solutions are more desirable than analog ones in practice, we attempt to show the theoretical linkages of analog and digital orthogonal transmultiplexers in order to highlight their commonalities and possible extensions. Then, we focus on the specific DFT based OFDM transceiver and show its inherent mathematical features which make it a very attractive solution, especially in adverse communication environments such as for DAB. An OFDM transceiver is separated into its transmitter and receiver in order to make the following discussions easier to understand.

2.2 MULTICARRIER MODULATION

Throughout the following analysis, the matrix operator $(\cdot)^t$ is used to denote transposition, $(\cdot)^*$ denotes complex conjugation and $(\cdot)^H = ((\cdot)^t)^*$.

An OFDM system multiplexes the incoming services (or subchannels) consisting of discrete data streams, modeled here for sake of simplicity by a unique sequence $I(n)$ with an initial rate of $1/T_u$, into several (say K) substreams $I_n(k) = I(kK+n)$, $0 \leq n \leq K-1$. Each bit stream is transmitted over its own orthogonal subcarrier, $g_k(t)$, where K subcarriers form an *orthogonal* set. The bit substreams $\{I_n(k)\}$ are first packed into symbols, by means of a given constellation scheme, e.g. quaternary phase-shift keying (QPSK) for the DAB or a quadrature amplitude modulation (QAM) for the Digital Video Broadcasting (DVB) application. Then, they are transmitted via subchannel k. Therefore, K QPSK/QAM subsymbols are simultaneously transmitted by the subcarriers. All the subsymbols transmitted during the same time duration T_s (symbol period) at time k (block k) are combined into a length-K vector constituting what is denoted as the OFDM symbol $\mathbf{s}(k) = (s_0(k), \cdots, s_{K-1}(k))^t$.

In the ideal case, the orthogonality property of the carriers, $\{g_k(t)\}$, ensures perfect recovery of the transmitted subsymbols at the receiver. The filter bank theory provides modulators of this kind that can be implemented as lossless (LL) perfect reconstruction (PR) filter banks (FB) [337]. The orthogonality is therefore defined not only between each pair of filters but also with their time shift versions by:

$$\begin{aligned}\langle g_k(t-iT_s), g_{k'}(t-i'T_s)\rangle &= \int_{-\infty}^{+\infty} g_k(t-iT_s)g_{k'}^*(t-i'T_s)\,dt \\ &= \delta_{i,i'}\delta_{k,k'},\end{aligned}$$

where $k, k', i, i' \in \mathbb{Z}$, $\langle\cdot,\cdot\rangle$ denotes the canonical scalar product and $\delta_{i,i'}$ is the Kronecker symbol defined as

$$\delta_{i,j} = \begin{cases} 1 \text{ for } i = j; \\ 0 \ \text{ otherwise.}\end{cases}$$

Traditional systems use the following set of orthogonal filters:

$$g_k(t) = u(t)e^{2j\pi\frac{kt}{T_s}} \tag{2.1}$$

where $u(t)$ is the normalized time domain rectangular window function of duration T_s

$$\begin{aligned} u(t) &= \frac{1}{\sqrt{T_s}}\text{Rect}\left(\frac{t}{T_s}\right) \\ \text{Rect}(t) &= \begin{cases} 1 & 0 \leq t < 1 \\ 0 & \text{otherwise.}\end{cases}\end{aligned} \tag{2.2}$$

It can be checked that this set of filters $\{g_k(t)\}$ is orthogonal, i.e.

$$\begin{aligned}\langle g_k(t-iT_s), g_{k'}(t-i'T_s)\rangle &= \int_{-\infty}^{+\infty} u(t-iT_s)u(t-i'T_s)e^{j2\pi\frac{(k'-k)t}{T_s}}\,dt \\ &= \frac{1}{T_s}\int_{[iT_s,(i+1)T_s]\bigcap[i'T_s,(i'+1)T_s]} e^{j2\pi\frac{(k'-k)t}{T_s}}\,dt \\ &= \delta_{i,i'}\delta_{k,k'}.\end{aligned}$$

The property of orthogonality is often misinterpreted in terms of spectral considerations. Actually, since filters $g_k(t)$ are all issued from a unique function $u(t)$ by simple frequency translations, the orthogonality is sometimes justified assuming that the frequency shift is large enough so that the $g_k(t)$ spectra do not overlap (as in FDM). But it is well known that the discrete time Fourier transform of a time limited signal (just as $u(t)$ in the present case) is a signal with infinite frequency bandwidth. Thus, since g_k satisfies (2.1), it is clear that perfect orthogonality cannot rely on simple spectral considerations. By the way, most orthogonal filter families overlap both in the frequency and time domain (as is the case for spreading codes used in code division multiple access (CDMA) schemes).

We show below that this modified orthogonality property and a discrete modeling of the OFDM modulator relies on a commonly used formalism in filter bank theory — the polyphase matrices. Moreover, the discrete equivalent of the OFDM modulator can be viewed as the synthesis part of a transmultiplexer [95].

2.3 A GENERAL FRAMEWORK: THE TRANSMULTIPLEXER APPROACH

After the modulation of $\mathbf{s}(n)$ by the set of $g_k(t)$, the transmitted time domain signal, $s(t)$, can be expressed as (cf. Figure 2.3)

$$s(t) = \sum_{k=0}^{K-1}\sum_{i\in\mathbb{Z}} S_k(i)\delta(t-iT_s) * g_k(t)$$

$$s(t) = \sum_{k=0}^{K-1}\sum_{i\in\mathbb{Z}} S_k(i)g_k(t-iT_s). \tag{2.3}$$

A straightforward implementation of this scheme would require K analog filters forming an orthogonal set. Since this implementation would not be very efficient, one usually first computes samples of the transmitted signal, $s(t)$. The resulting samples are then put through a digital-to-analog converter (DAC).

In discrete-time implementation, the continuous-time signal $s(t)$ transmitted during the time interval T_s is formed by a linear combination of K discrete-time orthogonal subcarriers $\{g_k(n)\}$. Since modulation is a linear and causal operation, at least $N \geq K$ samples of $s(t)$ must be available during the same time interval in order to perfectly retrieve the subsymbols embedded on the

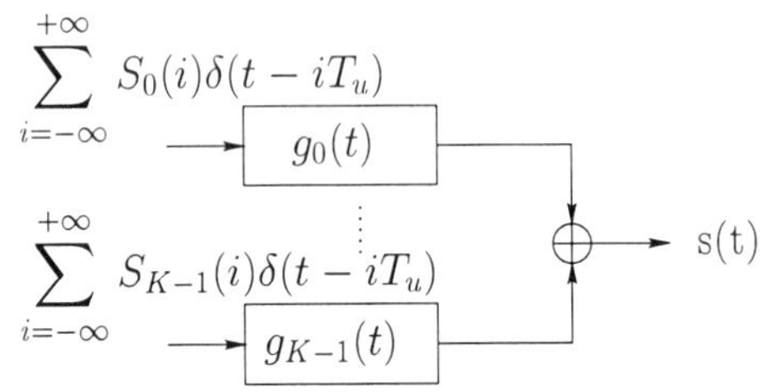

Figure 2.3 Continuous modeling of the OFDM modulator

orthogonal discrete-time subcarriers $\{g_k(n)\}$. Otherwise, some loss of information is unavoidable. More simply, the system is designed such that it performs a fixed block processing of the incoming data stream. The sampling rate T is chosen so that $NT = T_s$, $N \in \mathbb{Z}$ and $N \geq K$. The N time domain samples $s_n(k)$ to be emitted during block k can thus be expressed as

$$s_n(k) = s(kT_s + nT) = s[(kN + n)T], \qquad 0 \leq n \leq N - 1.$$

By construction, the sequences $s_n(k)$ form the Type-I polyphase components of the time domain sequence, s_n, to be transmitted [337]. According to the analog modulator in (2.3), we get

$$s_n(k) = \sum_{m=0}^{K-1} \sum_{i \in \mathbb{Z}} S_m(i) g_m \left[(k - i)NT + nT\right].$$

Applying the notations

$$G_m^l(z) = \sum_{n \in \mathbb{Z}} g_m \left[(nN + l)T\right] z^{-n}, \qquad 0 \leq l \leq N - 1, \tag{2.4}$$

$$G(z) = \left[G_m^k(z)\right]_{0 \leq k \leq N-1, 0 \leq m \leq K-1} \tag{2.5}$$

and defining $\mathrm{TZ}[x_n] = \sum_{n \in \mathbb{Z}} x_n z^{-n}$ as the Z-transform of any series $(x_n)_{n \in \mathbb{Z}}$, we obtain for $n \in \mathbb{N}_N$ fixed (where $\mathbb{N}_N$ is the set of integers $\mathbb{N}_N = [0, N - 1] \cap \mathbb{N}$)

$$s_n(z) = \mathrm{TZ}[s_n(k)] = \sum_{k \in \mathbb{Z}} s_n(k) z^{-k}$$

and

$$\begin{aligned} s_n(z^N) &= \sum_{k\in\mathbb{Z}} s_n(k) z^{-kN} \\ &= \sum_{m=0}^{K-1}\sum_{i\in\mathbb{Z}} S_m(i) \sum_{k\in\mathbb{Z}} g_m\left[(k-i)NT + nT\right] z^{-kN} \\ &= \sum_{m=0}^{K-1}\sum_{i\in\mathbb{Z}} S_m(i) z^{-iN} G_m^n(z^N) \\ &= \sum_{m=0}^{K-1} G_m^n(z^N) S_m(z^N) \\ &= (G_0^n(z^N), \cdots, G_{K-1}^n(z^N))\mathbf{S}(z^N) \end{aligned}$$

where $\mathbf{S}(z) = [S_0(z), \cdots, S_{N-1}(z)]^t = \mathrm{TZ}[\mathbf{S}(n)]$.

Denoting $\mathbf{s}(z) = [s_0(z), \cdots, s_{N-1}(z)]^t = \mathrm{TZ}[\mathbf{s}(\mathbf{n})]$, this proves that

$$\mathbf{s}(z) = [s_0(z), \cdots, s_{K-1}(z)]^t = G(z)\mathbf{S}(z). \tag{2.6}$$

Note that $s(z)$ is simply obtained from vector $\mathbf{s}(z)$ by performing the following operation:

$$s(z) = (1, z^{-1}, \cdots, z^{-N+1})\mathbf{s}(z^N)$$

which yields

$$\begin{aligned} s(z) &= \sum_{k=0}^{N-1} z^{-k} s_k(z^N) \\ &= \sum_{k=0}^{N-1} z^{-k} (G_0^k(z^N), \cdots, G_{K-1}^k(z^N))\, \mathbf{S}(z^N) \\ &= \left[\sum_{k=0}^{N-1} G_0^k(z^N) z^{-k}, \cdots, \sum_{k=0}^{N-1} G_{K-1}^k(z^N) z^{-k}\right] \mathbf{S}(z^N) \\ &= (G_0(z), \cdots, G_{K-1}(z))\mathbf{S}(z^N). \end{aligned}$$

Therefore the expression of the digital filters of the synthesis bank producing $s(z)$ are

$$g_m(z) = \sum_{l=0}^{N-1} G_m^l(z^N) z^{-l}, \qquad 0 \le m \le K-1. \tag{2.7}$$

A way of modeling the generation of $s(z)$ is to use the polyphase matrix formalism by the matrix filtering of $\mathbf{S}(z)$ by $G(z)$ as illustrated in the center diagram of Figure 2.4. This amounts to using the synthesis filter bank associated with $G(z)$ [337] with subband filters $g_m(z) = \sum_{l=0}^{N-1} G_m^l(z^N) z^{-l}$ as illustrated at the bottom of Figure 2.4. Regarding (2.7), (2.4) and (2.5),

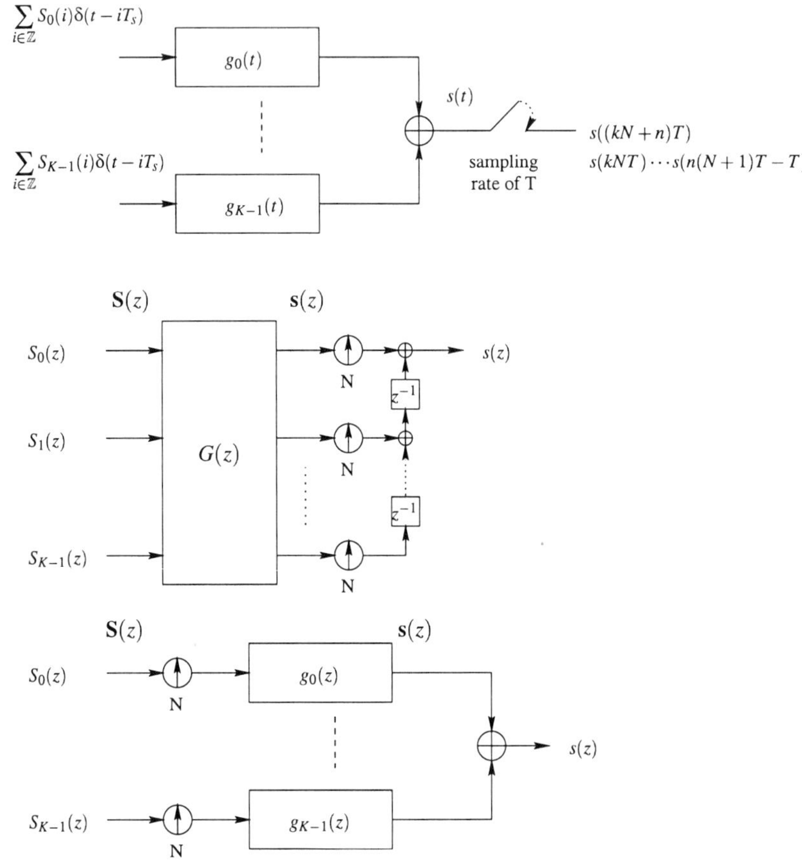

Figure 2.4 Equivalent models of the OFDM modulator

- $G(z)$ is denoted as the polyphase matrix [337] associated with the synthesis filter bank (SFB) performing the modulation
- $g_m(z)$ is the mth SFB filter
- $G_l^m(z)$ is the Type-I l^{th} polyphase component of $G_m(z)$

In the ideal case (no distortion nor noise added by the channel) let us focus on the perfect reconstruction condition (allowing to perfectly recover the data at the receiver). Denoting ˜ as the transconjugation operator whose argument is a Z scalar polynomial or a polynomial matrix ($\tilde{P}(z) = P^H(z^{-1})$), the condition that filters $g_m(t)$ should be orthogonal can be summarized in the discrete domain by the following condition on $G(z)$:

$$G(z)\tilde{G}(z) = I_K$$

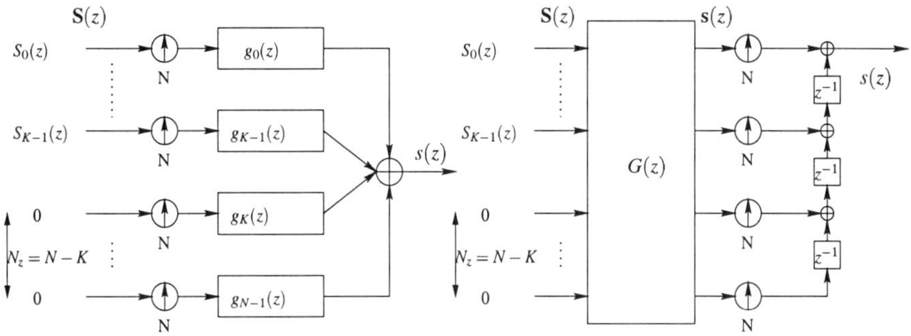

Figure 2.5 Discrete modeling of the oversampled OFDM modulator

where I_K denotes the $K \times K$ identity matrix. This orthogonality condition is known in the filter bank field as the lossless perfect reconstruction (LL PR) condition and is known to be fulfilled by LL PR filter banks (LL PR FB). However, usual LL PR FB form square polyphase matrices. A good solution can be obtained by taking a polyphase LL PR FB, and removing K filters from the bank. This amounts to increasing the number of columns of matrix $G(z)$ in order to obtain a square matrix, and feeding the additional filters by null symbols. In other words, we add $N_z = N - K$ orthogonal filters to the system, each of the new filters modulating null symbols (cf. Figure 2.5).

Since additional filters are used in parallel, one could guess that the computational cost for obtaining the filter bank output has increased. In fact this is not the case. Indeed, the "fast" transform algorithms are often provided for a number of subbands equal to the downsampling factor (they are referred to as critically sampled) [207] and, in general, these schemes cannot be significantly simplified when the last subbands are constrained to be null (this is the case at least for the discrete cosine transform (DCT) and the DFT). Hence the computational load is logically linked to N rather than to $N - K$ in any case.

2.3.1 Particular Cases of Transmultiplexers

DFT modulated filter banks. To facilitate our understanding let us focus on a particular OFDM system in which the transmitted signal $s(t)$ is obtained by modulating $\mathbf{S}(k)$ by a bank in which all orthogonal analog filters are derived from a single prototype filter $u(t)$ by regular frequency shift of its spectrum. Accordingly, we have

$$g_m(t) = u(t)e^{2j\pi f_m t},$$

where $f_m = f_0 + \frac{m}{T_s}$, $0 \leq m \leq K-1$ and the prototype window function, $u(t)$, is properly chosen such that the set of filters $\{g_m(t)\}$ is orthogonal. The DAB application belongs to this class of OFDM systems.

Define again $s_n(k) = s[(kN+n)T]$, $0 \leq n \leq N-1$ as the K time domain discrete samples (forming block k) to be sent through the channel. Assuming

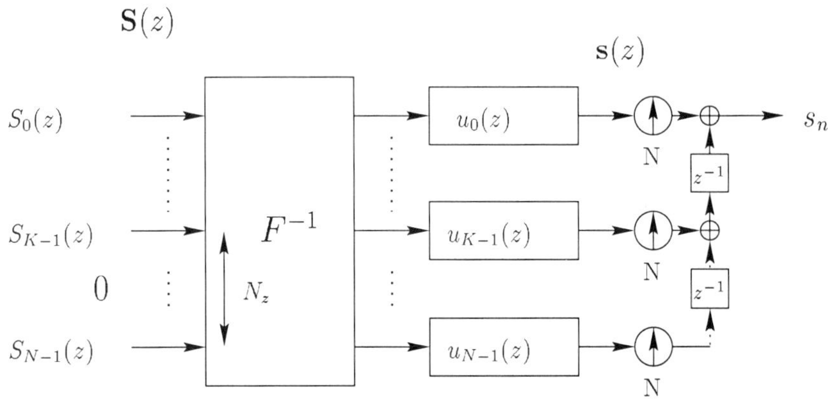

Figure 2.6 Discrete modeling of DFT modulated filter banks OFDM modulators

a baseband model (i.e. $f_0 = 0$) the previous definition yields

$$
\begin{aligned}
s_n(k) &= \sum_{i \in \mathbb{Z}} u[(k-i)T_s + nT] \sum_{m=0}^{K-1} S_m(i) e^{2j\pi f_m (k-i)T_s + nT} \\
s_n(k) &= \sum_{i \in \mathbb{Z}} u[((k-i)N + n)T] \sum_{m=0}^{K-1} S_m(i) e^{2j\pi nm/N}. \qquad (2.8)
\end{aligned}
$$

The IDFT of the vector $\mathbf{S}(k)$ enlarged by $N_z = N - K$ null components can be recognized in (2.8). As depicted in Figure 2.6, the first summation corresponds to filtering each output of the IDFT by the Type-I polyphase component, $u_n(z) = \sum_{k \in \mathbb{Z}} u[(kN + n)T]z^{-k}$, of the discrete sampled version $u(z)$ of $u(t)$. Therefore $\mathbf{s}(k)$ can thus be interpreted as the output of the cascade of a size N IDFT and the synthesis part of a filter bank. This specific scheme belongs to the DFT modulated filter banks class [207, 335, 96]. One can prove that the only $u(t)$ providing exact orthogonality is a rectangular window of length T_s. This is detailed in the next section.

What if the filter bank reduces to a transform?

Now consider the case where the prototype filter $u(t)$ reduces to the time domain rectangular window function of duration T_s as in (2.2). It is shown below that the samples of the channel signal, $s_n(k)$, can be obtained from $S_n(k)$ by applying a simple orthogonal transformation.

In the summation of (2.8), $u[((k-i)N + n)T] \neq 0$ if and only if $i = k$. In other words, the time-orthogonality is ensured by the simple fact that the time-extent of the carrier function is equal to the time-shift in which the inner product is zero. Hence, the polyphase components only have a single term thus yielding

$$s_n(k) = \frac{1}{\sqrt{NT}} \sum_{m=0}^{K-1} S_m(k) e^{2j\pi nm/N}, \qquad 0 \leq n \leq K-1 \tag{2.9}$$

Note that this expression is the IDFT F_N^{-1}, of the subsymbol sequence $\{S_n(k)\}_n$ enlarged by $N_z = N - K$ null components. Furthermore, with

$$F_N = \frac{1}{\sqrt{N}} (W_N^{ij})_{0 \leq i \leq N-1, 0 \leq j \leq N-1},$$

where $W_N = e^{-j\frac{2\pi}{N}}$, (2.9) can be restated in vector form as

$$\begin{aligned} \mathbf{S}(k) &= (S_0(k), \cdots, S_{K-1}(k), \overbrace{0, \cdots, 0}^{N_z = N-K})^t \\ \mathbf{s}(k) &= \frac{1}{\sqrt{T}} F^{-1} \mathbf{S}(k). \end{aligned}$$

As a result, the discrete-time modulator can be implemented as shown in Figure 2.4 with $G(z) = F_N^{-1}$.

In this case of Fourier modulated filters, the orthogonality between carriers is ensured by the orthogonality of the Fourier transform basis, $F_N F_N^H = I_N$. Note that there is no special significance to be attributed to the DAB due to the fact that the transform at the transmitter is an *inverse* DFT rather than a forward transform. If the baseband signal was chosen with respect to the highest frequency, a forward DFT would appear at the transmitter.

Oversampling and results based on the discrete model of the modulator

Mathematically, adding zero inputs to the DFT does not modify the output signal spectrum, provided that we assume a perfect DAC (here we intend by perfect DAC a perfect time interpolator). Indeed, sampling the analog OFDM modulated signal, $s(t)$, at a higher rate than the one imposed by the symbol duration only results in using a larger DFT transform and adding null components to the vector to be modulated in the equivalent discrete emitter model (it is equivalent to the interpolation formula used for band limited signals). Since all these versions of the discrete signal correspond to the same analog signal transmitted, they share the same spectrum.

Since this operation also increases the arithmetic complexity of the emitter, one could wonder why anybody would choose a higher sampling rate. This operation is appropriate when

- operating with non-ideal DACs (because it facilitates the discrete to continuous time interpolation since it gives a higher resolution to the resulting analog signal)
- at least partially implementing emitter filters in the discrete domain in order to comply with the out-of-band spectrum mask specification imposed by the standard.

Note that in what has been presented, the filters $g_m(t)$ have never been assumed to have non-overlapping frequency spectra. Therefore overlapping is allowed provided that the orthogonality condition is fulfilled. This could seem surprising at first glance but it is part of the "black magic" of the filter banks/transmultiplexer theory. Compared to FDM access schemes, this explains why OFDM is often referred to as a maximum spectral efficiency transmission technique.

The receiver (analysis filter bank). In the ideal channel case (no distortion and noiseless context), information can be perfectly recovered at the receiver as follows. If the received signal, $r(t) = s(t)$, is sampled at the same rate T as the transmitter, (2.6) still holds. Since the filter bank is lossless (orthogonal), we have:

$$G(z)\tilde{G}(z) = I_N.$$

Therefore, $\mathbf{S}(k)$ is obtained without any error at the output of the anti-causal analysis filter Bank (AFB), $\tilde{G}(z)$, where the modulator filters are assumed to be of order $KN-1$ as depicted in Figure 2.7.

Since it is not feasible to implement an anti-causal system, we will concentrate on a means for inverting the modulation operation in the ideal channel case. Bearing in mind that the demodulator filters are of memory $KN-1$, the order of each component of polyphase matrix $G(z)$ is $K-1$. Therefore the causal demodulator can be performed by a multiplication with matrix $z^{-K+1}\tilde{G}(z)$. A causal blocking structure can be derived from the first scheme of Figure 2.7 and is detailed in successive subfigures of the same figure observing that

$$r(z) = \sum_{i=0}^{N-1} r_i(z^N)z^{-i},$$

$$r(z) = z^{-N+1} \sum_{i=0}^{N-1} r_i(z^N)z^{N-1-i},$$

and

$$r(z) = z^{-N+1} \sum_{i=0}^{N-1} r_{N-1-i}(z^N)z^{i}.$$

Note the surprising presence of a delay z^{-1} in the causal modeling of the transmission system (detailed in the bottom figures of Figure 2.7). This comes from the fact that the reconstruction by LL PR FB of a signal is only possible up to a delay of $KN-1$. Thus this delay differs by one from a multiple of the block size. If this offset is not compensated, this results in a permutation and a delay in the outputs of the AFB as illustrated in [337]. In order to avoid representng this delay in the schemes, we will assume in the following that this delay is taken into account in the channel impulse response.

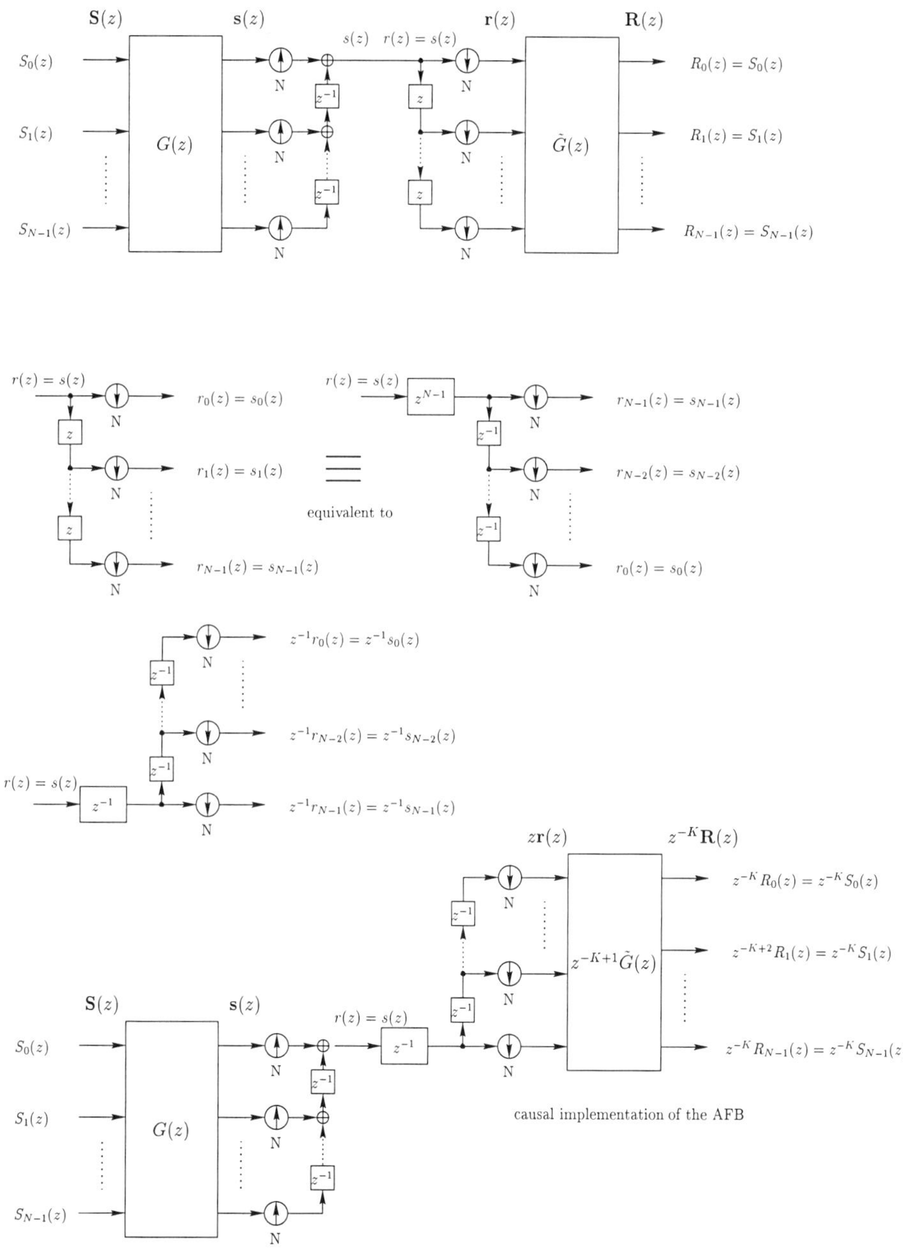

Figure 2.7 Causal modeling of the OFDM system and demodulator blocking

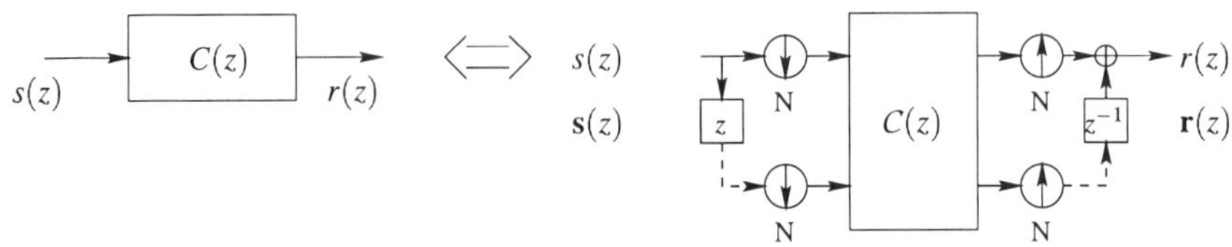

Figure 2.8 Equivalence between a scalar filtering and polyphase subband matrix filtering

2.4 EQUALIZATION OF DISCRETE MULTITONE SYSTEMS: THE GUARD INTERVAL TRICK

When the filter bank reduces to a block transform (a rectangular window $u(t)$ is used as prototype filter), the individual equivalent subband filters have a $\sin(\omega)/\omega$-like shape in the frequency domain, which is not very selective. Under this condition, dispersive channels have a strong influence on system performance, and some channel equalization has to be performed. This issue is investigated below.

2.4.1 *Block Channel Modeling*

Since an OFDM system basically operates on blocks of data, it is convenient to use a discrete model of the transmission channel.

Subband Filtering Matrix Expression. Here, we obtain the polyphase components of the output $r(z)$ of a scalar linear filter function, $C(z)$, of the input $s(z)$ as depicted in Figure 2.8. With the input, output and filter function represented as

$$s(z) = \sum_{k=0}^{N-1} s_k(z^N)z^{-k}$$

$$r(z) = \sum_{k=0}^{N-1} r_k(z^N)z^{-k}$$

$$C(z) = \sum_{k=0}^{N-1} C_k(z^N)z^{-k},$$

linear convolution is given by

$$r(z) = C(z)s(z) = \sum_{k,k' \in \mathbb{N}_N^2} s_k(z^N)C_{k'}(z^N)z^{-k-k'}.$$

One can easily check that the expression for the $r(z)$ polyphase components is

$$r_l(z) = \sum_{k=0}^{l} s_k(z)C_{l-k}(z) + \sum_{k=l+1}^{N-1} s_k(z)C_{N-1-k+l+1}(z)z^{-1}$$

which results in

$$\mathbf{r}(z) = \underbrace{\begin{bmatrix} C_0(z) & z^{-1}C_{N-1}(z) & \cdots & z^{-1}C_2(z) & z^{-1}C_1(z) \\ C_1(z) & C_0(z) & \searrow & & z^{-1}C_2(z) \\ \vdots & \searrow & \searrow & \searrow & \vdots \\ \vdots & & \searrow & \searrow & z^{-1}C_{N-1}(z) \\ C_{N-1}(z) & C_{N-2}(z) & \cdots & C_1(z) & C_0(z) \end{bmatrix}}_{\mathcal{C}(z)} \mathbf{s}(z),$$

where $\mathcal{C}(z)$ is $N \times N$, $\mathbf{r}(z) = [r_0(z) \cdots r_{N-1}(z)]^t$ and $\mathbf{s}(z) = [s_0(z) \cdots s_{N-1}(z)]^t$.

Scalar vector model of the channel effects. Assume that the channel is modeled as a linear time-invariant and causal filter of length M. In practice, the OFDM systems are designed such that the duration of the channel response is shorter than the transform size, N, i.e. at the utmost $D \approx N/4$ in DAB and DVB systems. For convenience, assume that a discrete-time filter is used to model the channel and is represented as the N-dimensional vector $\mathbf{C}$; note that most components of $\mathbf{C}$ are zero valued, i.e.

$$\mathbf{C} = (c_0, \cdots, c_{M-1}, 0, \cdots, 0)^t, \quad \text{with } c_k = 0 \text{ for } k \leq M.$$

Since the channel order is smaller than N, $\mathcal{C}$ simplifies to

$$\begin{aligned} \mathcal{C}(z) &= \begin{bmatrix} c_0 & c_{N-1}z^{-1} & \cdots & c_2 z^{-1} & c_1 z^{-1} \\ c_1 & c_0 & \searrow & & c_2 z^{-1} \\ \vdots & \searrow & \searrow & \searrow & \vdots \\ \vdots & & \searrow & \searrow & c_{N-1}z^{-1} \\ c_{N-1} & c_{N-2} & \cdots & c_1 & c_0(z) \end{bmatrix}_{N \times N} \\ &= C_0(N) + z^{-1}C_1(N) \end{aligned}$$

where both $C_0(N)$ and $C_1(N)$ are square scalar matrices of dimension N. A straightforward computation shows that the block vector of received signal samples, $\mathbf{r}(k)$, can be expressed as the product of the Sylvester matrix, $\tau(\mathbf{C})$, of dimension $N \times 2N$ and the vector of the previously transmitted samples yielding

$$\begin{aligned} \mathbf{r}(k) &= C_0(N)\mathbf{s}(k) + C_1(N)\mathbf{s}(k-1) \\ &= [C_1(N), C_0(N)] \begin{bmatrix} \mathbf{s}(k-1) \\ \mathbf{s}(k) \end{bmatrix} \\ &= \begin{pmatrix} 0 & c_{N-1} & \cdots & c_1 & c_0 & 0 & \rightarrow & 0 \\ \downarrow & \searrow & \searrow & \vdots & c_1 & \searrow & \searrow & \downarrow \\ \downarrow & & \searrow & c_{N-1} & \vdots & \searrow & \searrow & 0 \\ 0 & \rightarrow & \rightarrow & 0 & c_{N-1} & \cdots & c_1 & c_0 \end{pmatrix} \begin{bmatrix} \mathbf{s}(k-1) \\ \mathbf{s}(k) \end{bmatrix} \\ &= \tau(\mathbf{C}) \begin{bmatrix} \mathbf{s}(k-1) \\ \mathbf{s}(k) \end{bmatrix}. \end{aligned}$$

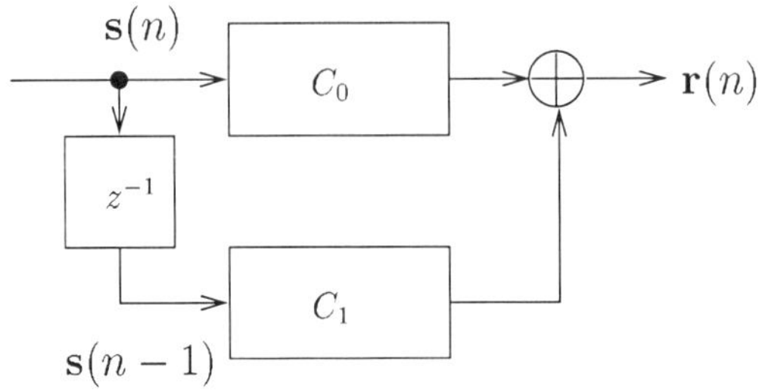

Figure 2.9 Block based discrete modeling of the transmission channel

Note that the two square matrices $C_0(N)$ and $C_1(N)$ form respectively the right and left halves of $\tau(\mathbf{C})$.

The corresponding discrete channel model block diagram is displayed in Figure 2.9. This model is always valid and leads to simple equalization schemes when the transmission filter, $u(t)$, is a rectangular window.

A simple equalizer. Fast filtering operations utilizing the fast Fourier transform (FFT) have been known for a long time. This procedure makes use of the cyclic convolution property of the DFT to transform a linear convolution into part of a cyclic one where the latter is computed by a single multiplication in the frequency domain. In order to reduce the complexity of channel equalization in a discrete multitone (DMT) system some approaches propose using a DFT as the demodulator. A smart solution has been proposed in [354]; it uses a simple equalization scheme in which the loss is minimal in spectral efficiency.

Since this scheme is efficient for a large number of subcarriers, it has been chosen as the standard for Terrestrial DAB (T-DAB standard) and the transmission of Video signals (T-DVB standard). Due to a smaller number of carriers, ADSL communication systems are much more open to other possible solutions.

The equalization, which would be computationally quite demanding for the cases of 256 or 2048 carriers, is made easy by using a simple trick. It can be interpreted as the dual of an overlap-save algorithm for fast filtering.

Its main feature is to force the channel to perform a cyclic convolution rather than a linear one by inserting a cyclic prefix between successive blocks, denoted as a guard interval. This operation consists of appending a block of redundant samples of length D to each block of the transmitted signal, $\mathbf{s}(n)$, where D is larger than the channel memory M. This assures that no intersymbol interference between OFDM symbols from different blocks will occur. This is easily understood, since the channel impulse response vanishes before the next OFDM symbol arrives. Furthermore, the GI is designed such that it wisely duplicates the tail of the block to its head so as to cyclically convolve the transmitted signal with the channel.

When the cyclic prefix is removed at the receiver [354], the effects of channel convolution are observed in the frequency domain, after the demodulation operation, as M complex gains, one per subcarrier or subchannel. The equalization task is thus performed by simple scalar division as illustrated in Figure 2.10. In order to better understand this process, the underlying operational steps are detailed below. The variables obtained after appending the guard interval are denoted by the superscript ^{gi} in the following.

Since the block size is increased by D samples, the discrete channel model given previously is still valid provided that matrices C_0 and C_1 are enlarged to dimension $P = N + D$. Thus, the following relations hold:

$$\begin{aligned}
\mathbf{s}^{\mathrm{gi}}(k) &= (s_{N-D}(k), s_{N-D+1}(k) \cdots, s_{N-1}(k), s_0(k), s_1(k), \cdots, s_{N-1}(k))^t, \\
\mathbf{r}^{\mathrm{gi}}(k) &= C_0(N+D)\mathbf{s}^{\mathbf{gi}}(k) + C_1(N+D)\mathbf{s}^{\mathbf{gi}}(k-1).
\end{aligned}$$

When $D \geq M$, suppressing the first D components of $\mathbf{r}^{\mathrm{gi}}(k)$ at the receiver leads to the simplified expression

$$\mathbf{r}(k) = (r_D^{\mathrm{gi}}(k), \cdots, r_{N+D-1}^{\mathrm{gi}}(k))^t.$$

Folding the structures of matrices $C_0(N+D)$ and $C_1(N+D)$ into a single circular matrix $C_c(N)$ of dimension N shows that $\mathbf{r}$ is obtained through linear combinations of $\mathbf{s}$ components. Therefore, after some calculations, it is easily shown that

$$\mathbf{r}(k) = C_c(N)\mathbf{s}(k),$$

where

$$C_c(N) = \begin{pmatrix}
c_0 & c_{N-1} & c_{N-2} & \cdots & c_1 \\
c_1 & c_0 & \searrow & \searrow & \vdots \\
\vdots & \searrow & \searrow & \searrow & c_{N-2} \\
c_{N-2} & & \searrow & \searrow & c_{N-1} \\
c_{N-1} & c_{N-2} & \cdots & c_1 & c_0
\end{pmatrix}_{N\times N} = C_0(N) + C_1(N).$$

Moreover, every I-circular matrix is diagonal in the Fourier basis, i.e.

$$F_N C_c F_N^{-1} = \sqrt{N}\mathrm{diag}(F_N \mathbf{C}).$$

Since the OFDM demodulator also includes DFT computation, after demodulation we get

$$\begin{aligned}
\mathbf{R}(k) &= F_N \mathbf{r}(k) = F_N C_c F_N^{-1}\mathbf{s}(k) \\
\mathbf{R}(k) &= \sqrt{N}\mathrm{Diag}(F_N^{-1}\mathbf{C})\mathbf{S}(k) = \sqrt{N}[C_0 S_0(k), \cdots, C_{N-1}S_{N-1}(k)]^t
\end{aligned}$$

where $(C_0, \cdots, C_{N-1})^t = F_N^{-1}(c_0, \cdots, c_{M-1}, 0, \cdots, 0)^t$. This shows that after demodulation the transmitted symbol is recovered up to a complex gain. Thus, the equalization is simply performed by dividing the output of each subchannel, or subcarrier, by the corresponding spectral gain of the channel (cf. Figure 2.10). This spectral gain can be chosen in either

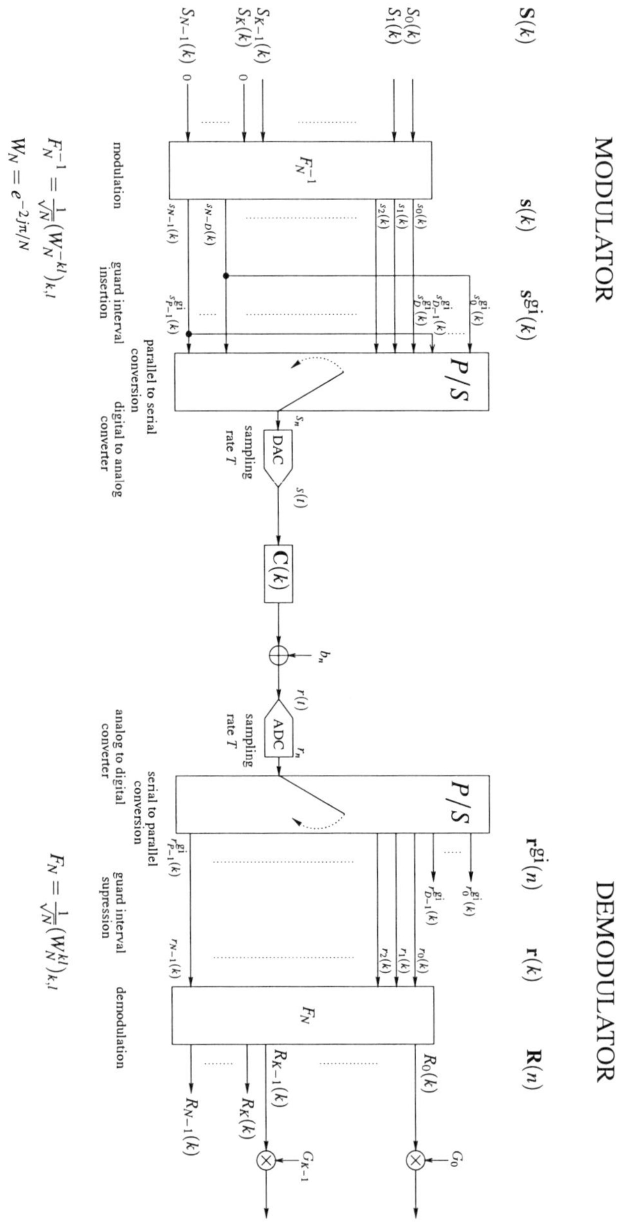

Figure 2.10 DAB equalization scheme based on the use of a GI

- a Zero Forcing (ZF) spirit $\Longrightarrow G_i^{ZF} = 1/C_i$

- or a Minimum Mean Square Error (MMSE) manner

$$\Longrightarrow G_i^{MMSE} = \frac{C_i^*}{|C_i|^2 + \sigma_{B_i}^2/\sigma_S^2} \quad \text{where } \mathbf{B}(k) = F_N \mathbf{b}(k).$$

Note that this is not an approximate relationship. The analog channel, which usually performs a linear convolution, has been "tricked" by the guard interval and instead performs a cyclic convolution.

It is important to realize that an MMSE equalization scheme is only justified in OFDM systems using other constellations than QPSK (e.g. QAM). Indeed, it is noted that the ratio $G_i^{MMSE}/G_i^{ZF} = \frac{|C_i|^2}{|C_i|^2+\sigma_{B_i}^2/\sigma_S^2}$ is a positive real number. Therefore, since the QPSK symbols are distributed on a circle and discriminated using an angle based criterion, the two different scalings resulting from the ZF and MMSE strategies will not affect the decision and lead to the same bit error rate.

It is clear now that a tremendous advantage of OFDM systems is the low complexity cost of the combined demodulation and equalization stages. As expected, the equalization scheme requires an estimation of the channel frequency response. This estimate is periodically obtained by arranging the transmitted symbols in frames containing known reference symbols. For a given channel bit-rate, this implies that the bit-rate available for the data is lowered by both the GI and the reference, or pilot, symbols. Usually, the system is tuned in such a way that it is a very small overhead. This corresponds to a slight loss of spectral efficiency and the effect of the GI insertion can be observed on the channel spectrum by noticing that it oscillates (while the regular system does not). This comes from the fact that the *sinc* shapes that perfectly overlap with full spectral efficiency are now "pushed away" proportionally to the ratio $(D+N)/N$.

This technique is very specific to the DFT-based OFDM systems. More generally, OFDM systems can be modeled by lossless perfect reconstruction transmultiplexers, as outlined in the previous section. In this case, the GI equalization scheme can no longer be used. Therefore, other channel equalization schemes must be used. Recently, it was shown that multirate signal processing theory can be used for designing more sophisticated precoder and post equalizer structures which are expected to find their use in the future [70, 186, 4].

2.4.2 A Unified Filter Bank Approach to the Guard Interval Insertion

To provide a general framework for modeling the adjunction of the cyclic prefix based on the polyphase formalism, first consider the general system in Figure 2.11. This figure represents a system equivalent to the one in Figure 2.10 in which the polyphase representation has been adopted. Note that the modulation matrix has been replaced by a general filtering matrix $F(z)$ (not necessary scalar) which can model modulators of length larger than the number of subbands (i.e. more selective filters).

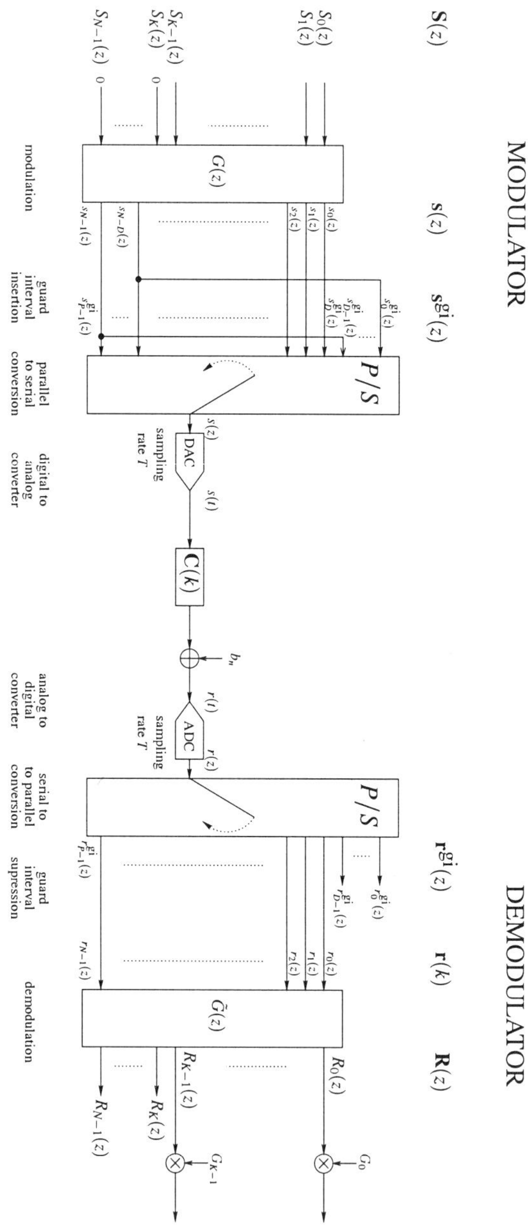

Figure 2.11 General OFDM transmission system

Let $H(z)$ be the $P \times N$ matrix defined by $H(z) = [G^{\mathrm{gi}}(z)^T, G(z)^T]^T$ where the $D \times N$ matrix $G^{\mathrm{gi}}(z)^T$ denotes the last D rows of $G(z)$. This modulator can be represented as illustrated in Figure 2.12 where matrix $H(z) = [G^{\mathrm{gi}}(z)^T, G(z)^T]^T$ simultaneously performs modulation and cyclic prefix adjunction. Another view of the system can also be derived using a classical filter bank. Indeed, the signal at the output of the "joint" modulator, $\mathbf{s}^{\mathrm{gi}}(k)$, can be expressed as

$$\mathbf{s}^{\mathrm{gi}}(z) = H(z)\mathbf{S}(z)$$

and the signal at the input of the digital analog converter becomes

$$\begin{aligned} s(z) &= \left(1, z^{-1}, \ldots, z^{-(P-1)}\right) \mathbf{s}^{\mathrm{gi}}(z^P) \\ &= \sum_{k=0}^{P-1} z^{-k} s_k^{\mathrm{gi}}(z^P) \\ &= \sum_{k=0}^{P-1} z^{-k} \left(H_0^k(z^P), \ldots, H_{N-1}^k(z^P)\right) \mathbf{S}(z^P) \\ &= \left(\sum_{k=0}^{P-1} z^{-k} H_0^k(z^P), \ldots, \sum_{k=0}^{P-1} z^{-k} H_{N-1}^k(z^P)\right) \mathbf{S}(z^P). \end{aligned} \tag{2.10}$$

The system can thus be depicted by the filter bank of Figure 2.12 where the oversampling rate is now P and the digital filters of the synthesis bank are defined as

$$H_m(z) = \sum_{l=0}^{P-1} z^{-l} H_m^l(z^P), \qquad 0 \leq m \leq N-1.$$

These schemes are known as oversampled filter-banks [57].

It has been reported recently [136] that it is possible to use non-trivial DFT modulated filter banks when the critical sampling (lossless) constraint is removed at a cost of spectral efficiency loss. The advantage is that the resulting prototype filters can be much more selective than in the critically sampled case. This reduces the overlap between bands and makes approximate, fast equalization schemes easy. However, it has been proven in this section that oversampling is exactly equivalent to inserting a cyclic prefix (guard interval). The selectivity improvement of such schemes is a result of the additional redundancy in conjunction with specific design criteria for the subband filters. In all cases, the anticipated equalization scheme inherently assumes the presence of a guard interval.

2.5 EXTENSIONS: CURRENT AND FUTURE RESEARCH

A general view of the previous sections already emphasizes the role of the selectivity of the filters and of the density of the mapping of the time-frequency

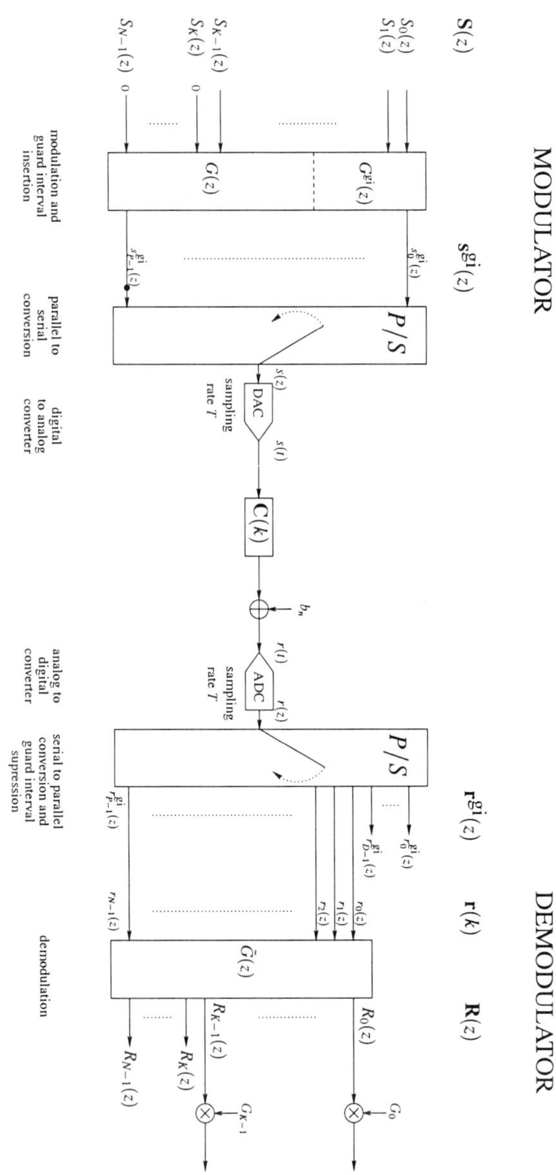

Figure 2.12 Polyphase representation of the OFDM system

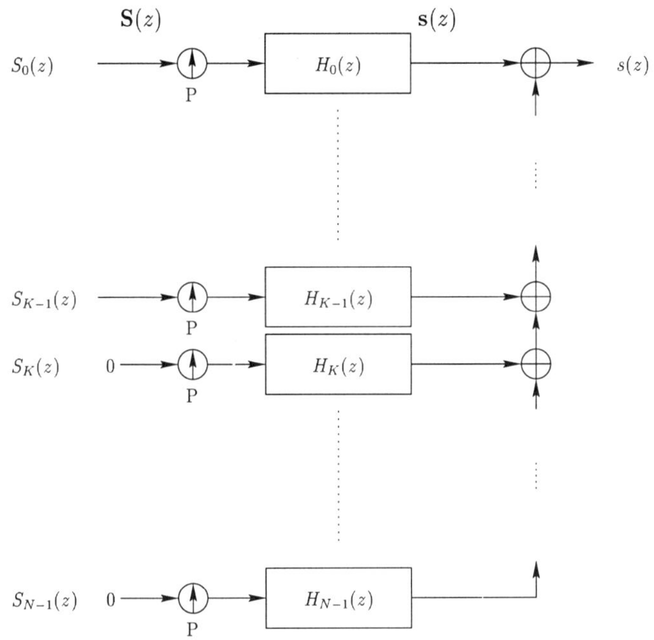

Figure 2.13 Filter bank representation of the OFDM system

plane performed by the filter bank. Many studies are being carried out along these lines. Previous results related to the cyclic prefix and the oversampled case can be understood as a way of increasing selectivity by allowing a smaller density of the mapping. The following paragraphs provide connections to other works related to this approach as well as to improved equalization strategies.

Carrier optimization. Investigating other modulation schemes enables improved channel separation and immunity to impulse noise. Thus, more recently, better frequency localized function sets, namely subband (wavelet) transforms, have been forwarded for OFDM applications [367, 192]. Note that in these schemes, the guard interval equalization trick does not apply since we cannot rely on the cyclic convolution mathematical property of the demodulator. Therefore, at the receiver, the orthogonality property between the subcarriers is destroyed via convolution with the channel.

If "wavelet-packet" based systems are used, however, the subcarriers can be precisely tuned according to the unevenness of a channel's power levels in the frequency domain — steep transitions would require narrow bands while constant amplitude zones could be handled within a single large subband yielding high efficiency bandwidth utilization. An allocation strategy could be based on the "waterfilling" procedure for maximum efficiency. Adaptive tiling of the time-frequency plane, depending on channel selectivity, can lead to unequal bandwidth subcarriers.

Other modulation basis. Another way of obtaining selective filters is to change the modulation scheme to a DCT. In this case, many solutions which allow long filters to be used as prototypes in a DCT-based OFDM modulation exist. Such solutions include the well-known modulated lapped transforms (MLT) or the extended lapped transforms (ELT) first proposed by Malvar [207].

The corresponding solutions would result in filters with improved approximations to the ideally shaped rectangular spectral window. Also note that the natural stacking of the channels corresponds to evenly stacked filter banks where linear phase solutions exist [193]. Other improved solutions include modified DFT filter banks [171] which map the DFT solution, oversampled by factor of two, to a critically sampled filter bank obtained by removing all the redundancies.

Another possibility is to relax the perfect reconstruction property, and to tolerate some amount of aliasing between subchannels (provided that it remains below the residual inter-symbol interference (ISI) introduced by the channel after equalization).

Oversampling. As explained below, DAC is easier when dealing with oversampled versions of the signal to be converted. Thus, another idea is to make full use of the inherent amount of oversampling (redundancy) present in OFDM systems. In that case, more selective synthesis banks than the classical rectangular time window modulated by the DFT can be found. Such a solution intrinsically results in some loss of spectral efficiency. Hence, a fair comparison should be made with systems sharing the same spectral efficiency, e.g. the ones using a guard interval. Such a comparison is detailed in [136]. Further improvements can also be obtained by considering jointly an Offset-QAM-type modulator and an oversampling factor of two. In this case, the full spectral efficiency is recovered. One such systems is known as the Isotropic Orthogonal Transform Algorithm (IOTA) modulation [186, 4] and is shown to be a good approximation of the densest frequency-time tiling known.

Equalization. The *guard-interval trick*, as explained earlier, is computationally efficient but clearly introduces some loss in spectral efficiency since part of the transmission time is not used for emitting useful data. This loss is on the order of the channel's memory length (given in number of samples) divided by the number of sub-carriers, D/N. This explains why in OFDM applications like DAB many carriers are used; that way the loss remains low ($\approx 25\%$). However, when the number of channels decreases, this technique naturally becomes less efficient.

Recent studies aimed at reducing the length of the ISI seen at the receiver and, therefore, also the redundancy introduced at the emitter, have yielded improved efficiency. In the context of ADSL communication systems, in order to increase the available bit-rate, it has been proposed in [39] that the duration of the GI be lowered, thus introducing some residual ISI. This ISI, coming from the difference in size between the channel memory and the GI length, is reduced by an adaptive filter with a small number of taps placed in front of the demodulator. Its function is to "confine" the channel impulse response so that the combined channel-equalizer response is seen as a smaller size by the receiver.

Due to the presence of a guard time at the emitter in classical DFT OFDM systems, it seems also that the strategy of artificially inducing cyclostationarity (transmitter induced cyclostationarity [330]) is particularly well suited to OFDM systems. Several works have been undertaken along these ideas and published around the same time [68, 106]. The resulting equalization methods based on the cyclocorrelations of the received signal provide robust estimation/identification of the channel impulse response without any constraints on its zeros.

Another research challenge is to find efficient and simple equalization strategies for the filtered OFDM case where the guard interval trick is not valid. Recent results have characterized how in the specific case of an OFDM system an oversampling of the received signal introduces some cyclostationarity and have derived a blind equalization method exploiting this property [69, 70]. In this case, both the guard interval and the pilot symbols required for estimating the frequency domain channel impulse response are useless and can be suppressed, leading to a higher useful bit-rate. Simpler algorithms for this approach are the focus of current and future studies.

2.6 CONCLUSION

This chapter has explained the current status of OFDM systems as used in digital broadcasting and more specifically in the DAB application. By proposing a discrete modeling of these systems based on the polyphase filter bank formalism, the flexibility brought by the multicarrier concept has been shown to allow many different tunings. This freedom is intrinsically linked to the way multicarrier systems transmit symbols in the time-frequency plane and to the selectivity of the (de-)modulator filters, allowing the use of simple equalization schemes.

Such flexibility makes OFDM a good candidate for many future broadband systems providing multimedia services to mobile and portable terminals. Though this technology is currently being deployed in Europe for terrestrial digital broadcasting (DVB, DAB, and Digital Radio Mondiale (DRM): dealing with digital broadcasting in the AM bands), OFDM is also used in the United States for high bit-rate point-to-point communications over twisted pairs (xDSL) and has been recently adopted for future broadband wireless radio local area networks (e.g. in the ETSI Broadband Radio Access Network: BRAN standardization effort for the air inteface of HIPERLAN II systems).

3 INTERFERENCE EXCISION IN SPREAD SPECTRUM COMMUNICATIONS

Michael J. Medley[1], Mehmet V. Tazebay[2] and Gary J. Saulnier[3]

[1]Air Force Research Laboratory
Rome, NY
medleym@rl.af.mil

[2]Sarnoff Corporation
Princeton, NJ
mvt@sarnoff.com

[3]Rensselaer Polytechnic Institute
Troy, NY
saulng@rpi.edu

3.1 SPREAD SPECTRUM SIGNALING

Since its inception circa the mid-1940's, the term *spread spectrum* (SS) has been used to characterize a class of digital modulation techniques for which the transmitted signal bandwidth is much greater than the minimum bandwidth required to fully represent the information being sent. Despite what might seem to be an inefficient utilization of resources, the combined processes of "spreading" and "despreading" the information bearing signal offers potential improvement in communications capability that more than offsets the cost incurred in using additional bandwidth. Indeed, SS offers such benefits as

- interference rejection
- low power spectral density
- resistance to multipath fading
- selective addressing capability for multiple access communications

Interference rejection rather broadly describes the SS system's ability to operate in a congested or corrupted environment, one that would most likely compromise the utility of conventional narrowband modulation techniques. This characteristic, combined with low power spectral density, have fostered military interest and investment in SS technology since World War II. In defense applications, the low power spectral density associated with SS waveforms is often exploited to support low probability of intercept and detection (LPI/D) communications. Since such capabilities can be utilized to meet user demands for transmission privacy and reliability over jammed channels, SS signaling represents a natural communications paradigm capable of meeting stringent military requirements for robust, covert communications.

The need for a means to combat multipath fading in mobile communications has been a primary catalyst for transitioning interest in SS techniques into commercial markets and applications. Robustness to multipath is realized as a result of the SS waveform's similarity to white noise, namely the autocorrelation of the spread waveform closely approximates an impulse function. As a result, multiple time-delayed replicas of the original signal plus noise can be resolved and coherently combined at the receiver to effectively raise the input signal-to-noise power ratio (SNR). This noiselike quality of the spread signal also facilitates the design and implementation of multi-user/multiple-access communications systems in which each user is assigned a unique signature code and allowed to transmit simultaneously. At the receiver, the desired signal is extracted from the composite sum of all the users' data plus noise through correlation with the appropriate signature sequence. Such an approach essentially delineates the underpinnings of code-division multiple-access (CDMA) systems in use today.

3.1.1 Spreading the Spectrum

In practice, signal spreading and despreading are accomplished using a data-independent pseudo-random, or pseudo-noise (PN), sequence. Maximal sequences (m-sequences), which consist of a series of 0's and 1's, are often used as spreading codes because of their ease of generation and good randomness properties. As an example, Figure 3.1 illustrates a simple length seven m-sequence and its autocorrelation function; specific details regarding the origin and implementation of such sequences are left to other texts and references [50, 77, 251, 295, 381]. As shown in this figure, the spreading sequence, $\{1,1,1,0,0,1,0\}$, converted to $\{1,1,1,-1,-1,1,-1\}$ for transmission, produces the following cyclic autocorrelation response

$$R[k] = \begin{cases} +L & k = nL \\ -1 & \text{otherwise} \end{cases} \tag{3.1}$$

where n is any integer. Here, L denotes the number of samples, or *chips*, in the spreading code. Each chip has a duration of T_c seconds. Hence, the rate of the PN sequence, called the *chip rate*, is $R_c = \frac{1}{T_c}$ chips/sec. Note that (3.1) is valid for all m-sequences independent of the value of L.

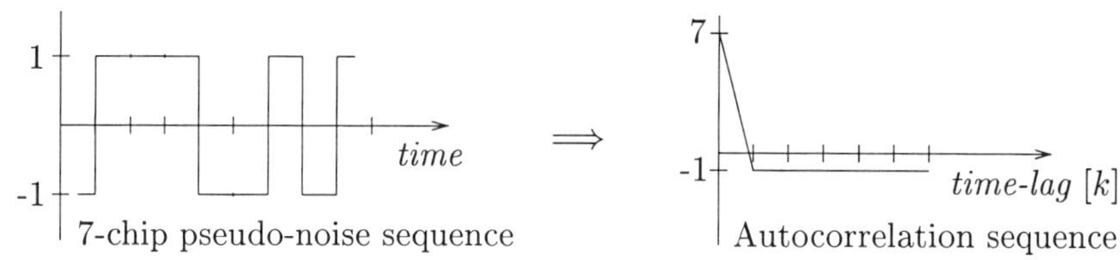

Figure 3.1 Length $L = 7$ m-sequence and corresponding cyclic autocorrelation response.

In direct-sequence (DS) spread spectrum, spreading is accomplished by multiplying the input data bits (± 1's) by the m-sequence as shown in Figure 3.2, producing a high rate sequence of ± 1's. Demodulation, or despreading, which is also depicted in this figure, is performed by correlating the received data sequence with the known spreading code and sampling at the appropriate instance in time. In practice, the bit duration, T_b, is typically much greater than T_c, i.e. $T_b \gg T_c$. Consequently, the chip rate is often orders of magnitude larger than the original bit rate $R_b = \frac{1}{T_b}$, thus necessitating the increase, or *spread*, in transmission bandwidth. The frequency response of the spread waveform has a $\sin(x)/x$ shape with main lobe bandwidth of $2R_c$. The magnitude-squared frequency response of such a waveform is depicted in Figure 3.3; pulse shaping is often used to suppress the sidelobes and effectively reduce SS bandwidth when necessary.

Alternatives to DS modulation include frequency-hop (FH), time-hop (TH) and related hybrid combinations. In FH-SS, a broadband signal is created by shifting the carrier of a narrowband waveform over a wide range of frequencies in a pseudo-random manner. In this case, the pattern through which frequencies are hopped is determined by a pseudo-noise sequence known (ideally) only to the transmitter and approved receivers. TH-SS is essentially the *dual* of FH-SS in that the transmitted waveform consists of short data pulses pseudo-randomly shifted in time. Although these techniques effectively produce wideband signals, the instantaneous bandwidth of FH-SS signals and the instantaneous time-spread of TH-SS waveforms are relatively small compared to the total spread bandwidth and bit duration, respectively. As a result, FH-SS is potentially vulnerable to powerful narrowband interference as is TH-SS to impulsive noise. To address these concerns, hybrid modulation techniques incorporating a combination of DS, FH and/or TH are frequently utilized.

3.1.2 Processing Gain and Jamming Margin

Regardless of which spreading mechanism is employed, the SS receiver must know the spreading sequence used at the transmitter in order to despread and recover the data. At the receiver, correlating (DS) or de-hopping (FH/TH) the received signal using a locally generated copy of the spreading sequence simultaneously collapses the spread data signal back to its original bandwidth

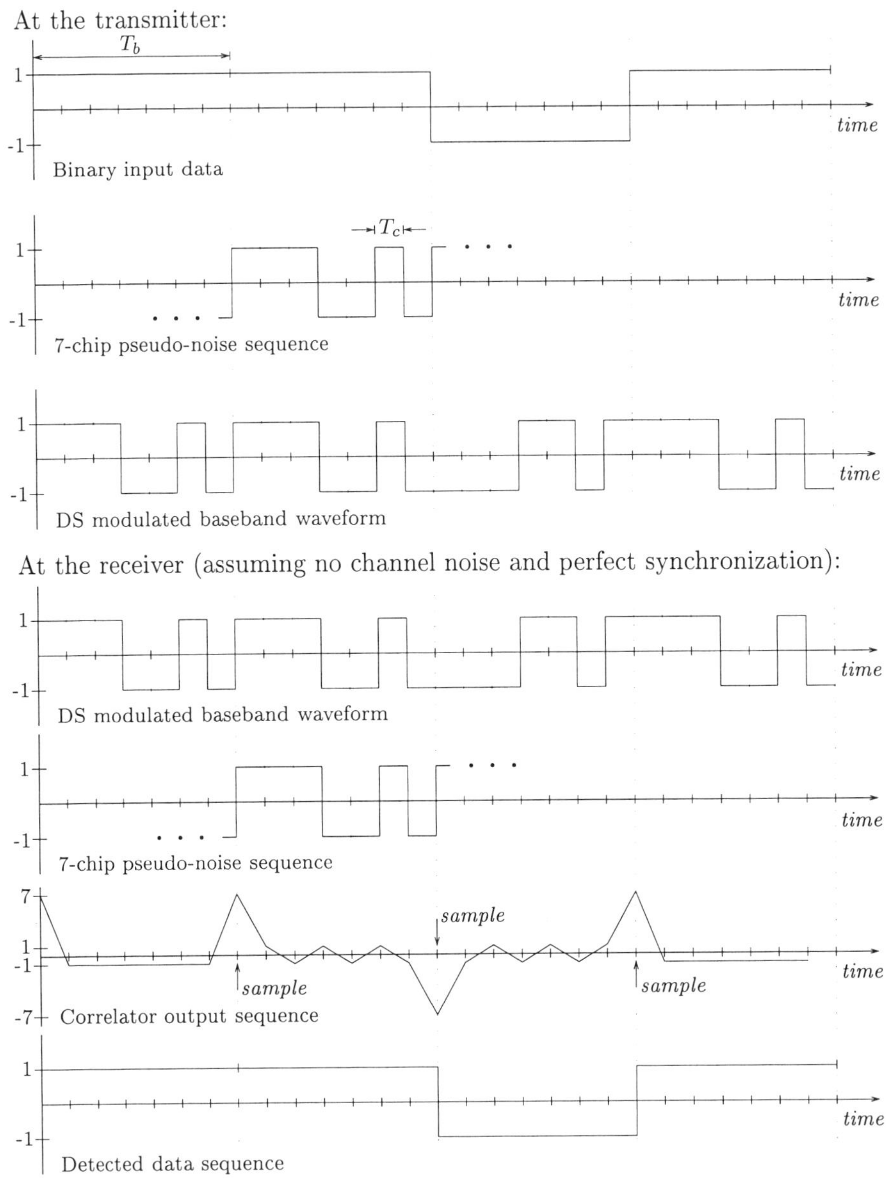

Figure 3.2 Direct-sequence spread spectrum modulation and demodulation.

while spreading any additive noise or interference to the full SS bandwidth or greater. A low-pass filter with bandwidth matched to that of the original data is subsequently used to recover the data and reject a large fraction of the spread interference energy. The ratio of the signal-to-noise ratio after despreading, SNR_o, to the input signal-to-noise ratio, SNR_i, is defined as the *processing*

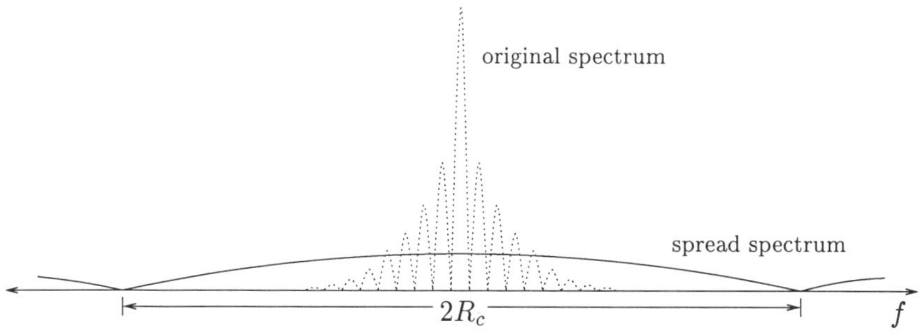

Figure 3.3 Magnitude-squared frequency response of a DS-SS waveform.

gain, G_p, i.e.

$$G_p \triangleq \frac{SNR_o}{SNR_i}.$$

Note that in both SNR_i and SNR_o the noise term implicitly denotes the sum of additive white Gaussian noise (AWGN) *plus* any additional interference. Given an input data rate of R_b bits/sec, G_p can be approximated in DS-SS systems by the ratio of the chip rate to the data rate,

$$G_p \approx \frac{R_c}{R_b} = \frac{T_b}{T_c} = N$$

where N corresponds to the number of chips per spread data bit; $N = L$ when individual data bits are modulated by the entire spreading sequence. In essence, G_p roughly gauges the improvement in the anti-jam capability and LPI/D quality associated with SS signaling.

System performance is ultimately a function of SNR_o, which determines the bit-error-rate (BER) experienced by the communications link. For a given data rate, spreading the transmitted signal energy over a larger bandwidth allows the receiver to operate at a lower value of SNR_i. The range of SNR_i for which the receiver can provide acceptable performance is determined by the *jamming margin*, M_J, which is expressed in decibels (dB) as

$$M_J = G_p - [SNR_{o_{min}} + L_{sys}] \tag{3.2}$$

where $SNR_{o_{min}}$ is the minimum SNR_o required to support the maximum allowable BER and L_{sys} accounts for any losses due to receiver implementation. Whereas G_p conveys a general idea of the effectiveness of SS, M_J represents an even more useful metric to system designers indicating how much interference can be tolerated while still maintaining a prescribed level of reliability.

The jamming margin represents a fundamental limit on the robustness of any SS system with respect to interference and is independent of the time-frequency distribution of the jammer energy within the DS-SS signal's bandwidth [251].

In other words, provided that the jammer energy is evenly distributed over the symbols in time, M_J is a function only of the jammer power (as reflected in SNR_o) and remains constant regardless of whether the jammer in question is characterized as a single-tone, an impulse in time, white noise or anything else. If the power of the interfering signal is below the jamming margin, the SS receiver can operate as desired without assistance. Interference power in excess of the limit set by (3.2), however, significantly degrades system performance and inevitably renders simple SS receivers useless unless auxiliary means of interference mitigation are employed.

As previously mentioned, when the jammer power exceeds M_J, narrowband and impulsive interference pose particularly effective threats to FH-SS and TH-SS system reliability, respectively. Although such interference also degrades DS-SS performance, the fact that DS modulation spreads data bit energy uniformly in *both* time and frequency enables and encourages the use of transform domain processing. Since such processing is the primary focus of this chapter, only DS-SS systems are considered henceforth.

3.1.3 Interference Excision

Despite the numerous benefits and applications of SS signaling touched upon earlier and comprehensively documented in the literature, the primary interest in SS as regarding this chapter relates to its capacity for interference rejection. As is evident from (3.2), any level of interference protection can be obtained by designing the signal with sufficient processing gain. The price of greater protection, however, is an increase in the bandwidth of the transmitted signal for a given data bandwidth. Since practical considerations such as transmitter/receiver complexity and available frequency spectrum limit reasonably attainable processing gains, it is often necessary to employ signal processing techniques which augment anti-jam capability without increasing bandwidth.

In general, interference rejection schemes exploit structural differences between the spread waveform and the interference and work to suppress the interference [220]. Processing can be performed in the time domain (e.g. adaptive transversal filtering) [138, 140, 173, 281] the spatial domain (e.g. adaptive array antennas) [224], or the transform domain [105, 217, 221, 222, 272, 322]. In keeping with the theme of this text, the focus in this chapter is limited to transform domain signal processing algorithms, in particular, transform domain excision.

The fundamental objective of transform domain excision is to represent the received time domain waveform, which is assumed to consist of the spread data, interference and AWGN, in another domain wherein the desired signal and the interference are readily distinguished. Under ideal conditions, the transform domain representation of the interference appears as an impulse function while the pseudo-noise spread data and AWGN appear relatively "flat," i.e. their energies are uniformly spread throughout the spectrum. The portion of the received signal's spectrum deemed to be "jammed" is then identified and eliminated through excision without significant loss of the desired signal energy. Figure 3.4 illustrates this process although not for the ideal case; note that in this

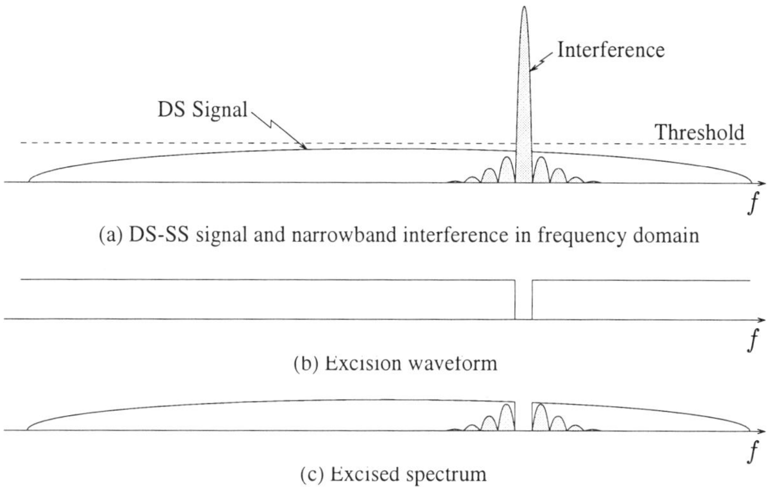

Figure 3.4 An illustration of the transform domain excision process.

figure, the relationship between the interference main lobe and adjacent sidelobes are exaggerated for illustration purposes only. The remaining transform domain components are subsequently inverse transformed, or re-synthesized, to produce the nearly interference-free desired signal [221].

From the preceding description and Figure 3.4 it is clear that in the transform domain an exciser essentially acts as a gating function, setting all spectral components with energy greater than an external threshold to zero while allowing those remaining to pass undistorted. Due to this binary nature, the excision process typically yields poorer BER performance than more sophisticated transform domain Wiener filtering schemes [215] unless all of the interference energy is contained in a sufficiently small number of transform domain bins. Nevertheless, in many cases, the simplicity of the receiver structure and the ability of the exciser to react rapidly to changes in the interference make it a prime choice in many narrowband interference suppression applications [62, 76, 104, 137, 265].

Block processing, as is typically performed in transform domain processing applications, inherently necessitates the partitioning of the input data sequence into finite-length intervals suitable for successive manipulation. Such segmentation is often accomplished through the use of windowing functions [243] which, unfortunately, introduce undesired spectral sidelobes. These sidelobes alias interference energy in frequency and thus hinder the exciser's ability to discriminate between the desired signal and the interference.

Sidelobe energy levels are directly related to the windowing function and can be reduced by increasing, if allowed, the window's main lobe bandwidth [243]. Given a fixed block length, this relationship results in a fundamental trade-off between efficient transform domain signal representation and spectral resolution. When the main lobe bandwidth is large, sidelobe energy levels are reduced at the expense of the transform's ability to provide adequate frequency resolution. Consequently, when the interference bandwidth is much less than that of the main lobe, the excision process often removes much more of the desired signal energy than necessary. On the other hand, when the main lobe bandwidth is small, narrowband jammers are more readily resolved but the sidelobes tend to be large, containing greater amounts of aliased interference energy.

Although non-rectangular windows produce smaller sidelobes, they increase computational complexity due to the use of overlapping segments of the input signal (as required for accurate reconstruction of the time domain waveform) [258]. To maintain consistency and equity with respect to the different transformation techniques under consideration in this chapter, the windowing functions used henceforth are assumed to be rectangular.

Regardless of windowing effects, the transformation technique, whether implemented via subband transform, analysis filter bank or some other form of signal decomposition, significantly impacts how well the spread data signal and the interfering waveform are resolved in the transform domain. In particular, its selection unequivocally establishes the relationship between the amount of interference energy and desired signal energy removed and thus the overall system BER. Ultimately, the goal of excision is to optimize the ratio of excised jammer energy to excised data signal energy such that the BER is minimized. In practice, however, suboptimal performance is typically conceded due to the complexity involved with determining the optimal transformation technique, which varies in accordance with changes in the interference, as well as the difficulty in establishing the excision protocol in real-time. Although tractable excision models are often employed to facilitate mathematical analysis, minimum BER performance is not guaranteed by maximizing the excised jammer energy to excised signal energy ratio. In practice, analytical results are experimentally verified and validated through numerical simulation and hardware emulation.

3.1.4 Chapter Overview

As alluded to earlier, this chapter is dedicated to the review and exposition of transform domain excision algorithms based on subband transforms and adaptive hierarchical filter banks. In it, recently developed transform domain approaches to interference rejection are compared to conventional excision techniques and are evaluated on the basis of BER performance and robustness with respect to jammer frequency, power and bandwidth. Since the study of optimal excision protocols is essentially independent of the underlying transform, its examination is omitted.

In the following section, conventional transform domain excision BER performance is analyzed using generic block transforms in narrowband interference environments. Results of this analysis are postponed until Section 3.3 wherein they are compared and contrasted with those obtained through the use of lapped transforms. Section 3.3 reveals the signal processing structures as well as the corresponding performance results associated with perfect reconstruction cosine-modulated filter banks. In Section 3.4, adaptive hierarchical filter banks are shown to provide improved robustness over conventional block transform domain excision approaches with respect to time and frequency domain interferers. An assessment of these state-of-the-art excision techniques are summarily expressed Section 3.5.

3.2 BLOCK TRANSFORM DOMAIN EXCISION

In 1980, transform domain filtering using real-time Fourier transforms and surface acoustic wave devices was originally introduced as a means of suppressing narrowband interference in continuous-time spread spectrum receivers [221]. Further development of this work eventually led to discrete-time transform domain excision algorithms based on the FFT [105, 222, 272]. Since then, advancements in digital signal processor technology have encouraged the use and refinement of excision algorithms that are compatible with commercially available chip sets and digital signal processing (DSP) hardware. Due in part to the ubiquitous popularity of the FFT, this relationship between algorithm development and implementation using state-of-the-art technology has resulted in the preponderance of excision algorithms based on the FFT. Consequently, for several years, excision algorithms have typically, almost implicitly, relied on the use of the FFT and have varied only in the type of windowing function employed, transform size, and the overall system processing gain.

In contrast to the analysis presented in [221], many analytical approaches invoke assumptions regarding the despread signal's statistical characteristics in an effort to simplify analysis. In particular, performance analysis is often facilitated by assuming that the despread narrowband interference is Gaussian distributed with a relatively flat spectrum, thus, making it essentially equivalent to an AWGN source [173, 256, 381]. Although this approximation works well in the presence of narrowband Gaussian waveforms, it is not necessarily appropriate when the narrowband interferer is characterized as a single-tone sinusoid.

Here, as in [221], the despread narrowband interference is not assumed to be Gaussian distributed. Instead, BER performance is analyzed for discrete-time systems as an explicit function of single-tone interference and is then generalized for a broader class of narrowband signals using the Gaussian approximation. Analytical and simulation results are presented in subsequent sections. The performance of conventional block transform domain excision algorithms serves as a benchmark to which lapped transform and hierarchical filter bank techniques are compared.

3.2.1 Binary Hypothesis Testing

In the following sections, transform domain filtering techniques are used to improve the detection of spread spectrum signals in the presence of narrowband interference and AWGN. Before proceeding with the analysis, however, it is worthwhile to briefly review the fundamental principles of detection theory derived from the binary hypothesis test [340]. For clarity, the modulation and demodulation operations between the transmitter and receiver are assumed to be transparent. Accordingly, the receiver has perfect knowledge of the carrier frequency and phase and thus processes real data. With the understanding that later sections are dedicated to the analysis of BER performance in the presence of narrowband interference, a basic review of the binary hypothesis test for real data vectors in Gaussian noise is presented here. Although this discussion is limited to binary signaling, these results can be straight-forwardly extended for higher level signaling formats.

Assuming antipodal signaling with the data bit energy normalized to unity, the received Gaussian random vector is given by

$$\underline{r} = d[n]\underline{c} + \underline{\eta},$$

where $d[n]$ is the transmitted data bit at time n, $\underline{c}$ is the spreading code, and $\underline{\eta}$ denotes a Gaussian random vector generated from samples of a zero-mean Gaussian random process with variance σ_η^2. The two hypotheses under consideration are thus given as

$$H_0 : \begin{cases} d[n] = -1 \\ \underline{r} = +\underline{c} + \underline{\eta} \end{cases}$$
$$H_1 : \begin{cases} d[n] = +1 \\ \underline{r} = -\underline{c} + \underline{\eta}. \end{cases}$$

Consistent with the assumption of antipodal signaling, the above hypotheses differ only in the polarity of the current data bit.

In practice, data bit decisions are typically based on the polarity of the correlation between the received signal and the reference spreading code. The necessary test statistic is thus obtained by sampling the output of the correlator or matched filter, namely

$$\xi = \underline{c}^T \underline{r}.$$

Given ξ, the data bit decision rule becomes

$$\begin{array}{lll} \text{choose} & H_0 \text{ if } & \xi < 0 \\ \text{choose} & H_1 \text{ if } & \xi > 0. \end{array} \tag{3.3}$$

Accordingly, a decision of H_0 yields the data bit estimate, $\hat{d}[n] = -1$, while choosing H_1 produces $\hat{d}[n] = +1$. Denoting the mean and variance of ξ as μ_ξ and σ_ξ^2, respectively, the corresponding probability of bit error can be expressed as

$$P_e = Q\left(\frac{\mu_\xi}{\sigma_\xi}\right) \tag{3.4}$$

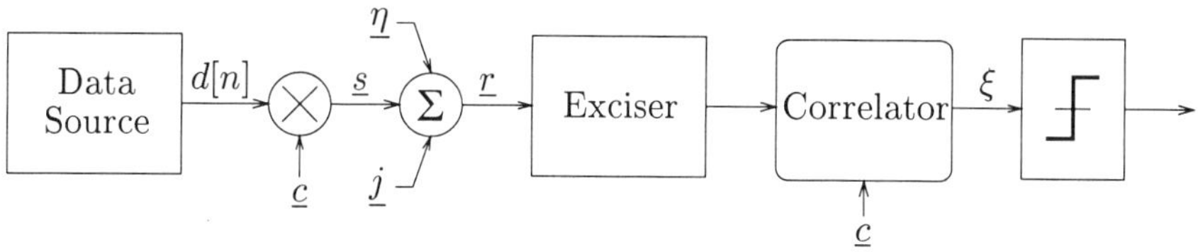

Figure 3.5 Block diagram of a DS-SS communication system.

where

$$Q(x) = \frac{1}{\sqrt{2\pi}} \int_x^{\infty} e^{-(y^2/2)} dy$$

is the Gaussian tail probability or Q-function. Note that the binary hypothesis test, as discussed above, is often referred to as the general Gaussian problem and has been thoroughly treated in the literature [340].

3.2.2 DS-SS Communication System Model

Figure 3.5 illustrates a rudimentary DS-SS communication system corrupted by AWGN and interference. As discussed earlier, the transmitted data corresponds to data bit samples modulated by a length L spreading sequence. Here it is assumed that each data bit is modulated by a single full-length PN code sampled once per chip, i.e. $N = L$. Thus, each sample of the spread data bit may be expressed as

$$s_i = \sqrt{P_s} d[n] c_i \qquad\qquad i = 0, 1, \ldots, N-1 \tag{3.5}$$

where $\sqrt{P_s}$ is the signal power at the receiver input and $d[n] \in \{+1, -1\}$ and $c_i \in \{+\frac{1}{\sqrt{N}}, -\frac{1}{\sqrt{N}}\}$ for $i = 0, 1, \ldots, N-1$ represent the random binary data bit sequence and PN code samples, respectively. Note that the value of $d[n]$ is assumed constant over the bit duration. Therefore, if a positive (negative) data bit is sent, $d[n] = +1$ (-1) for $i = 0, 1, \ldots, N-1$. Without loss of generality, the signal power is henceforth normalized to unity such that (3.5) becomes $s_i = d[n]c_i$ for $i = 0, 1, \ldots, N-1$.

In the channel, the thermal noise samples, η_i, represent AWGN samples with two-sided power spectral density $N_0/2$. The interference samples are generated from a single-tone interferer, represented as $j_i = A_j \cos[\delta\omega i + \theta]$, where A_j is a constant denoting amplitude, $\delta\omega$ is the offset from the carrier frequency and θ is a random phase uniformly distributed in the interval $[0, 2\pi)$. Accordingly,

$$r_i = d[n]c_i + A_j \cos[\delta\omega i + \theta] + \eta_i \tag{3.6}$$

with $i = 0, 1, \ldots, N-1$.

At the receiver, the input signal is partitioned into disjoint length-N data segments corresponding to individual data bits. Thus, in accordance with (3.6),

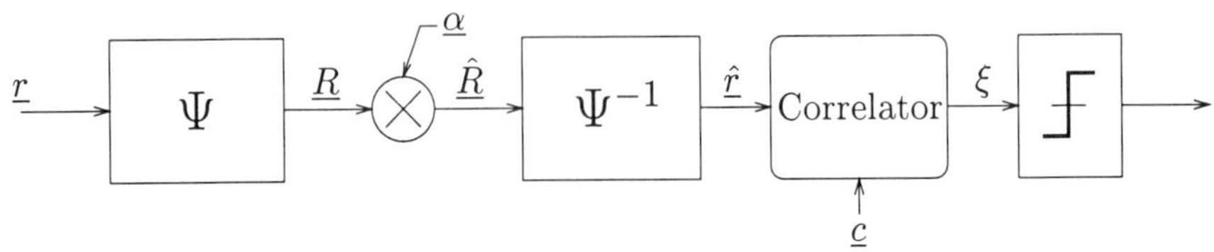

Figure 3.6 Discrete-time receiver employing transform domain filtering.

the $N \times 1$ received input vector consists of the sum of N samples from the message bit with those from the additive noise and interference, expressed vectorially as

$$\underline{r} = \underline{s} + \underline{j} + \underline{\eta}, \tag{3.7}$$

where $\underline{r} = [r_0 \; r_1 \; \cdots \; r_{N-1}]^T$. This waveform is processed by the exciser and subsequently input to a detection device, a hard limiter in Figure 3.5, for data bit decisions. Note that under the assumptions of one sample per chip and $N = L$, each input vector represents a single data bit spread by N chips; accordingly, processing is performed on a bit-by-bit basis.

3.2.3 Analyzing the Excision Process

Transform domain excision is typically performed as shown in Figure 3.6. Here, blocks Ψ and Ψ^{-1} represent the forward and inverse $N \times N$ block matrices associated with the N-point transform. The block transform domain coefficients generated from the input vector, $\underline{r}$, are expressed as

$$\underline{R} = \Psi \underline{r}. \tag{3.8}$$

In accordance with the properties of unitary block matrices, specifically $\Psi^{-1} = \Psi^{\dagger}$, the associated inverse transform operation is given by

$$\underline{r} = \Psi^{\dagger} \underline{R}. \tag{3.9}$$

Note that when strictly real block transforms are used, the † notation, which denotes complex conjugate transposition, can be replaced with the simple transpose operation, denoted by ^{T}.

The elements of the excision vector, $\underline{\alpha}$, are limited to values from the binary set $\{0, 1\}$ and thus determine which spectral coefficients are removed and which are passed without modification. Accordingly, the excised spectral coefficients are given by

$$\hat{\underline{R}} = \text{diag}(\underline{\alpha})\underline{R} = \begin{bmatrix} \alpha_0 R_0 \\ \alpha_1 R_1 \\ \vdots \\ \alpha_{N-1} R_{N-1} \end{bmatrix},$$

where diag $(\cdot)$ denotes an $N \times N$ matrix with diagonal elements corresponding to the components of its $N \times 1$ argument. Combining this expression with (3.9) and (3.8) yields

$$\underline{\hat{r}} = \Psi^{\dagger}\text{diag}(\underline{\alpha})\Psi\underline{r} \tag{3.10}$$

which is the re-synthesized, or filtered, time domain exciser output.

Assuming synchronization, the bit decision variable can be obtained after excision by correlating $\underline{\hat{r}}$ with the reference spreading code, yielding

$$\xi = \underline{c}^T\underline{\hat{r}}. \tag{3.11}$$

Since the system is linear, $\underline{\hat{r}} = \underline{\hat{s}} + \underline{\hat{j}} + \underline{\hat{\eta}}$. Considering that ξ corresponds to the correlation between the filtered signal and the reference waveform, its polarity indicates the value of the transmitted data bit. In practice, ξ is typically put through a threshold device with the decision boundary set to zero to determine the bit decision.

Although not a necessity, it has been assumed that the length, N, of the data blocks is equal to the order of the transform as well as the number of chips in the spreading code. Clearly, for many systems the length of the spreading code may not be equal to the dimensionality of the block or subband transform. In cases where the transform dimensionality exceeds the data bit length, it is often sufficient to zero-pad the received data bit vector to the appropriate length or, if possible, use augmented spreading codes. If, on the other hand, the data bit duration is greater than N, each bit can be processed N samples at a time. Under these conditions, overlap algorithms may be necessary to reconstruct the filtered data signal. Throughout this chapter, it is assumed that the former condition exists, i.e. the spreading code length is equal to that of the data block, with the realization that the subsequent analysis can also be applied to longer spreading codes when the appropriate reconstruction algorithms are utilized.

In accordance with (3.10), $\underline{\hat{j}} = \Psi^{\dagger}\text{diag}(\underline{\alpha})\Psi\underline{j}$. Due to the inherent dependency of $\underline{j}$ on phase, ξ is also a function of the uniformly distributed random variable θ. Thus, in order to evaluate the corresponding BER, as originally expressed in (3.4), the probability of bit error must first be rewritten in terms of the bit error probability conditioned on phase, $P_{e|\theta}$, as

$$P_e = \frac{1}{2\pi}\int_0^{2\pi} P_{e|\theta}\,d\theta \tag{3.12}$$

where

$$P_{e|\theta} = Q\left(\frac{\mu_{\xi|\theta}}{\sigma_{\xi|\theta}}\right) \tag{3.13}$$

and $\mu_{\xi|\theta}$ denotes the expectation of ξ conditioned on phase, $E\{\xi \mid \theta\}$.

By determining the conditional mean and variance of ξ, the BER expressions of (3.13) and (3.12) can be evaluated for arbitrary $\underline{\hat{r}}$. Since $\underline{\hat{r}}$ depends explicitly on the transform or filter bank used, evaluation of these expressions directly relates the receiver BER to the analysis/synthesis techniques employed.

3.2.4 Decision Variable Statistics

Given that the jammer phase is unknown, the conditional mean of ξ must be determined for a fixed value of θ. Assuming that the transform domain exciser coefficients and transformation matrix are fixed, the conditional expectation of (3.11) can be expressed as [216]

$$\mu_{\xi|\theta} = \underline{c}^T \Psi^\dagger \mathrm{diag}(\underline{\alpha}) \left(\Psi \underline{c} + \Psi \underline{j}_\theta \right), \tag{3.14}$$

where $\underline{j}_\theta = E\{\underline{j}|\theta\}$ reflects the conditioning on phase. Assuming that the two-sided power spectral density of the zero-mean noise term is $N_0/2$, the conditional variance of ξ is given by

$$\sigma^2_{\xi|\theta} = \frac{N_0}{2} \underline{c}^T \Psi^\dagger \mathrm{diag}(\underline{\alpha}) \mathrm{diag}(\underline{\alpha}) \Psi \underline{c}. \tag{3.15}$$

Note that since this expression is independent of θ, $\sigma^2_{\xi|\theta} = \sigma^2_\xi$. Inserting these quantities into (3.13) and (3.12), one can evaluate the BER performance of the receiver depicted in Figure 3.6 for an arbitrary set of basis functions and/or spreading codes.

3.2.5 The Gaussian Interference Approximation

The analysis presented thus far has focused specifically on receiver performance in the presence of single-tone interference. As alluded to previously, many studies of spread spectrum receiver performance in narrowband interference environments approximate the despread interference as a Gaussian distributed random variable. Likewise, in this section the received vector statistics are not conditioned on the jammer phase. Instead, the narrowband interference, whether single-tone or bandpass, is despread and approximated as an additive white Gaussian noise process. Since the resulting correlator output is a simple function of the desired signal and Gaussian noise, the probability of bit error as expressed in (3.4) determines the overall BER.

As before, the decision variable, ξ, is given in (3.11). In this case, however, the despread jammer energy is treated as zero-mean Gaussian noise, thus it no longer contributes to the mean, μ_ξ. Accordingly, the modified expression for the expected value of ξ becomes

$$\mu_\xi = \underline{c}^T \Psi^\dagger \mathrm{diag}(\underline{\alpha}) \Psi \underline{c}. \tag{3.16}$$

In the absence of phase conditioning, the modified variance must take into account contributions of noise from both the AWGN and the interference. Denoting the autocovariance matrix of the jammer as

$$\mathbf{K}_{\underline{jj}} = \begin{bmatrix} R(0) & R(1) & R(2) & \cdots & R(N-1) \\ R(1) & R(0) & R(1) & \cdots & R(N-2) \\ R(2) & R(1) & R(0) & \cdots & R(N-3) \\ \vdots & \vdots & \vdots & \ddots & \vdots \\ R(N-1) & R(N-2) & R(N-3) & \cdots & R(0) \end{bmatrix} \tag{3.17}$$

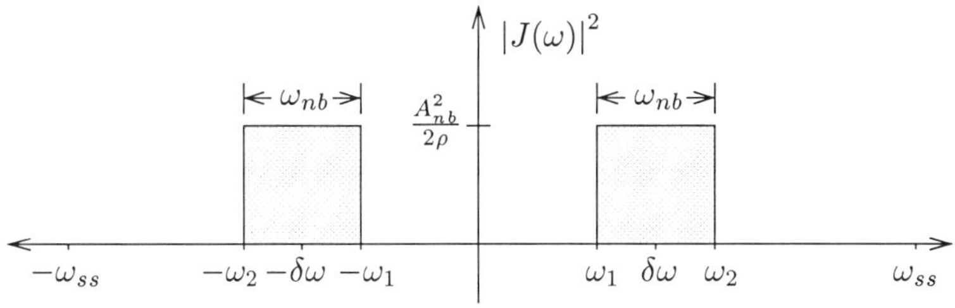

Figure 3.7 Narrowband Gaussian interference power spectral density.

with autocorrelation components

$$R(k) = E\{j[n]j[n+k]\},$$

the variance of ξ is given by

$$\sigma_\xi^2 = \underline{c}^T \Psi^\dagger \text{diag}(\underline{\alpha}) \Psi \left(\frac{N_0}{2} \mathbf{I} + \mathbf{K}_{\underline{jj}} \right) \Psi^\dagger \text{diag}(\underline{\alpha}) \Psi \underline{c} \tag{3.18}$$

where $\mathbf{I}$ is the $N \times N$ identity matrix. Notice that this expression is identical to (3.15) when $\mathbf{K}_{\underline{jj}}$ is equal to the null matrix. Insertion of (3.16) and (3.18) into (3.4) yields the associated bit error probability.

Single-Tone Interference. If the narrowband interference is characterized as the real single-tone jammer,

$$j[n] = A_j \cos[\delta\omega n + \theta],$$

whose power is defined relative to that of the desired signal by the jammer-to-signal ratio (JSR),

$$JSR \triangleq \frac{A_j^2}{2P_s},$$

the autocorrelation sequence used in (3.17) is given by

$$R(k) = \frac{A_j^2}{2} \cos(\delta\omega k).$$

The use of these functions in determining (3.16) and (3.18) is typically sufficient provided that the Gaussian interference approximation models the single-tone interference with acceptable accuracy. When more precise analysis is required, (3.14) and (3.15) must be used.

Narrowband Gaussian Interference. In many applications, the narrowband interference can be approximated by an equivalent narrowband Gaussian noise source, an example of which is shown in Figure 3.7. For a given amount of interference power, the narrowband Gaussian interferer affects a larger frequency range than a single-tone jammer but with lower power spectral density. Denoting the percentage of spread signal bandwidth jammed as ρ and the total interference power as A_{nb}^2, the two-sided power spectral density of $j[n]$ is simply $\frac{A_{nb}^2}{2\rho}$. The corresponding JSR is thus given by

$$JSR = \frac{A_{nb}^2}{P_s}.$$

In this case, $\mathbf{K}_{\underline{jj}}$, can be expressed as in (3.17) but with elements derived from the narrowband Gaussian source. With the spread spectrum signal bandwidth denoted as ω_{ss} and the jammer bandwidth as given by $\omega_{nb} = \rho\omega_{ss}$, the lower and upper cutoff frequencies of the narrowband spectral response are defined as $\omega_1 \triangleq \delta\omega - \frac{\omega_{nb}}{2}$ and $\omega_2 \triangleq \delta\omega + \frac{\omega_{nb}}{2}$, respectively. The autocorrelation sequence associated with this interference model is

$$R(k) = \frac{A_{nb}^2}{\rho} \frac{\sin\left(\frac{\omega_{nb}}{2}k\right)}{\pi k} \cos\left(\delta\omega k\right).$$

Assuming that the discrete-time sampling frequency is twice the spread spectrum bandwidth, i.e. $f_s = 2\omega_{ss}$, ω_{ss} can be normalized to π. As ρ approaches 100%, ω_{nb} and $\delta\omega$ thus approach π and $\pi/2$, respectively. In the limiting case, wherein $\omega_{nb} = \pi$ and $\delta\omega = \pi/2$, $R(k) = A_{nb}^2\delta(k)$ as is consistent with the definition of white noise.

3.2.6 Notes

In various laboratory environments, the analytical approach exercised here has been experimentally validated through computer simulation and hardware emulation using surface acoustic wave devices, real-time DSPs and floating point gate-arrays (FPGAs). In the following section, this analysis, is extended into the lapped transform domain. Therein, analytical results demonstrating the efficacy of block and lapped transform domain excision algorithms in the presence of various narrowband interference sources are comparatively presented.

3.3 LAPPED TRANSFORM DOMAIN EXCISION

Recent trends in signal processing have rekindled both commercial and academic interests in the application of multirate filter banks as well as related forms of time-frequency decomposition techniques such as wavelets, subband transforms and cosine-modulated filter banks to the interference excision problem [165, 216, 218, 248, 278, 279]. Although the new time-frequency analysis methods promise an opportunity to address historic weaknesses of FFT-based techniques, such as their susceptibility to impulsive wideband jammers, such expectations have not yet been fully realized due to their latent dependence on the interfering signal's structure and statistics. Nevertheless, novel approaches to interference suppression incorporating lapped transform (LT) domain or adaptive time-frequency excision offer improvement and added robustness in BER performance as compared to algorithms based on the FFT.

Excision using filter banks [217] and "spectrally-contained orthogonal transforms" (SCOT) [278, 279] has been shown to be very effective at mitigating narrowband interference. The analysis presented in [217] focuses on the development and analysis of transform domain filtering and data demodulation/detection schemes using orthonormal block and lapped transforms and forms the basis of this section. In a similar manner, the work presented in [278] essentially addresses the application of narrowband excision using time-weighted discrete Fourier transforms (DFT) to data demodulation and closely examines its corresponding BER performance. In contrast, the excision analysis offered in [279] focuses primarily on the use of time-weighted DFTs and cosine-modulated filter banks for radiometric signal detection. The primary purpose of this section is to present a fundamental analysis and evaluation of the LT domain excision process as a means of mitigating narrowband interference.

With LTs, the basis vectors are not restricted in length as they are in conventional block transforms like the DFT and the discrete cosine transform (DCT). Indeed, whereas the lengths of the block transform basis vectors are limited to the number of transform domain cells, or bins, the LT basis vectors have length, $\mathcal{N}$, that is equal to some even integer multiple of the number of bins, i.e. $\mathcal{N} = 2KN$, where N is the number of bins and K is the *overlapping factor* [207]; the inputs to successive transforms are produced by overlapping segments of the received signal. Thus, in comparison to traditional length-N basis vectors, the basis vectors associated with LTs typically yield improved stopband attenuation in the frequency domain for a given subband bandwidth. Fortunately, there exist efficient filter bank structures that allow these longer basis vectors to be used while only moderately increasing the number of required arithmetic operations [207].

The following section introduces two special cases of the LT, modulated lapped transforms (MLT) and extended lapped transforms (ELT) [207], and demonstrates how they can be used in a transform domain exciser. A subsequent section then presents an analysis of the BER performance of these transform domain excisers in the presence of tone and narrowband Gaussian interference with the corresponding results illustrated thereafter. To demonstrate the performance improvement realized using LTs, these results are depicted alongside those obtained using traditional block transform domain excision techniques.

3.3.1 Lapped Transforms

Inherent in the design of LTs is the satisfaction of the *perfect reconstruction* (PR) criteria, which imply that for an input sequence, $r[n]$, the reconstructed signal samples, $\hat{r}[n]$, are equal to the original values to within a constant scale and delay adjustment, that is

$$\hat{r}[n] = cr[n - n_0],$$

where c and n_0 are constants. Like block transforms, LTs can be represented by a transformation matrix, Ψ, whose rows are the individual basis vectors. Whereas the transformation matrix for a block unitary transform is square, the LT transformation matrix, as considered here, is real with dimensions $N \times 2KN$.

Viewing the LT transformation matrix as an infinite-dimensional block diagonal extension of Ψ, namely $\tilde{\Psi} = \mathrm{diag}\,(\cdots \ \Psi \ \Psi \ \Psi \ \cdots)$, and considering only real signals and transforms, PR can be achieved if and only if [207]

$$\tilde{\Psi}\tilde{\Psi}^T = \mathbf{I} \qquad \Longleftrightarrow \qquad \tilde{\Psi}^T\tilde{\Psi} = \mathbf{I}. \tag{3.19}$$

In this case, $\mathrm{diag}\,(\cdot)$ denotes a matrix whose block diagonal sub-matrices correspond to those given in the argument. These equations indicate the necessary conditions for orthogonality between the LT basis vectors and the conditions required for orthogonality of their overlapping "tails."

3.3.2 *Modulated Lapped Transforms*

To be consistent with the development of LTs as presented in [206, 207], modulated lapped transforms are considered here as a subset of general LTs with $K = 1$. The basic premise of the MLT is to use a $2N$-tap lowpass filter as a subband filter prototype which is shifted in frequency to produce a set of orthogonal bandpass FIR filters spanning the frequency domain. Denoting the lowpass prototype as $h[n]$, the sinusoidally modulated basis vectors can be expressed as [207]

$$\psi_k[n] = h[n]\sqrt{\frac{2}{N}}\cos\left[\left(n + \frac{N+1}{2}\right)\left(k + \frac{1}{2}\right)\frac{\pi}{N}\right], \tag{3.20}$$

where $0 \le n \le 2N - 1$ and $0 \le k \le N - 1$.

The design of the lowpass prototype, $h[n]$ is of central importance to the development of the MLT. To meet the PR requirements imposed by (3.19), $h[n]$ must satisfy the following requirements [207],

$$h[2N - 1 - n] = h[n] \qquad 0 \le n \le N - 1 \tag{3.21}$$

and

$$h^2[n] + h^2[n + N] = 1 \qquad 0 \le n \le N/2 - 1. \tag{3.22}$$

Note that the range over n in the last equation is limited to $N/2 - 1$ due to the symmetry of $h[n]$ as indicated by (3.21). Although there are many solutions to the above equations, the half-sine windowing function [207],

$$h[n] = -\sin\left[\left(n + \frac{1}{2}\right)\frac{\pi}{2N}\right],$$

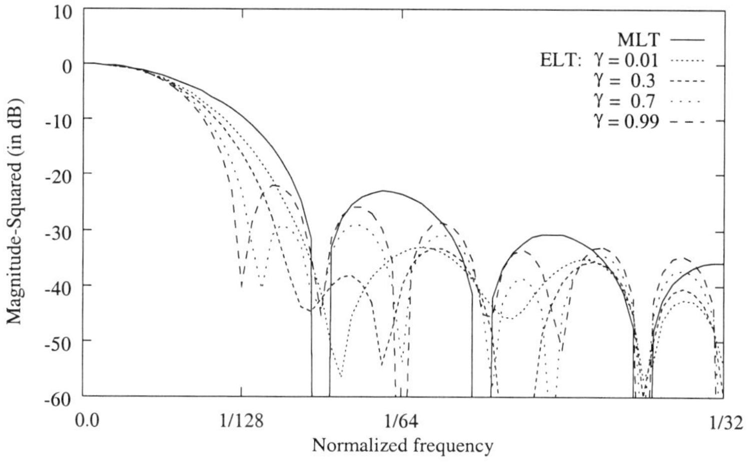

Figure 3.8 MLT and ELT lowpass filter prototype frequency responses.

is used throughout this chapter as the MLT lowpass filter prototype. This windowing function is commonly used since it satisfies both the polyphase normalization, ensuring efficient implementation, and the PR criteria. Figure 3.8 illustrates the frequency response associated with the half-sine windowing function for $N = 64$ - note that for clarity the frequency responses of the lowpass filter prototypes are plotted only on the range $[0, 1/32]$ instead of $[0, 1/2]$. Whereas non-windowed FFT basis vectors yield sidelobes that are roughly 13 dB down from the main lobe, the level of attenuation in the sidelobes associated with the MLT basis vectors is approximately 23 dB.

3.3.3 MLT Domain Processing

With $K = 1$, the $N \times 2N$ transformation matrix Ψ contains the LT basis set, $\{\psi_k[n],\ 0 \le k \le N-1,\ 0 \le n \le 2N-1\}$; thus there are N basis vectors with length $2N$. Transform domain signal processing using the MLT is illustrated in Figure 3.9. As in (3.7), by segmenting the input data stream samples into contiguous N-length data blocks with $N = L$, successive data vectors are produced which, assuming synchronization, represent independent data bits, $d[n]$.

Assuming that the data block of interest at time n is the $N \times 1$ vector $\underline{r}_n$ the governing LT analysis expression is given in terms of two adjacent data vectors as

$$\underline{R}_n = \Psi \begin{bmatrix} \underline{r}_n \\ \underline{r}_{n+1} \end{bmatrix} = \Psi_{II}\underline{r}_n + \Psi_{I}\underline{r}_{n+1} \tag{3.23}$$

where Ψ_I and Ψ_{II} denote $N \times N$ partitions of the the MLT transformation matrix according to the relationship $\Psi = [\Psi_{II} \mid \Psi_I]$. The corresponding inverse,

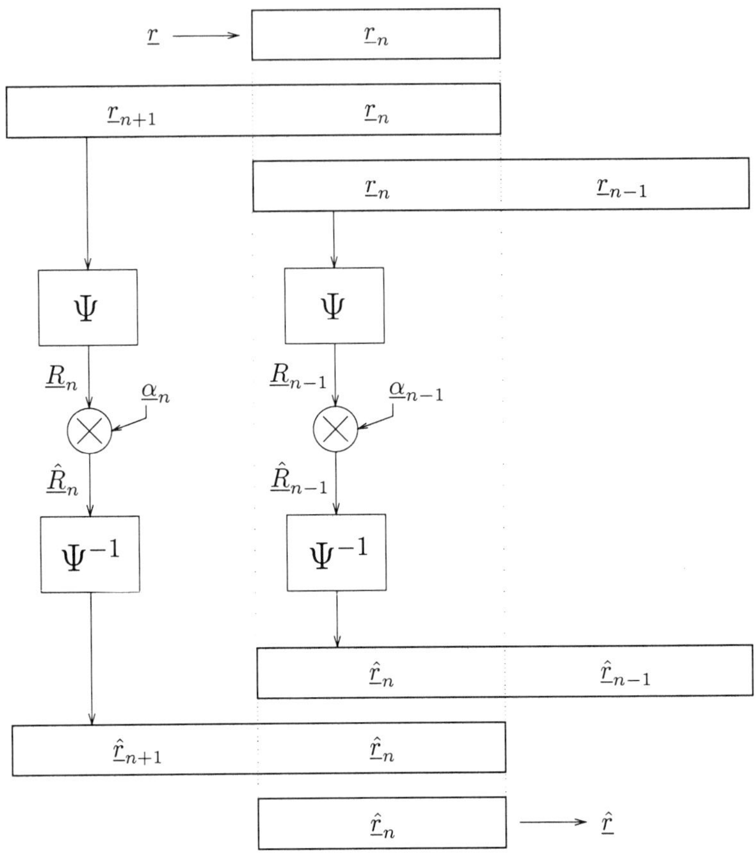

Figure 3.9 Signal processing using the MLT.

or synthesis, expression is given by

$$\underline{r}_n = \Psi_I^T \underline{R}_{n-1} + \Psi_{II}^T \underline{R}_n \tag{3.24}$$

where, as is consistent with (3.23), $\underline{R}_{n-1} = \Psi_{II}\underline{r}_{n-1} + \Psi_I\underline{r}_n$. Here, it is clear that both the current set of transform domain coefficients, $\underline{R}_n$, as well as the previous set, $\underline{R}_{n-1}$, are required to perfectly reconstruct the original input vector $\underline{r}_n$. Notice that as illustrated in Figure 3.9, in order to fully recover $\underline{r}_n$, input vectors $\underline{r}_{n-1}$, $\underline{r}_n$ and $\underline{r}_{n+1}$ are required.

In light of Figure 3.9 and 3.24), it is clear that MLT domain processing necessitates the use of two sets of transform domain excision coefficients, $\underline{\alpha}_{n-1}$ and $\underline{\alpha}_n$. Expressing the modified spectral coefficients as

$$\hat{\underline{R}}_n = \text{diag}(\underline{\alpha}_n)\underline{R}_n \qquad \text{and} \qquad \hat{\underline{R}}_{n-1} = \text{diag}(\underline{\alpha}_{n-1})\underline{R}_{n-1}, \tag{3.25}$$

the filtered time domain waveform can be written as

$$\hat{\underline{r}}_n = \Psi_I^T \hat{\underline{R}}_{n-1} + \Psi_{II}^T \hat{\underline{R}}_n. \tag{3.26}$$

Correlating this vector with the original spreading code yields the final decision variable,

$$\xi = \underline{c}^T \hat{\underline{r}}_n, \tag{3.27}$$

which is identical to that given in (3.11).

3.3.4 *Extended Lapped Transforms*

The ELT represents a subclass of lapped transforms with $K = 2$ and sinusoidally modulated basis vectors. As was the case with the MLT, the ELT basis vectors are generated by shifting the lowpass filter prototype, again denoted as $h[n]$, in the frequency domain such that the resulting N subband filters span the range of frequencies from zero to π. With an overlapping factor of 2, the length of $h[n]$ and, hence, the subband filters, is four times the number of subbands. By allowing the time-support of the basis vectors to be spread over four data blocks, (3.20), after modifying the range of n such that $0 \leq n \leq 4N - 1$, can be used to define the sinusoidally modulated subband filters.

Similar to (3.21) and (3.22), which were used in the previous section to represent the conditions placed on the MLT lowpass filter prototype by the PR constraints of (3.19), constraints for the ELT lowpass prototype are given by [206]

$$h[4N - 1 - n] = h[n]$$

$$h^2[n] + h^2[n + N] + h^2[n + 2N] + h^2[n + 3N] = 1$$

$$h[n]h[n + 2N] + h[n + N]h[n + 3N] = 0$$

where $0 \leq n \leq 2N - 1$ in the first equation and $0 \leq n \leq N/2 - 1$ in the last two. As presented in [206], one class of windows satisfying these equations is defined as

$$\begin{aligned}
h[\tfrac{N}{2} - 1 - i] &= -\sin\theta_i \sin\theta_{N-1-i} \\
h[\tfrac{N}{2} + i] &= \sin\theta_i \cos\theta_{N-1-i} \\
h[\tfrac{3N}{2} - 1 - i] &= \cos\theta_i \sin\theta_{N-1-i} \\
h[\tfrac{3N}{2} + i] &= \cos\theta_i \cos\theta_{N-1-i}
\end{aligned}$$

where $0 \leq i \leq N/2 - 1$ and the set of angles, θ_i, is given by

$$\theta_i = \left[\left(\frac{1-\gamma}{2N} \right) (2i+1) + \gamma \right] \frac{(2i+1)\pi}{8N}.$$

The additional parameter, γ, is a variable in the range $[0,1]$ that controls the roll-off of the prototype frequency response. Consequently, γ controls the trade-off between stopband attenuation and transition bandwidth of $h[n]$ and, thus, the ELT basis vectors [207].

Figure 3.8 demonstrates the effect of different γ on $h[n]$. Clearly, stopband attenuation is maximized as γ approaches zero whereas the bandwidth is minimized as γ heads towards unity. Depending on the value of γ, the level of sidelobe attenuation ranges from 22–34 dB.

For both simplicity and convenience, only ELT basis vectors derived from the above equations are considered in this chapter. For completeness, however, it should be noted that several techniques by which to develop optimal prototype filters have been presented in the literature [177, 234, 262]. A particularly appealing approach introduced by Vaidyanathan and Nguyen [234, 235, 338] utilizes eigenfilter design techniques to minimize the quadratic error in the passband and stopband of the prototype filter, $h[n]$. The resulting subband filters typically have very high attenuation in the stopbands and, thus, are characterized by low levels of inter-subband aliasing. Using the eigenfilter approach, lowpass filter prototypes can also be designed with complex coefficients [235].

3.3.5 ELT Domain Processing

The $N \times 4N$ ELT transformation matrix Ψ contains the ELT basis set, $\{\psi_k[n], 0 \leq k \leq N-1, 0 \leq n \leq 4N-1\}$. Accordingly, the ELT processes four data vectors per iteration with the resulting transform domain coefficients at time n defined as

$$\underline{R}_n = \Psi \begin{bmatrix} \underline{r}_n \\ \underline{r}_{n+1} \\ \underline{r}_{n+2} \\ \underline{r}_{n+3} \end{bmatrix} = \Psi_{IV}\underline{r}_n + \Psi_{III}\underline{r}_{n+1} + \Psi_{II}\underline{r}_{n+2} + \Psi_{I}\underline{r}_{n+3} \tag{3.28}$$

with the $N \times N$ matrix partitions defined relative to Ψ as

$$\Psi = [\Psi_{IV} \mid \Psi_{III} \mid \Psi_{II} \mid \Psi_{I}].$$

Appropriate adjustments to the time indices in (3.28) yields similar expressions for $\underline{R}_{n-1}$, $\underline{R}_{n-2}$ and $\underline{R}_{n-3}$. The reconstructed data vector, synthesized from its ELT spectral coefficients, is given by

$$\underline{r}_n = \Psi_I^T \underline{R}_{n-3} + \Psi_{II}^T \underline{R}_{n-2} + \Psi_{III}^T \underline{R}_{n-1} + \Psi_{IV}^T \underline{R}_n. \tag{3.29}$$

Thus, to fully recover $\underline{r}_n$, past and future input vectors from $\underline{r}_{n-3}$ to $\underline{r}_{n+3}$ are required. A physical description of an ELT domain signal processing system can be obtained by simply extending the MLT-based system shown in Figure 3.9 by two additional stages [215].

As expressed in (3.29), ELT domain processing necessitates the use of four sets of transform domain coefficients. As in the MLT case, the set of excised transform domain vectors, once again expressed as

$$\underline{\hat{R}}_k = \text{diag}(\underline{\alpha}_k)\underline{R}_k, \qquad k = n, n-1, n-2, n-3,$$

can be inverse transformed and recombined to yield the filtered time domain data vector,

$$\underline{\hat{r}}_n = \Psi_I^T \underline{\hat{R}}_{n-3} + \Psi_{II}^T \underline{\hat{R}}_{n-2} + \Psi_{III}^T \underline{\hat{R}}_{n-1} + \Psi_{IV}^T \underline{\hat{R}}_n. \tag{3.30}$$

Correlation of $\underline{\hat{r}}_n$ with the spreading code produces the bit decision variable according to (3.27).

3.3.6 *Analyzing LT Domain Excision*

In this section, the BER performance of lapped transform domain excisers in the presence of single-tone and narrowband Gaussian interference is examined. Although not considered here, the extension of the BER analysis to multiple-tone interference is straightforward and is discussed in some detail in [223]. An expression for the BER is first derived using the MLT and then modified to yield a similar expression for ELT domain excision. BER performance for block transform domain excision has been presented previously; results for block transforms are compared to those obtained via LT domain excision in the next section.

As in the block transform domain analysis, modulation and demodulation operations between the transmitter and receiver are assumed to be transparent. The receiver thus has perfect knowledge of the carrier frequency and phase and the corresponding received data signal is real. Assuming data bit synchronization, the boundaries of the transform input vectors are time-aligned with those of the data bit producing, as in (3.7), input vectors consisting of the sum of the transmitted signal, AWGN and single-tone interference, namely

$$\underline{r}_n = \underline{s}_n + \underline{j}_n + \underline{\eta}_n.$$

As in the previous analysis, $\underline{s}_n = d[n]\underline{c}$ denotes the spread data bit, antipodal signaling is used and each data bit is spread using one full length of the spreading sequence. It is once again assumed that the received signal is sampled once per chip with the length of the spreading sequence equal to the number of transform domain bins, N.

As expressed in (3.27), the bit decision variable associated with LT domain excision algorithms is given by

$$\xi = \underline{c}^T \underline{\hat{r}}_n$$

where $\underline{\hat{r}}_n$ is as defined in (3.26) for the MLT and (3.30) for the ELT. As suggested in (3.3), to make a bit decision, ξ is put through a threshold device with the decision boundary set to zero. The mean and variance of ξ once again allow the conditional and unconditional bit error rate expressions given in (3.13) and (3.12) to be evaluated for arbitrary input signals.

MLT Domain Excision. Assuming that the data bit energy is normalized to unity, the spread data bit at time n is simply given by

$$\underline{s}_n = d[n]\underline{c}. \tag{3.31}$$

Thus, consistent with (3.23), the MLT domain coefficients associated with the spread data signal and the interference are expressed as

$$\underline{S}_n = \Psi \begin{bmatrix} d[n]\underline{c} \\ d[n+1]\underline{c} \end{bmatrix} \quad \text{and} \quad \underline{J}_n = \Psi \underline{j}_{\theta_n} \tag{3.32}$$

where, since the single-tone jammer operates continuously, $\underline{j}_{\theta_n}$ denotes the $2N \times 1$ interference vector with phase θ_n.

Since the jammer phase, θ, is unknown, the conditional mean, $\mu_{\xi|\theta}$, of ξ must first be determined for a fixed value of θ. Assuming that the transform domain filter coefficients and transformation matrix are fixed and that a +1 data bit has been sent ($\underline{s}_n = \underline{c}$), the conditional expectation of ξ can be expressed as

$$\mu_{\xi|\theta} = \underline{c}^T \left\{ \Psi_I^T \left(\underline{\hat{S}}_{n-1} + \underline{\hat{J}}_{n-1|\theta_{n-1}} \right) + \Psi_{II}^T \left(\underline{\hat{S}}_n + \underline{\hat{J}}_{n|\theta_n} \right) \right\} \tag{3.33}$$

where $\underline{\hat{S}}_n$ and $\underline{\hat{J}}_{n|\theta_n}$ are related to $\underline{S}_n$ and $\underline{J}_{n|\theta_n}$ through (3.25) with $\underline{J}_{n|\theta}$ defined as the conditional expectation

$$\underline{J}_{n|\theta_n} \triangleq E\left\{ \underline{J}_n \,|\theta_n \right\}. \tag{3.34}$$

A similar expression holds for time $n-1$.

Given that the zero-mean noise term is characterized as AWGN with two-sided power spectral density $N_0/2$, the variance of ξ can be written as

$$\sigma_\xi^2 = \frac{N_0}{2} \sum_k \underline{C}_k^T \mathrm{diag}(\underline{\alpha}_k)\mathrm{diag}(\underline{\alpha}_k)\underline{C}_k \tag{3.35}$$

where the summation is over $k = n-1, n$ and where, for ease of notation, the MLT domain representations of the reference spreading code are given by

$$\underline{C}_{n-1} = \Psi_I \underline{c} \quad \text{and} \quad \underline{C}_n = \Psi_{II} \underline{c}. \tag{3.36}$$

Noting that the block transform domain representation of the spreading code is $\underline{C} = \Psi\underline{c}$, it is readily apparent that this expression is quite similar to that given in (3.15). As in the case of block transforms, this expression is independent of θ.

Using the expressions for $\mu_{\xi|\theta}$ and σ_ξ^2, as given by (3.33) and (3.35), one can evaluate the BER performance of the receiver depicted in Figure 3.9 for an arbitrary set of basis vectors and spreading code. As before, the probability of bit error expressions are presented in (3.12) and (3.13).

ELT Domain Excision. Similar to the preceding discussion, the ELT domain representations of the spread data signal and the interference waveform at time n are given by

$$\underline{S}_n = \Psi \begin{bmatrix} d[n]\underline{c} \\ d[n+1]\underline{c} \\ d[n+2]\underline{c} \\ d[n+3]\underline{c} \end{bmatrix} \quad \text{and} \quad \underline{J}_n = \Psi \underline{j}_{\theta_n}$$

where $\underline{j}_{\theta_n}$ now represents a $4N \times 1$ interference vector with phase θ_n. Replacing n with the appropriate time indices, $n-3, n-2, n-1$, yields the full set of ELT domain coefficients. Extending (3.33) and (3.35) to account for the four sets of transform domain coefficients yields

$$\mu_{\xi|\theta} = \underline{c}^T \left\{ \Psi_I^T \left(\hat{\underline{S}}_{n-3} + \hat{\underline{J}}_{n-3|\theta_{n-3}} \right) + \Psi_{II}^T \left(\hat{\underline{S}}_{n-2} + \hat{\underline{J}}_{n-2|\theta_{n-2}} \right) \right.$$
$$\left. + \Psi_{III}^T \left(\hat{\underline{S}}_{n-1} + \hat{\underline{J}}_{n-1|\theta_{n-1}} \right) + \Psi_{IV}^T \left(\hat{\underline{S}}_{n} + \hat{\underline{J}}_{n|\theta_{n}} \right) \right\},$$

which is the conditional mean of ξ with respect to $d[n] = +1$ and θ_n. The $\underline{J}_{n|\theta_n}$ terms are again given by (3.34). Furthermore,

$$\sigma_\xi^2 = \frac{N_0}{2} \sum_k \underline{C}_k^T \mathrm{diag}(\underline{\alpha}_k) \mathrm{diag}(\underline{\alpha}_k) \underline{C}_k$$

is the variance of ξ with the summation over $k = n-3, n-2, n-1, n$

$$\underline{C}_{n-3} = \Psi_I \underline{c}, \quad \underline{C}_{n-2} = \Psi_I \underline{c}, \quad \underline{C}_{n-1} = \Psi_I \underline{c} \quad \text{and} \quad \underline{C}_n = \Psi_{II} \underline{c} \tag{3.37}$$

denote the ELT domain spreading code coefficients. Together with (3.12) and (3.13), these expressions can be used to determine the probability of bit error associated with the ELT domain excision algorithm.

Narrowband Gaussian Interference. Treating the despread narrowband interference as an additional zero-mean AWGN source, the mean of the MLT-based decision variable becomes [217]

$$\mu_\xi = \sum_k \underline{C}_k^T \mathrm{diag}(\underline{\alpha}_k) \underline{C}_k \tag{3.38}$$

with the associated variance given as

$$\sigma_\xi^2 = \frac{N_0}{2} \sum_k \underline{C}_k^T \mathrm{diag}(\underline{\alpha}_k) \left(\frac{N_0}{2} \mathbf{I} + \mathbf{K}_{\underline{jj}} \right) \mathrm{diag}(\underline{\alpha}_k) \underline{C}_k \tag{3.39}$$

where $k = n-1, n$ with $\underline{C}_k$ given in (3.36). Likewise, expressions for the mean and variance associated with the ELT related decision variable are identical to (3.38) and (3.39), respectively, with $\underline{C}_k$ denoted in (3.37) and $k = n-3, n-2, n-1, n$. As in the earlier analysis, the BER associated with the MLT and ELT domain excision systems can be obtained by substituting the appropriate expressions for the mean and variance of ξ into (3.12) and (3.13) and evaluating.

3.3.7 *Modeling the Excision Process*

In practice, since transform domain excision decisions are typically a function of the spectral distribution of the interfering signal's energy, many applications evaluate the magnitude of each transform domain bin and remove those that exceed a preset threshold. Although this process, simply termed "threshold excision," is easily implemented in hardware, its stochastic nature complicates mathematical analysis and often necessitates (at least for analytical purposes) a simpler, more tractable process model. One such model, which allows objective evaluation of various linear transforms and filter banks, is considered here.

To obtain the average probability of bit error in this chapter, transform domain vectors corresponding to the received data signal with the k largest magnitude bins removed are determined for several different phases of the interferer - this technique is referred to as the "k-bin" excision algorithm. Using these excised vectors, the corresponding bit error probabilities conditioned on phase are calculated and averaged over a large number of phases to obtain the final value of the bit error rate. To simplify the analysis, the selection of the k bins to be excised is based solely on the ensemble average of the transform domain distribution of the narrowband interference energy; this approach is valid provided that relatively large JSR ratios are considered and that only a small percentage of bins are removed.

As mentioned earlier, spread spectrum receivers incorporating MLT or ELT domain excision are implemented as shown in Figure 3.9, where, for the ELT, two additional sections are added. Although not explicitly shown in the receiver diagram, the transform domain excision coefficients are based on the transform domain distribution of interference energy using the k-bin excision algorithm. Throughout the following sections, LT domain excision vectors are based on the entire set of transform domain coefficients and are evaluated simultaneously, thereby inherently taking into account the distribution of data bit energy across all sets of transform domain coefficients.

3.3.8 *Results*

In the following sections, analytical BER results illustrate the performance of spread spectrum receivers employing lapped transform domain excision in the presence of narrowband interference. In each of the interference scenarios considered, a 64-chip augmented spreading code is used to modulate binary data. Consequently, the dimensionality of the transformation matrix is 64×64 for block transforms, 64×128 for the MLT and 64×256 when using the ELT. The ELT lowpass filter prototype is generated using $\gamma = 0.5$. In each case, the number of bins excised is parenthetically indicated in the figure legend.

Single-Tone Interference. Based on earlier analysis, receiver performance using transform domain excision and a variety of orthonormal block transforms in the presence of a single-tone interferer with $\delta\omega = 0.127$ rad/sec and JSR = 20 dB is shown in Figure 3.10. As in [217], the block and subband transforms tested include the DFT, DCT and the Karhunen-Loève transform (KLT). As

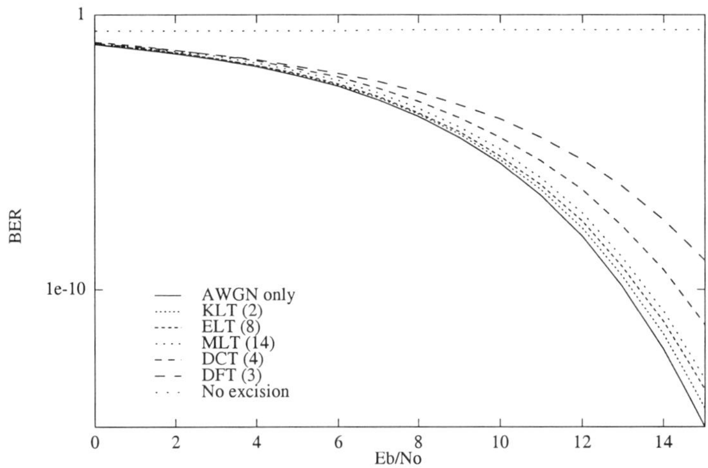

Figure 3.10 Excision in the presence of single-tone interference, JSR = 20 dB and $\delta\omega$ = 0.127 rad/sec.

expected, the KLT, which confines the interfering sinusoids to two basis vectors, yields the best BER performance [215]. Among the fixed transform techniques considered, both lapped transforms produce results comparable to the KLT and significantly better than those achieved by most of the block transform implementations. In fact, with respect to the KLT performance results, the ELT and MLT bit error rates are within 0.2 dB and 0.4 dB, respectively, whereas the DCT generates the lowest BER of all the block transforms considered yet is roughly 1.2 dB worse.

As suggested in the design of the ELT lowpass filter prototypes discussed earlier, there is always a trade-off between the bandwidth of the main lobe and the level of attenuation in the sidelobes. Due to windowing effects and, thus, relatively poor attenuation in the sidelobes, the performance obtained using the DFT is rather poor for the given interference environment. In contrast, the lower BER associated with lapped transform domain excision can be attributed to the relatively small bandwidth and high stopband attenuation associated with the LT subband filters, a direct result of the longer basis vectors.

To quantify algorithm robustness with respect to jammer power and frequency, Figures 3.11 and 3.12 illustrate the performance of the excision-based receivers as a function of the JSR and $\delta\omega$, respectively. In Figure 3.11, the number of bins excised is held fixed as the jammer power is allowed to increase. In this figure, the KLT basis vectors have not been recalculated for each JSR value tested. Thus, although the two excised KLT basis vectors optimally represent the interfering tone when the JSR = 20 dB [215], as it increases the interference energy present in the remaining bins increases almost uniformly. As a result, the residual interference acts as a white noise source with its power commensurately related to that of the single-tone jammer. Due to the high

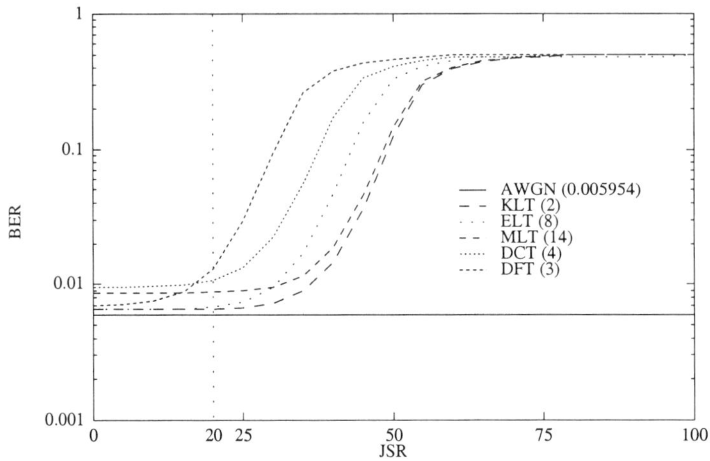

Figure 3.11 Excision in the presence of single-tone interference as a function of JSR, $E_b/N_0 = 5$ dB and $\delta\omega = 0.127$ rad/sec.

levels of stopband attenuation associated with the lapped transforms, the MLT and ELT are capable of tolerating larger power interferers than the block transforms without significantly compromising performance. In fact, for both the MLT and ELT, relatively low BER results, namely a BER of less than 9.5×10^{-2} (as compared to the AWGN level of 5.95×10^{-2}), are maintained up to a JSR of approximately 30 dB. As in practice, if the jammer power is sufficiently large, additional or alternative measures, such as coding or longer spreading codes, must be implemented.

Figure 3.12 illustrates the BER obtained as a function of jammer frequency offset, $\delta\omega$. In these figures, the KLT is not considered since it must be recalculated for each value of $\delta\omega$ and, thus, is not robust with respect to frequency. Here, the relative insensitivity of LT domain excision with respect to the jammer frequency is demonstrated as the performance of both LT-based algorithms do not deviate significantly from theoretical BER performance in AWGN alone at any frequency. In contrast, the block transform implementations are highly sensitive to frequency, again a result of the frequency responses of the transform basis vectors.

Narrowband Gaussian Interference. Excision-based receivers have also been evaluated in the presence of narrowband Gaussian interference. As discussed previously, the parameters characterizing this type of interferer are the JSR, the center frequency, $\delta\omega$, and the fractional bandwidth, ρ. Regarding the following results, $E_b/N_0 = 5$ dB, JSR $= 10$ dB, $\delta\omega = 0.127$ rad/sec and $\rho = 0.1$ unless otherwise noted.

Figure 3.13 illustrates receiver performance as a function of E_b/N_0. As in the case of the single-tone interference shown in Figure 3.10, the best per-

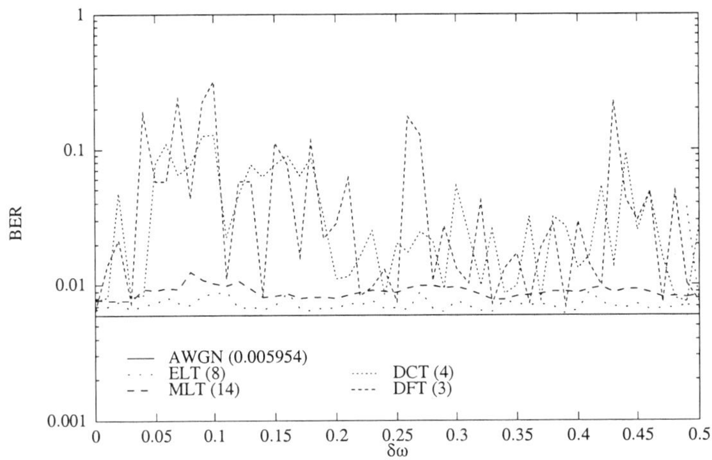

Figure 3.12 Excision in the presence of single-tone interference as a function of frequency, $E_b/N_0 = 5$ dB and JSR = 20 dB.

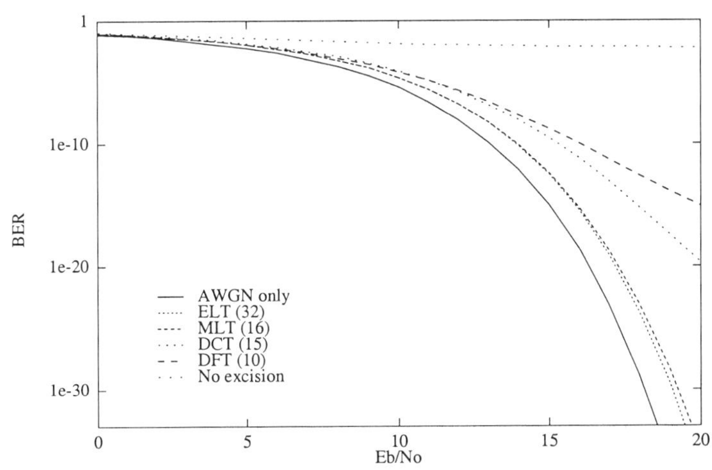

Figure 3.13 Excision in the presence of narrowband Gaussian interference, JSR = 10 dB, $\delta\omega = 0.127$ rad/sec and $\rho = 0.1$.

formance is obtained using lapped transforms, which consistently yield BER results within 1.0–1.5 dB of that obtained in AWGN alone; in contrast, all of the block transforms considered clearly produce substantially poorer results. Such results are rather dramatic considering that the removal of 10% of the spread spectrum signal energy results in an immediate loss of roughly 0.5 dB in E_b/N_0. Although not included here, additional laboratory results verify that similar performance is maintained over all $\delta\omega$.

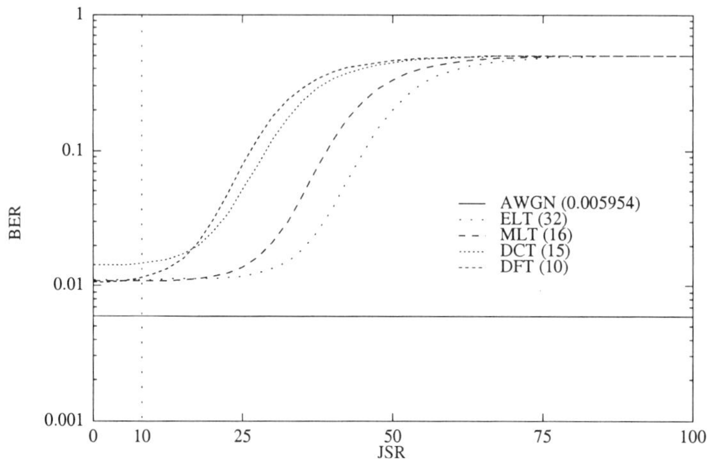

Figure 3.14 Excision in the presence of narrowband Gaussian interference as a function of JSR, $E_b/N_0 = 5$ dB, $\delta\omega = 0.127$ rad/sec and $\rho = 0.1$.

Figures 3.14 and 3.15 illustrate algorithmic performance as a function of the narrowband Gaussian JSR and ρ, respectively. Figure 3.14 shows that the lapped transform domain excision schemes again provide the best performance as the JSR increases. In Figure 3.15, the bandwidth of the interfering signal is allowed to vary from approximately zero, which approximates the single tone, to 100%, which is equivalent to white noise. To allow for such a broad range of bandwidths, the center frequency has been repositioned at $\delta\omega = 0.25$ rad/sec. From this figure, it is clear that for jammer bandwidths below 10%, the MLT and ELT yield similar values for the BER. As the bandwidth increases, all of the transform domain excision schemes yield unacceptable BER results.

3.4 ADAPTIVE TIME-FREQUENCY EXCISION

In contrast to block and lapped transform domain excision algorithms, adaptive time-frequency (ATF) excision is capable of tracking and suppressing time-varying, non-stationary interference. The novelty of the ATF exciser is two-fold: (1) it evaluates the time-frequency features of the received data vector in order to determine in which domain to perform excision, time or frequency, and (2) it adapts its subband decomposition structure so as to efficiently represent the interference energy in as small a time-frequency space as possible.

In determining the best domain for excision, the energy associated with each component of the received data vector is measured and compared to a preset threshold. Elements with energy levels in excess of this threshold are considered "captured" with the total number of such elements denoted as N_c. When N_c is less than or equal to a predetermined limit, N_l, the captured elements are set to zero [322]. Hence, if the interference is time localized, time domain

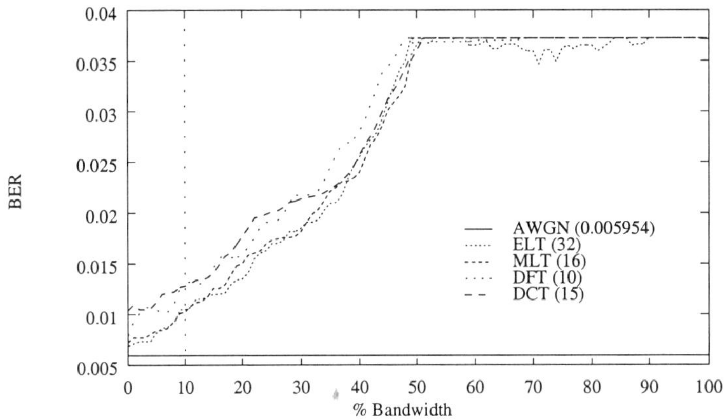

Figure 3.15 Excision in the presence of narrowband Gaussian interference as a function of ρ, $E_b/N_0 = 5$ dB, JSR = 10 dB and $\delta\omega = 0.25$ rad/sec.

excision is used to suppress it. On the other hand, when $N_c > N_l$, the energy distribution of the received data vector is deemed sufficiently widespread in the time domain such that the likelihood of time localized interference is small and, thus, transform domain techniques, if any, are required. Clearly, if the interference energy is not significant the processing gain associated with the SS waveform may offer sufficient protection. Figure 3.16 displays the flow diagram of the ATF exciser algorithm [322].

As explained in the following section, the adaptive tree structuring algorithm examines the spectrum of the received signal and determines the best subband tree structure for interference analysis and, hence, mitigation. Subbands containing significant amounts of interference energy are removed with the remaining components recombined via a synthesis filter bank to yield the filtered data vector. The subband tree structure changes in accordance with changes in the input spectrum thus tracking statistical variations in the received data.

3.4.1 *Adaptive Subband Tree Structuring (Wavelet Packets)*

Fixed block transforms typically assume that the input signal is stationary and, thereby, insensitive to statistical variations. The KLT on the other hand is optimal in that it completely decorrelates the input signal and maximizes its energy compaction in the transform domain. The set of basis vectors defining the KLT are derived from eigenanalysis of the covariance matrix of the received data vector. Unfortunately, however, in many practical situations only estimates of the data statistics are available. Such a liability, combined with the formidable computational complexity associated with eigenanalysis, limits the KLT's utility in practice and motivates the development of alternative signal dependent decomposition techniques.

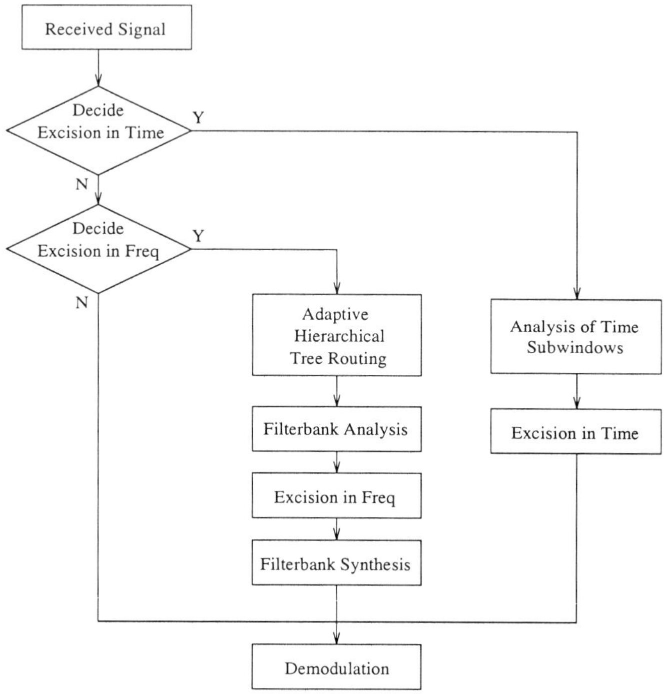

Figure 3.16 The flow diagram of the adaptive time-frequency exciser algorithm.

An adaptive subband transform inherently tracks the variations of the input spectrum by minimizing the spectral leakage between transform domain bins and, thus, reduces the detrimental effects of the exciser on the desired signal spectrum. A simple subband tree structuring algorithm (TSA) can be generated using either a bottom-to-top (tree pruning) or top-to-bottom (tree growing) approach. In tree pruning, the signal is decomposed into a full tree. Cost function analysis is then applied to every node in the tree with the cost of each node compared to its two child nodes. In contrast to tree pruning, tree growing generates a tree from top-to-bottom where at each node the decomposition is justified. Here, the cost function is applied to the parent sequence and the sequences of the child nodes. If the cost of the two child nodes is less than that of the parent, the child nodes, yielding a more efficient representation, are retained. If, on the other hand, the child nodes provide a less efficient representation, the parent node is terminated without further decomposition. Surviving child nodes are subsequently compared with their own offspring and are grown in a manner consistent with the previously defined rules. By applying such a signal adaptive process, the less efficient decompositions are avoided and the most suitable basis set for representing the input SS signal is determined.

In the ATF excision algorithm, the energy compaction measure quantifies the unevenness of the given signal spectrum. By utilizing this measure as a

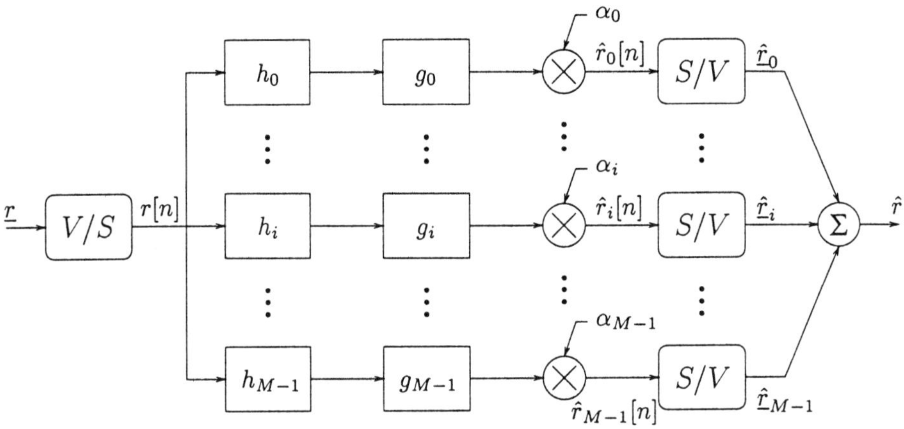

Figure 3.17 Filter bank-based interference exciser.

cost function, the adaptive TSA effectively tracks the variations of frequency-localized input signals. In each stage of the hierarchical filter bank, the subband output energy distribution is analyzed. Each node is decomposed if and only if the energy compaction at this node exceeds a predefined threshold.

3.4.2 Analysis of the Adaptive Time-Frequency Exciser

A modified version of the ATF exciser is displayed in Figure 3.17. In contrast to the originally defined receiver structure [322, 323], excision here is depicted as a post-synthesis filtering process. Such an alteration does not affect system performance and is only introduced to simplify the following discussion. Since processing is performed using discrete-time FIR filters, the input data vector $\underline{r}$ is converted from its parallel data format to a serial data sequence, $\{r[n]$ for $n = 0, 1, \ldots, N-1\}$, whose values are obtained from the input vector according to the relationship $\underline{r} = [r[0]\ r[1]\ \cdots\ r[N-1]]^T$. In this figure, vector-to-serial and serial-to-vector data conversions are denoted by the blocks labeled V/S and S/V, respectively. The set of M analysis and synthesis filters are represented as $\{h_i\}$ and $\{g_i\}$, respectively. Note that when performing time domain excision, they are effectively described as simple impulse functions, that is $h_i[n] = \delta[n-i]$ and $g_i[n] = \delta[n-i]$ for $i = 0 \ldots M-1$ with $M = N$ (N is the length of the input data vector).

In accordance with Figure 3.17, the ith branch output can be written in terms of its length-$(2N-1)$ composite analysis/synthesis filter response,

$$t_i[n] = (h_i[n] * g_i[n]) = \sum_{m=0}^{N-1} h_i[m] g_i[n-m],$$

as

$$\hat{r}_i[n] = r[n] * t_i[n] = \sum_{m=0}^{2N-2} t_i[l]r[n-l]$$

where $*$ denotes the discrete-time convolution operation. Assuming that the input data vector is of length N, this process yields $2N-1+N-1=3N-2$ output values. Neglecting the convolution tails, the $N \times 1$ filtered output vector associated with the ith branch is given by

$$\underline{\hat{r}}_i = [\hat{r}_i[N] \ \hat{r}_i[N+1] \ \cdots \ \hat{r}_i[2N-1]]^T .$$

Multiplication by the excision weights, $\alpha_i \in \{0,1\}$, and summing over i produces the filtered output

$$\underline{\hat{r}} = \sum_{i=0}^{M-1} \alpha_i \underline{\hat{r}}_i ,$$

which, when correlated with the reference spreading code, $\underline{c}$, yields the decision variable

$$\xi = \underline{c}^T \underline{\hat{r}}.$$

As in the previous sections, the mean and variance of this term determine the corresponding BER as expressed in (3.12). The derivation of and actual expressions for the mean and variance tend to be rather complicated and are developed in [323]. BER results associated with the ATF exciser are demonstrated in the following section.

3.4.3 Performance Evaluation of the Adaptive Time-Frequency Exciser

Figure 3.18 illustrates BER performance results obtained in the presence of single-tone interference with $\delta\omega = 0.306$ rad/sec and JSR $= 20$ dB. Results obtained via ATF excision as well as block transform domain excision using a 63-point KLT, 128-point DFT and 128-point DCT are depicted. A spreading sequence with length $L = N = 63$ is assumed.

Figure 3.19 displays the performance of ATF and fixed transform-based excisers for non-stationary pulsed (time-localized) wideband Gaussian interference. This jammer is randomly switched on and off with a 10% duty cycle and JSR $= 20$ dB. As demonstrated by these results, the ATF exciser is capable of identifying the domain in which processing should be performed and successfully mitigating time localized interference. As anticipated, none of the fixed transform-based excisers effectively suppress the interference. Thus, the ability to recognize and mitigate interference localized in either the time or frequency domain offers potential performance advantages over conventional fixed transform domain interference excision algorithms.

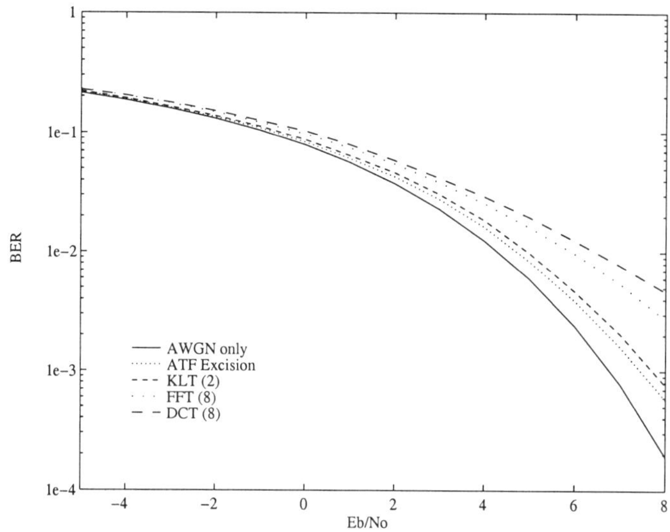

Figure 3.18 ATF excision in the presence of single-tone interference, JSR $= 20$ dB and $\delta\omega = 0.306$ rad/sec.

3.5 SUMMARY

As demonstrated in this chapter, excision performance is largely dependent on the transform's ability to compactly represent the interfering signal energy in the transform domain. Accordingly, narrowband interference is best removed using a uniform bank of bandpass filters with unity gain in the passband and infinite stopband attenuation. Of the transforms considered here and in [215], lapped transforms most closely approximate this ideal.

From the results shown, it is apparent that lapped transform domain excision algorithms are relatively insensitive to jammer frequency. Such a characteristic is often advantageous in practice since one is seldom guaranteed that the frequency of the interfering tone is known or constant. Considering that these algorithms are also relatively insensitive to jammer power and that their complexity using polyphase filter bank structures rivals that of conventional block transform algorithms [207], the MLT and ELT must be considered as viable transformation techniques for narrowband interference excision applications.

Despite the demonstrated efficacy of block and lapped transform domain excision techniques against narrowband interference, such algorithms are largely ineffective in aiding data transmission over channels corrupted by impulsive or wideband interference. In such cases, ATF excision has been shown as a potentially effective alternative. Adaptive TSAs tracking variations in the input signal's statistics yield hierarchical filter bank structures which adapt in time

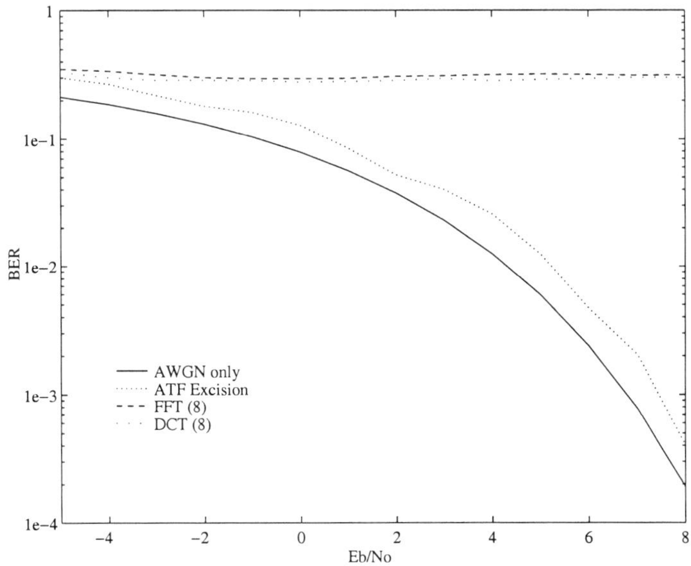

Figure 3.19 ATF excision in the presence of wideband Gaussian interference with 10% duty cycle and JSR = 20 dB.

to efficiently represent jammer energy and facilitate its removal. This approach offers the user the ability to dynamically redesign the subband transform such that its time-frequency plane partitions are consistent with the energy distribution of the undesired waveform. Hence, ATF excision is capable of mitigating a variety of types of interference including those originating from narrowband or wideband sources as well as those with other arbitrary time-frequency distributions.

4 TRANSFORM-BASED LOW PROBABILITY OF INTERCEPT COMMUNICATIONS

Richard S. Orr[1], Thomas C. Farrell[2] and Glenn E. Prescott[3]

[1]Stanford Telecom
Reston, VA
rorr@csi.stel.com

[2]TRW
Redondo Beach, CA
thomas.farrell@trw.com

[3]University of Kansas
Lawrence, KA
prescott@eecs.ukans.edu

4.1 INTRODUCTION TO SPREAD SPECTRUM COMMUNICATIONS

Spread spectrum originated from the need to protect signals from interference or intentional jamming. Voice and data communications employing spread spectrum have become staples in the inventory of governments around the world. Naturally, other properties of spread spectrum transmission became apparent to the investigators developing these systems, the key ones being multiple access (MA) and low probability of intercept (LPI) communications.

Spread spectrum signals are often used in the military environment to provide LPI, or covert, communications. Recently, spread spectrum has also been incorporated into such civilian applications as wireless local networks and cellular telephones, where its multiple access advantages are exploited along with low transmitter power and low probability of interference.

As the use of LPI systems becomes more widespread, so does the interest of people other than the intended receiver to detect and determine key features of the signals. In the military environment, the desirability of detecting and characterizing the enemy's signals is obvious. In the civilian world, there is a requirement to police the electromagnetic spectrum and a need for field engineers to determine how much traffic a spread spectrum band carries in a particular environment. This is of course a detection problem, and it falls within the scope of our discussion.

4.1.1 *Transform Constructs for LPI*

The focus of this chapter is LPI, more specifically, the application of time-scale transform domain techniques to both the synthesis and analysis of LPI signals. Those familiar with spread spectrum realize that traditional signal design is a matter of selecting an allocation of the available degrees of freedom in time and frequency (or scale). Research in two-dimensional signal representations has shown that transform domain techniques provide novel decompositions or *tilings*, of the time-frequency plane not seen before. A simple example convinces us that these new tilings should be exploited.

Consider a communication system that is subject to strong sinusoidal interference within its operating band. If the system is designed such that no one frequency location is critical to reliable transmission of the information, the interfering tone can in principle be filtered out with little loss of communication fidelity. Suppose now that the nature of the interference changes suddenly from tones to short duration bursts of energy. Mechanisms put in place to cope with tone interference become of little use, since the unwanted signal now fills the instantaneous spectrum when it is on. Were the transmitted signal designed such that no one time location is critical to reliable performance, the spots corrupted by the bursts could be located and time gated out of the subsequent processing, again with little loss. A time-frequency design can achieve this. More importantly, because the key operations involved in modulating and demodulating signals in such a system consist of well-known and efficient digital signal processing algorithms, adaptive modification of the signals and the transmission plan is possible [195].

Transmultiplexers, which map time-division multiplex (TDM) data into a frequency-division multiplex (FDM) format, have been employed for years in telephone networks and are a basic example of a time-frequency mapping. Time-scale mappings involve partitioning data into multirate streams and causing these streams to represent expansion coefficients over a multiscale basis. As one might expect, there are innumerable options for how to do this. Only a scant few of these are addressed in this chapter.

Time-scale processing has further benefits, especially as an analysis tool. In Section 4.4 of this chapter, we discuss one way a transform-based technique may be used to detect traditional hopped LPI signals.

4.1.2 *Chapter Overview*

Section 4.2 of this chapter introduces the key concepts at a level of specificity sufficient for our needs. Criteria for a signal to be considered LPI or LPD (low probability of detection) are presented, followed by a discussion of waveform design techniques one can use to achieve these criteria. The section concludes with a brief survey of traditional LPI spread spectrum types.

In Section 4.3 we develop transform-domain LPI constructs from the viewpoint of expansions of functions over basis sets. In a transform-domain setting the information to be communicated is impressed on the expansion coefficients, which are used to synthesize the transmitted signal. At the receiver the coefficients, and hence the information they represent, are recovered. Observe that the paradigm is the inverse of that in signal processing: the transmitter function is synthesis and the receiver function is analysis. The notion of *scale* is introduced as prelude to discussion of some wavelet-based constructs. Organization of resources at both single and multiple scales turns out to be useful. Because discrete wavelet transforms (and filter banks that generate and invert them) are linear, periodically time-varying systems, they represent an element requiring synchronization in practical communications system. Rather than being a burden, the receiver filter bank synchronization process can be organized to offer synchronization speed features not available in single-scale systems. A rather lengthy discussion is devoted to this novel and still developing field.

In Section 4.4 we present some material on the detection of LPI signals in the presence of white Gaussian noise (WGN) using transform based techniques. Included is a general mathematical development of the problem and a specific implementation using a quadrature-mirror filter (QMF) bank tree.

4.2 LPI AND LPD SIGNALS

This section introduces the operative LPI and LPD definitions, defines criteria for these, and surveys design principles that enable the criteria to be met.

4.2.1 LPI and LPD Defined

The terms "low probability of intercept" and "low probability of detection" are used interchangeably by many, without clear insight into what the subject of LPI/LPD is all about. It is useful to observe a distinction between the terms. When we speak of LPD in this chapter, we refer to signals whose presence is difficult to discern. A casual observer will certainly not discover that the signal shares some environment with him. The interesting technical challenge, of course, is to accomplish the same end when the observer is quite interested and not at all casual, and has at his disposal sophisticated means. By almost anyone's definition, *detection* is the LPD issue.

LPI, on the other hand, carries with it the notion that one who succeeds in observing should not be able to derive information from the signal environment that enables characterization of the signal in terms of signal type, modulation parameters, carrier frequency, bandwidth, et cetera, leaving its content open to subsequent exploitation. These parameters are called *features*, with the determination of their existence and value referred to as *features extraction.* The passage from feature extraction to LPE (low probability of exploitation), whose focus is the intelligent use of information derived from the signal — for example, vulnerability to decryption (the recovery and reading of the message plain text) or direction finding (locating the emitter) — is of considerable interest and depth, but would take us far afield. In this chapter, we are more interested in whether the extraction of features from a signal can be used as a weapon in the detection battle, and therefore stop short of addressing the complex topic of LPE. An orderly treatment of all three topics can be found in [237].

The LPI vulnerabilities suggested above are familiar to those having studied the theory of cyclostationarity. For example, it maybe possible to process a low energy density signal in a way that a narrowband component is produced, and if the signal persists long enough, the narrowband component can be integrated until whatever it represents becomes apparent at a signal-to-noise ratio (SNR) high enough to indicate signal presence. This example encapsulates our interest in LPI and in fact is the impetus for the creation of so-called *featureless waveforms* that attempt to elude all feature detectors. The intent of the following section is to present LPI/LPD criteria that have some measure of universal acceptance and have actually been used in the LPI/LPD designs based on wavelet-related principles.

4.2.2 LPI/LPD Criteria

Our criteria draw from the featureless waveform concept. Inasmuch as features tag a signal as man-made, the function of a featureless waveform is to take on the appearance of something created in nature. Radio engineers can immediately cite a number of natural disturbances that show up in communication channels, the ubiquitous one being receiver thermal noise.

Although statisticians know a remarkable number of criteria for Gaussian statistics, we emphasize only the few most important ones, the ones communication engineers have spent a great deal of effort to satisfy. In the list below, each criterion carries a label (LPI or LPD) to identify which characteristics it represents. Reasonable people might differ in how to assign these labels.

Gaussian Marginal Distribution (LPI). The experimental histogram of a series of observations should exhibit the familiar zero-mean, bell-shaped curve within some acceptable tolerance.

Complex Gaussian Marginal Distribution (LPI). If the observations are acquired at intermediate frequency (IF) by extraction of in-phase (I) and quadrature (Q) components, or complex down conversion, each component should be marginally Gaussian and zero mean and the I and Q components should be statistically independent and identically distributed. These properties can be said to hold if one is satisfied that the component means are zero, the variances are identical and the second-order cross moment (cross-correlation) is zero. These characteristics can be represented in polar coordinates as well, in which case the amplitude density should be Rayleigh and the phase density uniform. The complex plane presentation of the data brings home the point that there should be no naturally occurring axes that define I and Q; the rotational symmetry of the two-dimensional Gaussian density is required.

Higher Order Moments (LPI). All odd-order central moments of the zero-mean Gaussian density are zero. Those of even order follow the recursive law

$$E\left(x^{2n}\right) = (2n-1)\sigma^2 E\left(x^{2n-2}\right),$$

where the variance is σ^2. One can imagine moment matching tests to an arbitrarily high order, but it has been suggested that the ability to do practical work beyond or even at the eighth moment is questionable, primarily due to the adverse SNR characteristics that occur when taking high powers of signals well below the thermal noise level.

Impulsive Autocorrelation (LPD). Ideally the entire signal should present itself as a white Gaussian process, implying that the joint statistics of any set of samples must prove Gaussian. To check second-order statistics experimentally, the joint density of pairs having a common time spacing must be developed. Stationarity comes under test here; if for some reason points spaced by, let us say, 1 microsecond are highly correlated, it could escape notice in a histogram test but should be strongly evident in the autocorrelation function. Such tests become difficult and computationally expensive as the number of variables and spacings increases, and are not widely used beyond second order.

White Power Spectrum (LPD). The signal power spectrum should be bandpass white, so that no one part of it is more important than any other. The flat spectrum is of course the minimax solution to making the largest spectral value as small as possible, denying an interceptor any favored band for detection. Similarly, a jammer is offered no selectivity in his frequency allocation.

Low Power (LPD). The advantage of a low power transmission is obvious. One component of achieving this is through energy efficient transmission. The energy efficiency of a communication waveform refers to the required value of the bit SNR, E_b/N_0, at the receiver to achieve a specified performance measure (usually the bit error rate, BER). Efficient modulation supports the aim of LPD by keeping the required transmitter power near the theoretical minimum, denying precious power to an interceptor. As a reference point, assume a system requirement for BER$= 10^{-5}$ is in force for an additive white Gaussian noise (AWGN) channel. Using uncoded coherent binary phase-shift keying (BPSK) modulation, this is achieved at $E_b/N_0 = 9.6$ dB. Adding a rate-1/3 convolutional code drops the requirement by about 5 dB and turbo codes can operate within 1 dB or so of Shannon's theoretical limit of -1.4 dB [296]. In viewing this 10 dB spread between the mundane and the sophisticated, it is well to remember that the alternatives cannot meaningfully be compared without a concomitant processing complexity evaluation.

Low Power Spectral Density (LPD). Not only should the power be low, its spectral density should be low as well. Of course, any advantage accrued through low power transmission translates proportionately to power spectral density reduction. Further reduction comes at the price of increased bandwidth, which is why LPI/LPD transmissions use spread spectrum signals. A quantitative reference for "low" is the receiver thermal noise power spectral density. An informed link budget calculation indicates approximately where the received signal will lie relative to the noise floor at a given intercept receiver. Typical spread spectrum systems are designed to place the received signal at least 10 dB below the noise floor at any unintended site.

Since we can't count on correctly anticipating the lower limit of resolution in an observer's spectrum analyzer, the uniform spectral density may have to be achieved "instantaneously" (everywhere in the band at once) rather than "on the average." This would lead to use of direct sequence (DS) waveforms rather than, for example, frequency hopping (FH). In leading to comments about different manifestations of spread spectrum, the discussion has almost allowed two monsters to escape from behind doors we would rather keep closed: (1) the subject of simultaneous time and frequency resolution, i.e. the Heisenberg inequality, and (2) the comparison of DS and FH systems. Both subjects deserve much more space than we intend to explore.

Lack of Cyclostationary Features (LPD). Many signals contain what are described as "hidden periodicities," that is, aspects that, while not themselves periodic, are capable of generating a periodic component when subjected to some transformation. No linear time-invariant transformation can map that which is aperiodic into something periodic, so the choices are nonlinear or linear time-varying. The most obvious example of cyclostationarity is a real-valued bi-level waveform consisting of only ± 1 values. Squared, this waveform becomes a direct-current (DC) constant with a spectrum consisting entirely of an impulse at 0 Hz.

Communication system design has for years relied on the development of spectral lines from the received signal for synchronization and tracking purposes. An example is the extraction of a "$4\times$ carrier" line in balanced quadrature phase-shift keying (QPSK) to provide frequency or phase lock. LPI/LPD design principles directly deny this possibility in asking for suppression of cyclostationarity. Some interesting consequences of this philosophy are discussed later in the chapter when LPI receiver design issues are discussed.

The most general quadratic process to which a real-valued signal can be subjected is an integrated lag product with a weighting function, as in

$$\int\int s(u+\tau_1)K(u,v)s(v+\tau_2)\,du\,dv.$$

For present purposes we set the weighting function equal to the impulse $\delta(u-v)$. When a lowpass signal is squared it produces both a baseband and a double carrier lobe, each of which can be examined separately. Integrating the squared signal is simply energy detection, which manifests itself as a spectral spike at 0 Hz. If there are hidden periodicities such as a uniform chip rate in the signal, the Fourier transform of a lagged product might, depending on the modulation, show line components at the chip rate frequency in either the baseband or double carrier lobes. Carrier components not suppressed by the modulation will show up as a line at the double carrier frequency. Put through a 4th law device, signals such as QPSK can show "energy" ($\int s^4$) or chip rate lines at baseband and a carrier line at the quadruple carrier frequency. Additional examples are found in [101] and the references therein.

Lack of Repetition (LPD). Given that the practice of cyclostationarity is proscribed, it should be no surprise that its cousin, repetition, is equally undesirable. The two differ (and are almost opposite) in that the hidden repetition in cyclostationarity is periodic, whereas repetitions are not hidden, but need not be periodic. As an example, the sequence ($\underline{1234}94867\underline{1234}55\underline{1234}9706459\underline{1234}$ ...) shows no period, but contains four copies of the subsequence, 1234.

One threat against repetition is an autocorrelation attack. Certain correlation lags may show highly nonrandom behavior and tip off an observer. Periodic repetition is even more highly susceptible to disclosure via autocorrelation.

4.2.3 Techniques to Achieve the Criteria

How can the criteria discussed above be achieved? Some suggestions, as opposed to an exhaustive list of ways, are offered for each criterion.

Gaussian Marginal Distribution. Many spread spectrum signals start with binary pseudo-noise (PN) codes, the amplitude density of which is about as far from Gaussian as possible (impulses of area 1/2 at ± 1). Techniques to "Gaussianize" such a signal are mathematically non-rigorous at best. Consequently, experimental evidence becomes very important. At least one technique, in which some low-order moments of the signal distribution are forced to closely match those of the Gaussian, has been successfully employed [277].

Filtering clearly destroys a highly tailored amplitude distribution like the binary. The appeal to the central limit theorem is that a linear, time-invariant filter creates at its output a linear combination of many identically distributed, and possibly uncorrelated, random variables. Suppose we create a binary code by drawing $+1$s and -1s equiprobably from a good random number generator and use these as a sequence of uniformly spaced inputs to a digital filter with a rectangular impulse response. Since we have taken steps to assure that the input sequence power spectrum is white, the power spectrum of the output matches that of the filter, in this case a $(\sin Nx/N \sin x)^2$ function. If we pass the resultant through a second filter that has a bandpass characteristic lying within the main lobe of the spectrum, the output can be expected to have a roughly Gaussian distribution. Unless the bandpass filter bandwidth is small relative to the code rate, the output spectrum will not be close to white. This can be fixed by inversely tapering the spectral flatness; its disadvantage is that to achieve a specified output spread bandwidth, the input code rate must be considerably higher than the bandwidth, and code generators are known consumers of power (analog) or computation rate (digital). (The latter is a real issue for inexpensive consumer equipment, such as hand-held units, but relatively innocuous for a satellite-based transmitter or receiver.) Due to lossy bandlimiting, the original binary code cannot, of course, be recovered at the receiver. In this case, the receiver's best option is to run an identical code generator as the local correlation reference.

A second way to process binary into Gaussian is with a filter bank. This method is discussed in some detail later, so for the present it suffices to say that although the Gaussianization mechanism is the same — lots of linear combinations going on — the filter bank offers some flexibility that the single filter does not. In particular, the filter bank can perform invertible, even orthogonal, transformations that allow the receiver to work with either the Gaussian or binary signal; also, the code generator never needs to run at a rate exceeding the spread bandwidth.

The third method is more deterministic. One starts with a signal set having certain desired properties, e.g. orthogonality, and maps it into a new signal set via a linear transformation. The "right" transform will map the signal values into new ones showing a Gaussian histogram. The trick, naturally, is to find the transformation. In one technique [277], the ultimate transform is constructed piecemeal as the product of Givens rotations. (A Givens rotation is a 2×2 unitary coordinate rotation matrix that can be applied to any two coordinates in the higher-dimensional matrix.) Each rotation is selected to optimize the match between the selected moments of the signal matrix entries (viewed as a collection of sample values) and the moments of the desired Gaussian distribution. The cost function driving this optimization is weighted least squares match of the form

$$C = \sum_{n=n_1}^{n_2} w_n (M_n - \hat{M}_n)^2,$$

where the $\{M_n\}$ are the target moments and $\{\hat{M}_n\}$ are the sample moments. A nonlinear optimization routine performs the iterations [277].

Complex Gaussian Marginal Distribution. All the material discussed above applies here as well. All that needs to be added is that whatever signal generation mechanism is used to produce a real-valued white Gaussian process, a similar method should be used to generate a second such signal that is statistically identical but statistically independent. These two signals become the baseband components of a new signal synthesized by up-converting the two real components to IF via quadrature-phased oscillators. If the statistics are as prescribed above, there is no memory in the signal of the I and Q axes, at least not to those unaware of the signal construction details, and the amplitude and phase have Rayleigh and uniform densities, respectively. The trickiest part is to assure the independence of the quadrature components. See comments below in this regard.

Higher Order Moments. There is no theoretical basis beyond the central limit theorem — to whatever extent it may apply — for claiming much knowledge about the higher order moments of a process obtained by filtering a wideband BPSK or QPSK code. Currently available evidence is anecdotal and experimental. Of course, for signals obtained using the moment-optimized transform technique [277], there is reason to expect moment conformance, but only up the highest order moment included in the optimization cost function. Beyond that there is no guarantee.

Autocorrelation. A design technique that generates a white process has the opportunity to be given any desired (and realizable) autocorrelation function, simply by passing the process through the appropriate linear time-invariant filter. Generally, the only desired modification is bandlimiting, and this may be accomplished by applying the real- or complex-valued sequences (generated in the design) to an analog or digital pulse-shaping filter. By looking at or performing tests on the output of the filter one should not be able to decide whether that output was generated from a white noise sequence or some other structured process.

Power Spectrum. Due to the Wiener-Khintchine theorem, which identifies the power spectrum as the Fourier transform of the autocorrelation function, the power spectrum becomes no more than a second way to address the concerns raised in the preceding paragraph. Periodicity, however, shows up strongly in a spectral display as lines.

Low Power. Two classical techniques for energy efficient signaling are large alphabet orthogonal (or simplex or biorthogonal) and error-correction coding applied to almost any base modulation. Some of the complexities of bandwidth efficient modulation can be avoided by shunning these for LPI/D use, since there is plenty of bandwidth available by assumption. There are wavelet-based versions of almost all suitable modulations, some of which are discussed later on. It should be noted that the spread spectrum technique usually does not contribute to or detract from the energy efficiency of a modulation scheme; this holds whether the modulation and spread spectrum are applied separately or together.

Low Power Spectral Density. In light of the previous discussion, it should be clear that to meet low power spectral density requirements, one must (1) make a signal with a flat spectrum and as much bandwidth as possible, or at least as much as you need, whichever is the limiting factor, and (2) use spreading techniques that give wide instantaneous bandwidth rather than those that hop a narrowband process over the available bandwidth. In situations where intercept resolution is not a concern, hybrid DS/FH systems can be a quite suitable compromise.

Lack of Cyclostationary Features. Our model process, AWGN, is stationary, therefore *not* cyclostationary. The feature suppression ideas mentioned here are intended to deny the designed signal any appearance of cyclostationarity.

There are two common causes for cyclostationarity in a communication system. The first arises from a constant rate of symbol transmission, and the second is due to the carrier frequency. A BPSK data stream on a PN code, for example, will be cyclostationary unless certain precautions are taken. In particular, pulse signal characteristics such as highly rectangular envelope shapes tend to exaggerate hidden periodicities. Causing the pulse spectrum to go to zero at frequency offset $1/T$ from the carrier can be shown to eliminate second order cyclostationarity [263]. In general, cyclostationarity of order $2n$ is eliminated by spectral confinement to $\pm 1/nT$.

The special case of Gaussian statistics is exceptional in this regard. A mathematically Gaussian random process has a moment structure such that its central *cumulants* above second order are zero. (Cumulants are the characteristic functions of powers of the variable.) To the extent that a process has jointly Gaussian statistics, its central cumulants above second order will vanish. An implication of this turns out to be that higher-order cyclostationarity is avoided when second order cyclostationarity is made to vanish, hence the importance of achieving Gaussian statistics in LPI signal design.

Lack of Repetition. Repetition is eliminated or minimized by creating modulation formats having a large number of parameters that can be varied rapidly. If the control of that variation is a truly random source, the resulting signal is unlikely to repeat a segment. Of course for successful communication to occur, the variation cannot truly be random and unknown to the intended recipient. It must be pseudo-random, meaning it can be constructed from some algorithm once the algorithm has been properly initialized. Exchange of the initialization information, which with some abuse of terminology we call *key variables*, is a problem that must be solved between communicators prior to their session. This problem of *key distribution* is, however, well outside our scope of discussion. The whole area is known within Government circles as TRANSEC (transmission security). A natural consequence of pseudo-randomness is the requirement to synchronize the pseudo-random generators at the two ends of the link.

PN codes are a well-known source of random-looking numbers. Those that come from linear shift registers can be broken by linear algebraic techniques if the PN sequence is directly observed. Shift register codes are periodic, since the register contents must eventually repeat. This limits their use to either other applications where coding is used more for band spreading and/or identification than for security, e.g. CDMA cellular phones, Global Positioning System (GPS) acquisition and ranging codes, or cases where the duration of their use is short compared to the period (the GPS "P" code has an inherent period in excess of 38 weeks, only one week of which is used by any one satellite). Nonlinear registers and the nonlinear codes they generate can be much more difficult to crack.

The most familiar DS spread spectrum systems use PN codes in a remarkably straightforward manner, the code simply multiplies the modulated baseband data, creating a wideband pseudo-random signal. Although the code exhibits itself quite directly in doing so, an LPI/D application is typically designed such that the SNR at any unintended recipient will be so low that reading individual chip values of the code should be a virtual impossibility. In addition to DS spread spectrum, there are more subtle uses for a PN code. If the modulation contains some integer-valued parameters, multiple bits of the code can be grouped to yield a binary representation of the integer. Codes with longer periods can be generated from shorter ones by systematic combination. The most striking example is the family of Gold codes, which have pleasing auto- and cross-correlation properties and are created as the modulo-2 sum (exclusive or) of two shift register sequences.

In conclusion, adroit use of pseudo-randomness can either remove repetition or reduce its occurrence to a tolerable level. Pseudo-random behavior is a key component of LPI/D waveform construction.

4.2.4 Traditional Signal Types

Several types of "traditional" LPI/D spread spectrum signals are most often discussed in the literature:

Fast Frequency Hopping. For this signal, the transmitter rapidly hops a carrier (pseudo-randomly) among a large number of center frequencies. The bandwidth of each hopped portion of the signal is determined by the hop rate. We refer to each hop as a "cell." The energy distribution of each cell has a sinc-squared shape in the frequency dimension (assuming fairly sharp hops with constant energy output in the time dimension). If the hop duration is T, it is common to take the cell bandwidth to range from the hop center frequency $\pm 1/2T$. The cell has a time-bandwidth product of unity and includes most of the cell's energy.

Time Hopping (TH). TH is similar to fast FH, except the carrier frequency is constant and the transmitter only transmits during one of several pseudo-randomly selected time slots.

Fast FH/TH. This is a combination of the two techniques described above. In all three signal structures mentioned thus far, the time-bandwidth product of each cell is unity.

DS. For this signal, the transmitter modulates the information with a high frequency pseudo-random waveform (known to the intended receiver). The effect is to spread the signal over a much wider bandwidth than it would otherwise occupy. The pseudo-random waveform is often a random binary waveform, spreading the signal energy under a sinc-squared envelope.

Fast FH/DS. For this signal, each fast FH cell is further spread in frequency with a pseudo-random waveform, creating cells with time-bandwidth products much greater than unity.

TH/DS and Fast FH/TH/DS. These combination techniques have cell time-bandwidth products much greater than unity.

Slow FH. In this case, the carrier signal is first modulated by the information using multiple frequency shift keying (MFSK). The transmitter then hops the signal; relatively slowly such that each cell's bandwidth is determined by the MFSK waveform. Here, the cells' time-bandwidth products are greater than unity. Phase modulation such as differential phase-shift keying (DPSK) is also compatible with slow FH.

4.3 TRANSFORM-BASED LPI CONSTRUCTS

All transform domain systems are derived from an easily explained common paradigm. First of all, a set of basis functions spanning the signal space of interest is established. The basis need not be orthogonal. A few examples are:

- The Fourier harmonics over an interval of length T and their periodic extensions
- Classical polynomials such as Laguerre, Legendre or Hermite functions
- Time-frequency translates of a "window" function
- Time-scale translates of a "mother wavelet"

A basis is of course used to represent elements of the space it spans as a linear combination, as in

$$s(t) = \sum_n c_n x_n(t),$$

where $\{x_n(t)\}$ represents the basis and with expansion coefficients, $\{c_n\}$, given by

$$c_n = \int s(t)\tilde{x}_n(t)\,dt.$$

When the system is orthogonal, $\{\tilde{x}_n(t)\}$ are identical or proportional to $\{x_n(t)\}$; when nonorthogonal, $\{\tilde{x}_n(t)\}$ are defined relative to a set of functions called the *biorthogonal* set. As far as this chapter is concerned, our objects of interest reside in the Hilbert L^2 space of square-integrable functions or its discrete equivalent, l^2 (square-integrable sequences).

In a transform domain system, data is manipulated into a form where it can play the role of expansion coefficients, and the transmitter's main function is to synthesize the signal corresponding to those coefficients. Figure 4.1 illustrates the functional steps that occur at both ends of a transform domain system. The

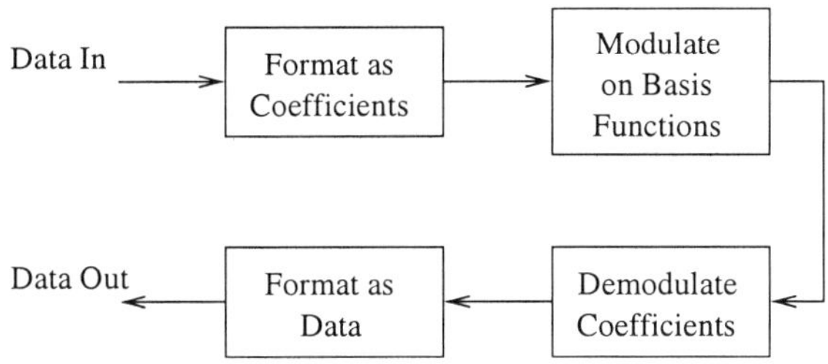

Figure 4.1 Transform domain transmit and receive functions.

transform domain construct describes many familiar modulation methods. As an example consider pulse position modulation (PPM), for which the defining equation is

$$s(t) = \sum_n c_n x(t - nT),$$

permitting the identification

$$\{x_n(t)\} = \{x(t - nT)\}.$$

Such sets have been called a *repetitive basis* [240]. In general, the functions $\{x_n(t)\}$ need not be a basis for the Hilbert space. They may instead span some subspace. This is certainly the case for the PPM repetitive basis. When the prototype function $x(\cdot)$ is the familiar sinc function, that is,

$$x(t) = \frac{\sin \pi t/T}{\pi t/T},$$

the basis spans the set of all functions bandlimited to $(-1/T, 1/T)$, but certainly not L^2.

A second example is a Gabor, or Weyl-Heisenberg basis. These are best described as a doubly-indexed set again generated from a prototype or *window* function $x(\cdot)$, namely

$$\{x_{m,n}(t)\} = \{x(t - nT) \exp(j2\pi t/T)\}.$$

Two orthogonal Gabor sets are known; $x(\cdot)$ is either a rectangular pulse or its Fourier dual, a sinc pulse. For other windows the set is nonorthogonal and may or may not be complete in L^2 [119].

As our final example we consider continuous-time wavelets, for which the set of functions consists of translations and dilations of a "mother wavelet,"

$$\{x_{m,n}(t)\} = \{x(a^m t - nT)\}.$$

When $x(\cdot)$ satisfies certain technical conditions that make it a wavelet, the function set, orthogonal or nonorthogonal, can be complete in L^2.

4.3.1 Single and Multiscale Systems

In some of the discussion that follows, we draw a distinction between single-scale and multiscale systems. This distinction is exactly what one should expect based on the nomenclature: a single-scale system contains signals all drawn from just one scale of the wavelet family, while a multiscale system has members representing more than one (possibly an infinite number of scales). The repetitive basis expansions mentioned above are a single-scale construct. A useful approach to multiscale systems is developed below, following some motivation by a comparable Fourier domain construct.

Beyond the exclusion of all but one scale in the former concept, there is no fundamental difference between the constructs. They simply represent alternatives for tiling the time-frequency plane. It is interesting to learn that wavelets, whose most highly advertised feature is scale, can be used to form valid communication construct when the scale variable is denied. Even when the wavelets are restricted to the two-band case, the function set consisting of all single-scale translates of the scaling function and the wavelet has more generality than pure PPM. The orthogonality of the scaling function and the wavelet might be likened to that of sine- and cosine-phased pulses, but in the wavelet case the scaling function and wavelet may appear much more dissimilar than the trigonometric pulses. For something drastically new to occur, the fundamental pulses must come from an M-band set, preferably for large values of M. The basis now consists of all single-scale translates of the $M-1$ wavelets and the scaling function. This idea is further developed in Section 4.3.3

4.3.2 A Fourier Transform Domain Communication System

The primary goal of this section is to illustrate the foundation logic of a Fourier transform domain (FTD) system, motivating a comparable architecture, based on wavelets and called a wavelet transform domain (WTD) system (developed in Section 4.3.4). The example illustrates how a Fourier transform domain approach can produce a featureless waveform.

Architecture. Consider the discrete-time communication system in Figure 4.2. Input bits presented serially are converted to parallel in groups of m and the information borne by each group is transmitted via a single channel symbol by selecting one of $M = 2^m$ signals. (Any error-correction coding that may have been previously applied is relevant to this discussion, beyond the observation that the FTD input then consists of encoded bits.) The orthogonal alphabet here is pulse position modulated (PPM), illustrated in Figure 4.3. In PPM, the signals are discrete impulse sequences of length M and taking on the value 1 in a single position and value 0 in the remainder.

An impulse has maximum energy concentration in the time domain but maximum spectral spread in frequency. Its spectrum is white with a linear phase slope proportional to the impulse delay. Denoting the time slots as 0 to $M-1$, the impulse at $t = 0$ has a spectrum of constant phase, but the one at $t = k$ accumulates $2\pi k$ radians of phase across its M Fourier coefficients,

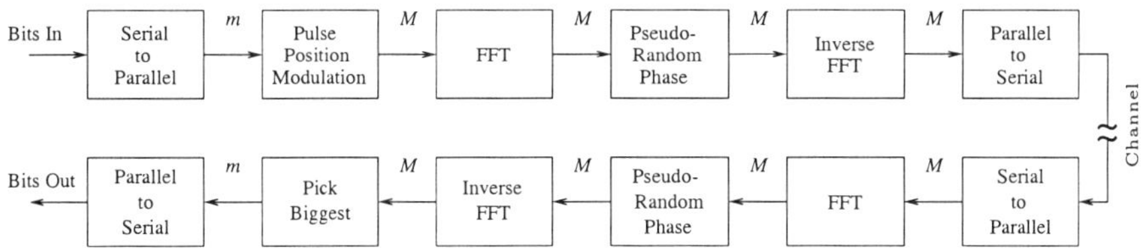

Figure 4.2 A Fourier transform domain communication system.

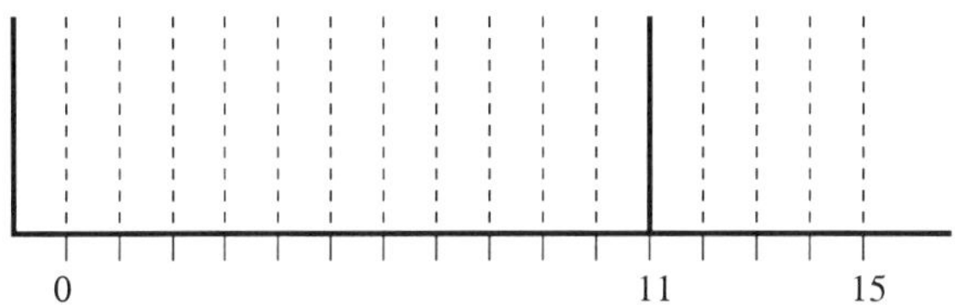

Figure 4.3 Example of a 16-ary PPM signal set. In this example, signal 11 is selected.

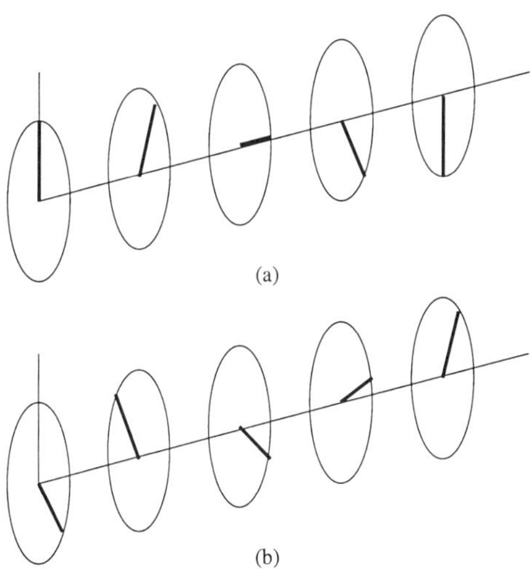

Figure 4.4 Phase randomization. (a) Uniform phase associated with Fourier coefficients of a pulse in a non-zero location and (b) phase after randomization.

or $2\pi k/M$ radians per coefficient. Next, we apply an independent pseudo-random phase shift to each of the M frequency components, such that the phase angles behave as though chosen at random from a uniform distribution on $[0, 2\pi)$. When the result has been returned to the time domain by an inverse finite Fourier transform (FFT), or IFFT, and converted to serial, it no longer resembles the original impulse.

After the waveform is passed through an AWGN channel, the communication receiver reverses the synthesis process (cf. Figure 4.2). The FFT of the sequence is taken and the phase pseudo-randomization is removed by application of the conjugate of the phasors impressed at the transmitter. Except for effects due to additive noise, the resultant has a linear phase function that causes its IFFT to collapse mainly into one location in time. Since the FFT is an orthogonal transform, the noise components of the Fourier coefficients are independent, identically distributed (i.i.d.) Gaussian variates. The optimum demodulation decision is made in favor of the time sample having the greatest algebraic value, and the corresponding serial bit stream may be reconstituted.

The phase randomization spreads the energy in time without disturbing the spectral flatness. When M is large, the high time-bandwidth product waveform out of the IFFT is noiselike, and its statistics become close to Gaussian according to the central limit theorem as induced by the linear combination of i.i.d. random variables formed by IFFT. Any two PPM waveforms corresponding to distinct M-ary symbols are orthogonal, as is easily concluded from the time domain representation, and they remain orthogonal under the FFT.

Observe that there is no need to actually originate the waveform in the time domain and carry out the FFT. One can start with the set of phase shifts appropriate to the pulse position, increment them by the pseudo-random sequence and take the IFFT. To achieve roughly 1° granularity in phase, 8 bits of PN sequence are needed per Fourier coefficient. For this function alone the PN code rate must exceed the chip rate by a factor of 8.

Computation. This FTD system exhibits the bandwidth expansion factor found in any M-ary orthogonal modulation, $M/\log_2 M$. One can add processing gain by using an FFT of length $N = KM$, for $K \gg 1$, and using for data transmission only a subset of size M of the possible PPM signals, e.g., those corresponding to pulses in positions kK where $0 < k < M-1$. The processing gain G (ratio of chip rate to bit rate, or number of chips per bit) is the product of the two bandwidth expansion factors

$$G = KM/\log_2 M \quad \text{chips/bit}. \tag{4.1}$$

Both the modulation and demodulation processing loads are dominated by the FFT. In this case, computational requirements are proportional to $KM \log_2 KM$ operations per transform, or $\log_2 KM$ per chip. Phase rotation accounts for one additional chip operation, a complex multiply. Denoting by R_c the chip rate, we have

$$R_{Tx}|_{FTD} = R_{Rx}|_{FTD} = (\log_2 M + \log_2 K + 1)R_c. \tag{4.2}$$

If we express the results in terms of the processing gain by eliminating K from (4.1) using (4.2), we have

$$R_{Tx}|_{FTD} = R_{Rx}|_{FTD} = [\log_2(G \log_2 M) + 1]R_c. \tag{4.3}$$

We later compare (4.3) to a corresponding result for WTD communications.

LPI/D Properties. An FTD system is classified as single scale. How does it measure up in terms of ability to achieve LPI/D properties? We have observed that as the number of coefficients increases, the time domain function takes on a Gaussian distribution, again in accordance with the central limit theorem. In the frequency domain the waveforms are white on a sample spectrum basis because the Fourier coefficients are all of unit magnitude. This falls into the "too good to be true" category; the transmitted signal would be more like natural white noise if the sample spectra were statistically white. For an observer to pick this up, however, he would have to be in an advantaged SNR posture and performing the IFFT on the correct block boundaries — not impossible, but usually difficult as M grows large. Notice that the autocorrelation function has a constant peak value (the phase conjugation makes the coefficients of equal magnitude and zero phase) and sidelobes that are as Gaussian as the waveform samples.

If the Fourier coefficients are created at baseband without the normal conjugate symmetry of a real-valued signal, the signal becomes complex valued, corresponding to independent quadrature components at RF. Each component should be Gaussian, and the statistical independence renders them jointly Gaussian. For there to be lack of repetition requires only that the PN generator producing the random phases of the spectral components have no repetition over the planned transmission interval. This brief survey finds the Fourier domain construct meeting most of the significant criteria.

4.3.3 A Single-Scale Wavelet Construct

The family of waveform designs reported here is responsive to the criteria of Section 4.2.1. A method that derives from genus-1 M-band wavelet coefficient matrices leads to a deterministic construction that supports all these criteria.

Waveform Selection. It is well-known that the most efficient, equi-energetic M-ary waveform set is the simplex set [9]. For BER around 10^{-5}, the 16– and 64-ary simplex are about 2 and 3 dB superior, respectively, to QPSK. Simplex modulation can be described as an orthogonal set with the mean of the signal vectors subtracted from each vector, i.e. a zero-mean version of orthogonality (cf. Figure 4.5). The simplex is slightly more efficient than orthogonal signals for equivalent performance, although for large alphabets this advantage asymptotically vanishes. More important for LPI is the fact that removal of the mean also removes a linear feature from the waveform; an interceptor who knows the signal format exactly cannot simply average the signal over successive transmission of the simplex words and expect to coherently accumulate energy that indicates signal presence.

Next we show how to develop such a set using wavelets. The procedure summarized below is also discussed in [277]. In this signal format, arbitrarily long M-ary simplex codewords are constructed from an M-band wavelet matrix by transposing the matrix and deleting the column that represents the lowpass wavelet, or scaling function. N copies of this new matrix are row-rotated

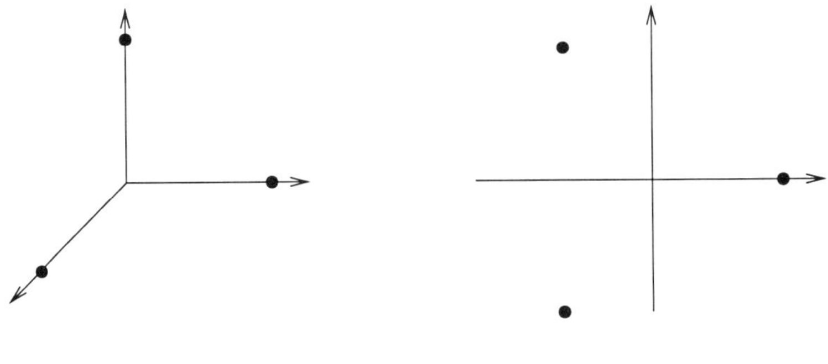

Figure 4.5 A simplex constellation as orthogonal signals with mean removed: (a) 3-D orthogonal constellation, (b) equivalent 2-D simplex constellation.

under control of a pseudo-random generator and concatenated to form an $M \times N(M-1)$ matrix whose rows are the codewords for one M-ary transmission. The next transmission is under control of a new portion of the PN sequence and correspondingly has different codewords. The following is a step-by-step description of the construction.

Mathematical Model of the Waveform. We begin with the definition of an M-ary wavelet coefficient matrix (WCM).

Definition 1 *A real-valued $M \times M$ matrix $\mathbf{A}$ is a WCM for orthogonal wavelets of compact support if its entries $\{a_{i,j}\}$ satisfy the following conditions:*

$$\sum_{j=0}^{M-1} a_{i,j} = M\delta_i \quad \text{and} \quad \sum_{j=0}^{M-1} a_{i,j}a_{i',j} = M\delta_{i,i'}.$$

$\mathbf{A}$ is thereby an orthogonal matrix satisfying $\mathbf{A}\mathbf{A}^t = M\mathbf{I}$, implying that the matrix $\mathbf{A}/\sqrt{M}$ is unitary. Although WCMs need not represent systems that are orthogonal or of compact support, we will use the designation WCM to imply both conditions throughout. The following theorem reveals the structure of $\mathbf{A}$.

Theorem 1 *The matrix $\mathbf{A}$ is a WCM if and only if it is of the form*

$$\mathbf{A} = \begin{bmatrix} 1 & \mathbf{0}_{M-1}^t \\ \mathbf{0}_{M-1} & \mathbf{U}_{M-1} \end{bmatrix} \mathbf{H}_M, \tag{4.4}$$

where $\mathbf{U}_{M-1}$ is an $(M-1) \times (M-1)$ unitary matrix, $\mathbf{0}_{M-1}$ is the length $M-1$ column vector of zeros, and $\mathbf{H}$ is the $M \times M$ canonical Harr matrix developed by the iteration

$$H_2 = \begin{bmatrix} 1 & 1 \\ 1 & -1 \end{bmatrix}, \quad H_{2^{m+1}} = \begin{bmatrix} H_{2^m} & H_{2^m} \\ H_{2^m} & -H_{2^m} \end{bmatrix}.$$

Proof: See [120].

We note that the matrix multiplying $\mathbf{H}_M$ in (4.4) is itself unitary and that $\mathbf{H}_M$ itself is orthogonal, satisfying $\mathbf{H}_M\mathbf{H}_M^t = M\mathbf{I}$. Partitioning $\mathbf{H}_M$ as

$$\mathbf{H}_M = \begin{bmatrix} 1 & \mathbf{1}_{M-1}^t \\ \mathbf{h}_{10} & \mathbf{H}_{11} \end{bmatrix},$$

where $\mathbf{1}_{M-1}^t$ is the $M-1$ vector of 1s, we can write

$$\mathbf{A} = \begin{bmatrix} 1 & \mathbf{0}_{M-1}^t \\ \mathbf{0}_{M-1} & \mathbf{U}_{M-1} \end{bmatrix} \begin{bmatrix} 1 & \mathbf{1}_{M-1}^t \\ \mathbf{h}_{10} & \mathbf{H}_{11} \end{bmatrix} = \begin{bmatrix} 1 & \mathbf{1}_{M-1}^t \\ \mathbf{U}_{M-1}\mathbf{h}_{10} & \mathbf{U}_{M-1}\mathbf{H}_{11} \end{bmatrix}.$$

Thus $\mathbf{A}$ results from a unitary mapping of $\mathbf{H}_M$ restricted to preserve the first row of $\mathbf{H}_M$.

Now consider the matrix $\mathbf{B}_M\mathbf{A}^t$. $\mathbf{B}$ is orthogonal, and its columns — except for the first — sum to zero. Deleting the first column of $\mathbf{B}$ to form $\mathbf{C}$ yields

$$\mathbf{C} = \begin{bmatrix} \mathbf{h}_{10}^t\mathbf{U}_{M-1}^t \\ \mathbf{H}_{11}^t\mathbf{U}_{M-1}^t \end{bmatrix}.$$

$\mathbf{C}$ is thus an $(M-1)\times M$ matrix, the sum of whose rows is the all-zero row vector. Clearly, each row of $\mathbf{C}$ has squared L^2 norm equal to $M-1$, since the row differs from that in the orthogonal $\mathbf{B}$ only by the deletion of a first entry which is a 1. Similarly, the inner product of any two distinct rows of $\mathbf{C}$ will equal -1, because these row vectors would have been orthogonal without deletion of the leading 1.

Definition 2 *For any positive number k, a collection of M vectors whose squared lengths equal $k(M-1)$, whose inner products of distinct vectors equal $-k$, and whose sum is the zero vector is called an M-ary simplex. A matrix whose rows are a simplex set is said to be simplex.*

Theorem 2 *The rows of matrix* $\mathbf{C}$ *constitute an M-ary simplex.*

Proof: The proof follows directly from the definition of a simplex and the observations preceding the definition.

We have constructed a simplex signal set whose detailed structure depends on the unitary matrix selected in (4.4). This is the basis for a more complicated, pseudo-randomly time-varying simplex set derived next. Construct a matrix $\mathbf{F}$ by concatenating N copies of $\mathbf{C}$ where N is any positive integer, i.e.

$$\mathbf{F} = [\mathbf{C}\ \ \mathbf{C}\ \ \cdots\ \ \mathbf{C}].$$

$\mathbf{F}$ is evidently simplex. Each $\mathbf{C}$-matrix in $\mathbf{F}$ is called a *block*. We now operate on the blocks of $\mathbf{F}$ as follows: (1) each $\mathbf{C}$-matrix may have its rows cyclically permuted any number of times between 0 and $M-1$, and (2) the matrix $\mathbf{C}$ may be replaced by its negative, $-\mathbf{C}$. These operations are performed independently on each copy of $\mathbf{C}$, and are pseudo-randomly driven. Each of the M possible permutations is taken to occur with probability $1/M$, and the algebraic sign is positive or negative with probability 1/2. Accordingly, $\mathbf{F}$ is rewritten as

$$\mathbf{F} = [\mathbf{C}_1 \ \mathbf{C}_2 \ \cdots \ \mathbf{C}_N], \tag{4.5}$$

where each $\mathbf{C}_k$, $k = 0, 1, \ldots, M-1$, represents a $\mathbf{C}$-matrix that has possibly been cyclically shifted and/or multiplied by -1.

Theorem 3 $\mathbf{F}$ *is simplex.*

Proof: Row permutation alters none of the three simplex properties, i.e., the row norm, the row inner product and the mean row vector. The same is true if any $\mathbf{C}_n$ is replaced by its negative. Thus the row sums of $\mathbf{F}$ become $(M-1)N$, the inner product of two distinct rows is $-N$, and the mean row is zero.

We extend the construction slightly by assuming that two such $\mathbf{F}$-matrices are generated. The matrices, denoted $\mathbf{F}_I$ and $\mathbf{F}_Q$, are each generated by the algorithm given above, but using *statistically independent* permutations and sign changes. The use of these matrices is seen in the conversion of the discrete sequences of matrix rows into an analog signal. $\mathbf{F}_I$ and $\mathbf{F}_Q$ are used to construct two analog waveforms as follows. Let $\{f_{I,k}\}$ and $\{f_{Q,k}\}$ be the sequences (I and Q row vectors) to be transmitted at a given time. Form the pulse-amplitude modulated waveforms

$$s_I(t) = \sum_{k=1}^{N(M-1)} f_{I,k}p(t-kT) \quad \text{and} \quad s_Q(t) = \sum_{k=1}^{N(M-1)} f_{Q,k}p(t-kT), \tag{4.6}$$

where $p(t)$ is a pulse waveform whose detailed properties are selected to meet certain conditions: (1) to suppress second-order cyclostationary behavior, the pulse should be bandlimited to less than half the chip rate [263], and (2) to prevent chip-to-chip intersymbol interference (ISI), the pulse autocorrelation function should be zero at all lags equal to a nonzero multiple of the chip spacing. These somewhat conflicting requirements are best resolved in favor of maintaining a desired pulse spectrum, inasmuch as the processing gain tends to average the ISI to a level below thermal noise. A good example to keep in mind is a pulse whose baseband bandwidth is approximately $1/2T$ Hz and whose spectral occupancy is rather uniform across the band, e.g. a $\sin x/x$ pulse.

The I and Q components are modulated onto in-phase and quadrature components, respectively, of an RF carrier. The complex baseband representation of the composite signal is

$$s(t) = s_I(t) + js_Q(t). \tag{4.7}$$

At IF, it can be regarded as

$$s(t) = s_I(t)\cos 2\pi ft + s_Q(t)\sin 2\pi ft.$$

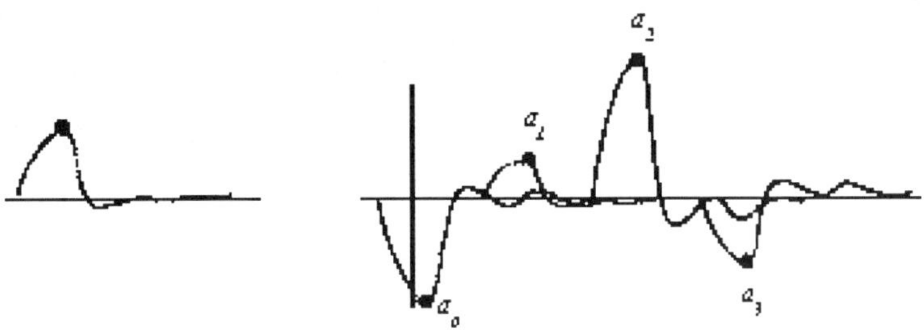

Figure 4.6 Example of a pulse and pulse superposition.

Because the I and Q channel signals are statistically independent, the net signal exhibits both amplitude and phase modulation and may be regarded as a form of quadrature amplitude modulation (QAM).

We have assumed M to be a power of two, $M = 2^m$. For each m bits of data to be transmitted, the randomization of the **F**-matrices is repeated. Thus the transmitted signal has virtually no chance of repeating a sequence if the period of the PN generator is great enough. The only constant feature of the waveform is the subsequence structure inherent in the use of a constant **C**-matrix throughout.

In the construction we talk of random permutations and sign changes. In the realization of the waveform, the permutations and signs are controlled by PN generators whose properties can be as close to ideal as desired, and whose periods can be arbitrarily long.

Waveform Detectability. Figure 4.7 shows an I/Q realization of a 1024-chip, $M = 16$ codeword corresponding to four input bits and, for comparison, an equivalent sample of bandlimited Gaussian noise. The appearance of the two waveforms has been compared to bandlimited noise in terms of its histogram, I/Q scatter plot, autocorrelation function and power spectrum; in none of these measures is a significant discrepancy between the two waveforms found. The histogram and scatter plots are found in Figures 4.8 and 4.9, respectively.

By limiting the bandwidth to less than half the chip rate, the waveform has inherent protection against second-order cyclostationarity. To the extent that Gaussian statistics are maintained, fourth- and higher-order detection attempts will fail, since cumulants of the Gaussian distribution higher than the second vanish. (Band limitation to less than one quarter the chip rate would also eliminate 4th order cyclostationarity irrespective of process statistics.) Figures 4.10 and 4.11 compare the wavelet waveform and a conventional QPSK waveform under conditions where the chip filtering is adjusted for both waveforms to suppress the carrier line at the doubled carrier frequency and the chip rate lines at the baseband lobe. Figure 4.10 shows the 4× carrier lobe out of a 4th law detector for both waveforms; a carrier line is evident in the QPSK but not in the wavelet signal. Similarly in Figure 4.11 we see QPSK rate lines protruding from the baseband lobe, but no comparable phenomenon of the wavelet.

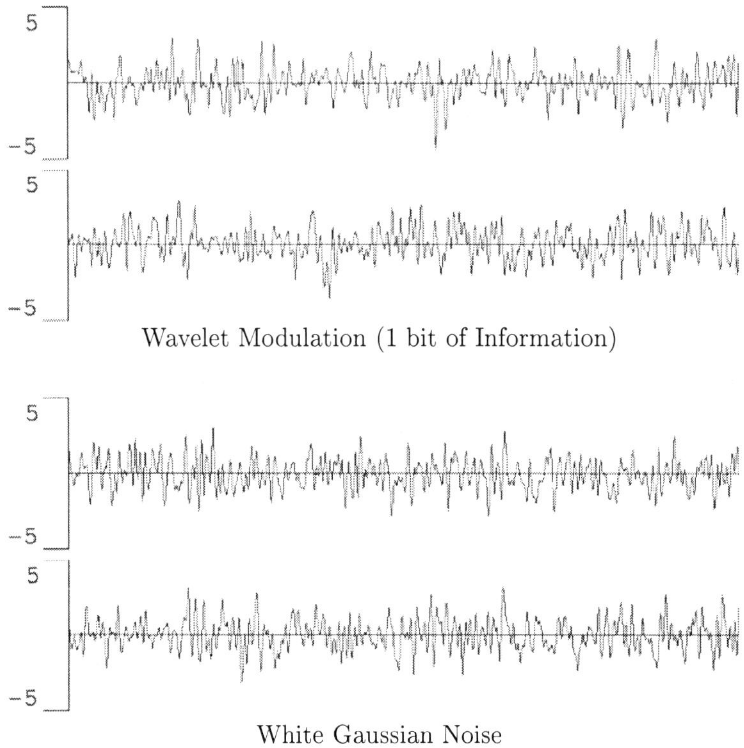

Figure 4.7 The wavelet waveform compared to bandpass Gaussian noise.

Attempts to detect the waveform based on its occasionally repetitive use of short wavelet coefficient sequences — length $M-1$ in an M-ary alphabet — are foiled by pseudo-random negation of the sequences in the waveform, preventing visible autocorrelation buildup on the chance occurrence of a too frequent repetition of any one segment.

4.3.4 Wavelet Transform Domain — A Multiscale Construct

We now introduce multiscale methods into communications system architecture. The developments outlined here are among the most promising techniques for the uses of wavelets in digital communications. This material is further discussed in [244].

Wavelet Domain Creation of Gaussian Sequences. In the FTD waveform, spectral whitening was accomplished by choosing a signal set whose Fourier coefficients have constant magnitude. A corresponding wavelet transform domain operation is to create constant magnitude wavelet coefficients by using PN sequences of ± 1s as the inputs to an orthogonal inverse discrete wavelet transform (IDWT), implemented as a discrete-time synthesis filter

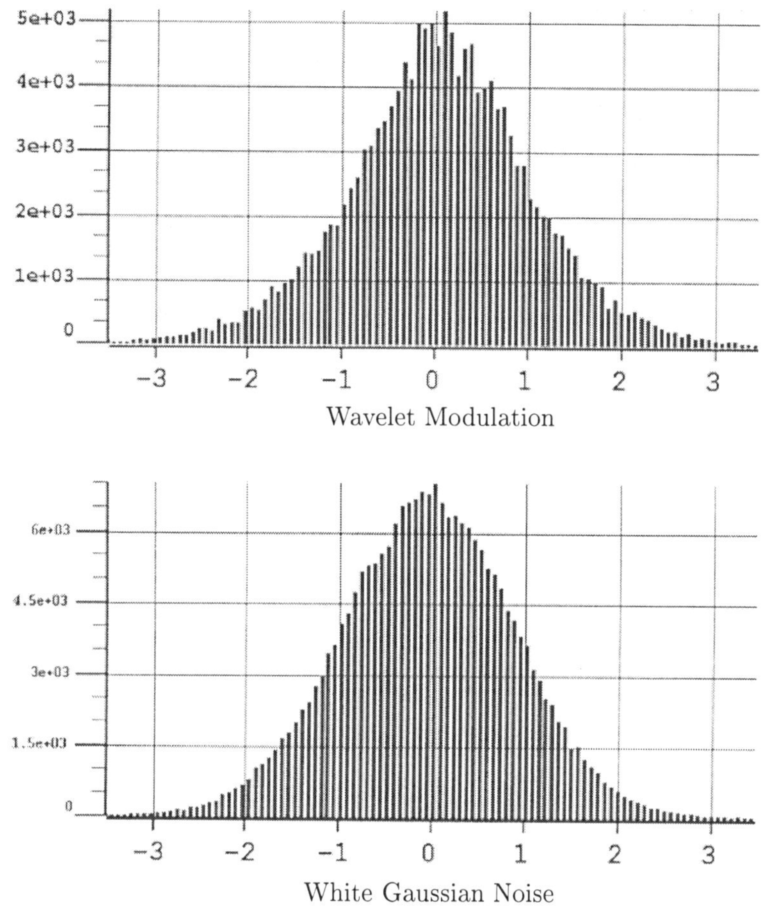

Figure 4.8 Comparison of amplitude histograms — wavelets and Gaussian noise.

bank. This operation is replicated for a quadrature-phase channel with an independent PN code. Alternatively, one can think of the PN input to a single filter bank as being complex-valued QPSK, $(\pm 1, \pm j)$. The driver sequences can be recovered by passing the signal through the corresponding analysis filter bank. The time domain waveform so generated exhibits the same key properties as the FTD waveform.

The PN sequence stream inputs that function as wavelet coefficients can be created from a single higher rate PN stream. Although demultiplexing a single high rate PN stream is not necessary simply for the creation of a Gaussian sequence, it proves important subsequently when we address the imposition of data modulation onto the sequence. Since coefficients at different scales must be supplied at different rates, the input PN sequence must be demultiplexed into streams at these rates. A J-scale IDWT demands J highpass subsequences and one lowpass subsequence, at J rates differing by factors of 2. Figure 4.12 shows

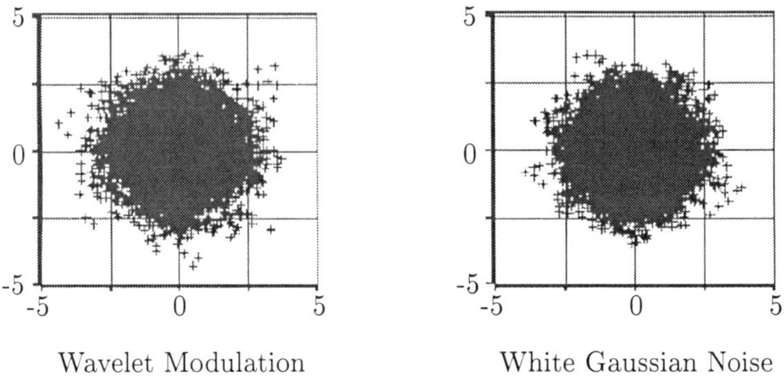

Figure 4.9 Comparison of I/Q scatter plots — wavelets and Gaussian noise.

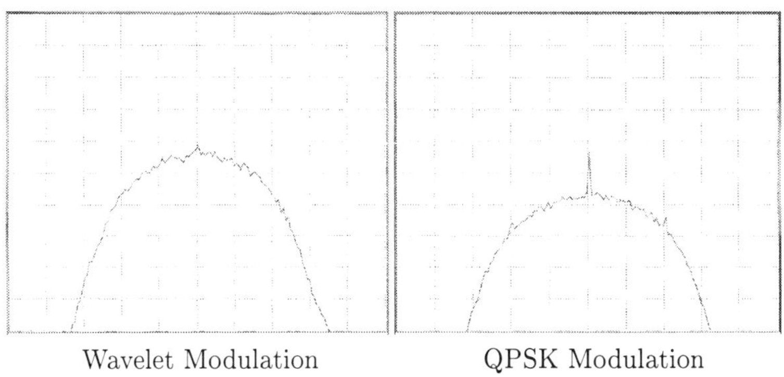

Figure 4.10 Comparison of fourth-power law detector results — $4\times$ carrier lobe.

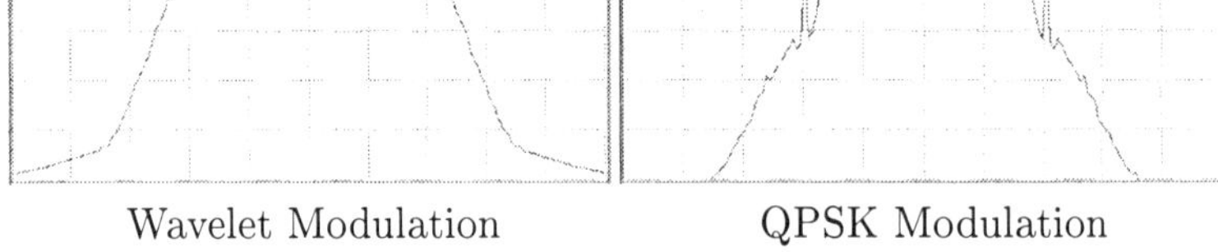

Figure 4.11 Comparison of fourth-power law detector results — baseband lobe.

a four-scale IDWT, or transmultiplexer, where each input bears a designation HP (highpass) or LP (lowpass) with an index denoting its relative rate. The input must be demultiplexed into 5 streams in this case. It can be shown that such a demultiplex is possible for a transform of any number of stages, the

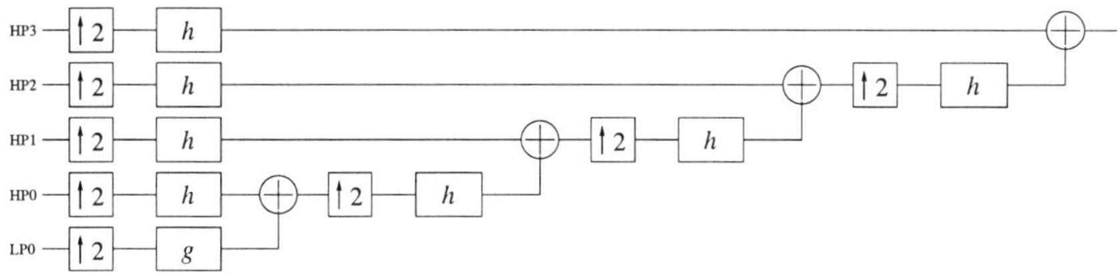

Figure 4.12 Four-scale inverse discrete wavelet transform.

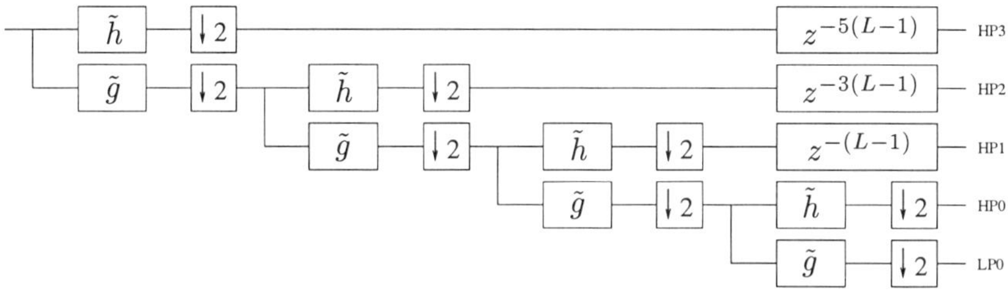

Figure 4.13 Realizable four-stage DWT.

result being streams having bits in the positions corresponding to the needed wavelet coefficient inputs. The general demultiplex has period 2^J in the input sequence, period 16 in the example. the signal at the IDWT output will have the characteristics of a white Gaussian noise process when the input is a white binary process. The Gaussian character comes via the central limit theorem phenomenon in the filtering stages, where the IDWT contains $2J$ filters of length L, each having $L/2$ nonzero numbers inside them at any time. Thus each IDWT output value is influenced by JL numbers.

Input PN sequence recovery requires careful application of the DWT. There are three areas of concern to discuss here: (1) realizability of the filter banks, (2) synchronization of the receive DWT, and (3) synchronization of the receive multiplexer. In conventional filter bank theory, a perfect reconstructing (PR) analysis-synthesis cascade reproduces the input at zero delay. Although the corresponding filter pair is unrealizable, delaying one filter to make its impulse response realizable results in a one-stage DWT/IDWT cascade that is simply an $(L-1)$ time unit delay. To compensate for this in a multi-scale transform pair, each DWT arm emanating from a highpass filter/down-sampler cascade must have a delay based on its symbol rate. Assuming, in Figure 4.13, that both $\tilde{g}$ and $\tilde{h}$ are realizable, it is not difficult to show that the branch delays are of the form $(2^j-1)(L-1)$ [267]. With these delays the alignment of data is correct for the subsequent multiplex.

Synchronization of the IDWT output relative to the timing of the down-samplers in the DWT is critical because the wavelet transform is not shift invariant. The output of an IDWT/DWT cascade, a linear and periodically time-varying (LPTV) system, may not match any delayed version of its input.

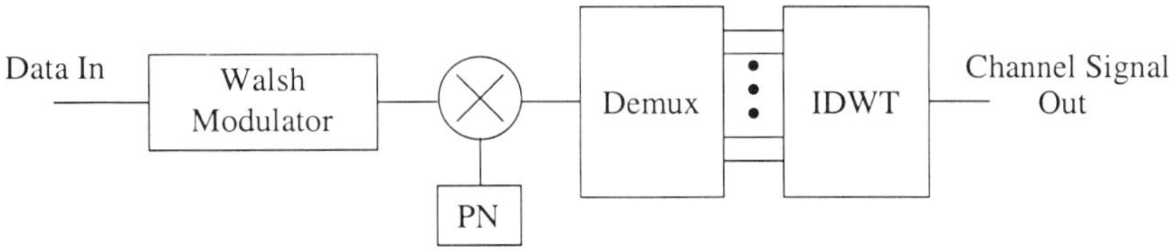

Figure 4.14 Walsh function modulator.

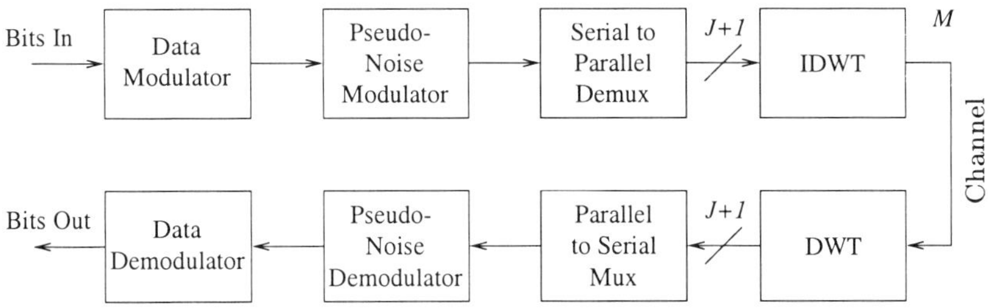

Figure 4.15 Wavelet transform domain communication system diagram.

One can always insert a delay before the receiver that will permit correct recovery. (See Section 4.3.5 for results showing how this synchronization might be achieved.) The other matter of attention is the multiplexer that reconstitutes the PN sequence. It, of course, requires the same synchronization as the rest of the receive processor, inasmuch as it is also periodically time-varying.

Efficient Modulation. The modulator transforms a white binary PN into a white "Gaussian" process, and the demodulator recovers the binary process. We can very efficiently impress information onto the Gaussian waveform without changing its statistical properties by employing modulation that retains the white binary characteristics of the IDWT input. For this purpose we use Walsh functions; other binary-valued modulation waveforms may be used as well.

Input data is modulated by grouping m consecutive bits and interpreting the resulting binary integer as the Walsh function index. The corresponding Walsh function is generated as a ± 1 sequence and point-by-point multiplied by the PN sequence, resulting in a spread spectrum processing gain of $G = \textit{chip rate/data rate} = M/\log_2 M = 2^m/m$. If the PN rate is increased to K chip slots, one achieves gain $G = KM/\log_2 M$. The product sequence is demultiplexed and transformed by the IDWT. Figure 4.14 illustrates the full modulator subsystem. In an RF transmission system, the *channel signal out* in Figure 4.14 would be upsampled, bandpass filtered digitally and D/A converted prior to frequency up-conversion, amplification and coupling to an antenna. Figure 4.15 shows the full system block diagram.

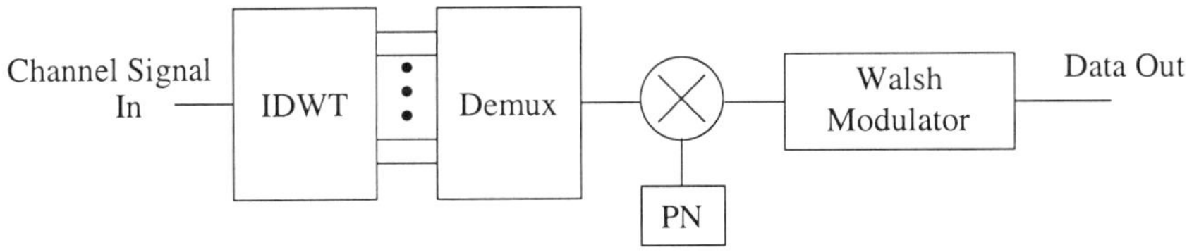

Figure 4.16 Walsh function demodulator.

Efficient Demodulation. Demodulation requires correct time, frequency and phase knowledge at the receiver. Here we ignore the acquisition and tracking procedures that achieve this.

The channel is modeled as an AWGN channel, for which the optimum demodulator (Figure 4.16) simply reverses the operations of the modulator. The input signal, after pulse matched filtering, sampling, A/D conversion and appropriate delay, enters a DWT module that creates a noisy version of the PN-spread Walsh sequence. After multiplication by the local reference PN code, the signal consists of the baseband Walsh function plus noise. The Walsh demodulator creates the minimum error probability version of the input bits and provides these as output. Alternatively, soft decision metrics for each Walsh word could be provided for subsequent decoding as required.

The Walsh demodulation process is subject to a fast algorithm implementation that makes its contribution to the overall receiver processing typically negligible. The first step is to reconstitute the elements of the Walsh function via a data chip matched filter, where "data chip" refers to an element of the modulation word, and *not* to the PN spreading sequence. Since the demodulator input is created by multiplying the PN sequence by the sequence of Walsh function chips, the matched filter sums the corresponding values out of the PN despreader, an integrate-and-dump implementation, reducing the data to one number per Walsh chip. Walsh chip values are then combined to realize a matched filter for each Walsh sequence. In this, the fast Walsh-Hadamard transform, one word is demodulated with exactly $M \log_2 M$ adds/word, or M adds/bit.

Computational Complexity. The elements of transmit signal generation are: (1) determining the Walsh sequence, (2) PN spreading, and (3) the IDWT. Setting up the Walsh sequence is negligible, e.g. reading one Walsh chip every K PN chips. PN multiplication is one operation/chip. Using a polyphase implementation of the IDWT filters [337], the computation for length L filters can be shown to be bounded by $2L$ multiply-accumulate operations (MACs) per chip. In total, the transmit computation rate is then approximately

$$R_{Tx}|_{WTD} = (2L+1)R_c. \tag{4.8}$$

Two real values of PN are needed at each chip, in contrast to the estimated eight for FTD.

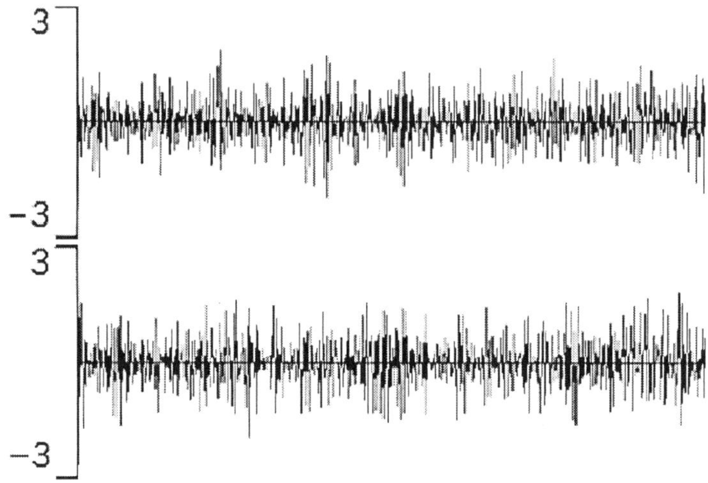

Figure 4.17 WTD waveform samples.

The PN and filtering results above apply equally well to the receiver. Demodulation takes more computing than modulation, but is still low-rate work because it proceeds at a pace related to the data rate, not the chip rate. We can overbound it with $M \log_2 M$ MACs/word, or M/G per chip,

$$R_{Tx}|_{WTD} = (2L + 1 + M/G)R_c. \tag{4.9}$$

One can see from this that the alphabet size can be quite large and still not appreciably contribute to the receiver load. For long filters fast convolution algorithms can weaken the dependence on L in (4.8) and (4.9).

Comparing (4.8) and (4.9) to (4.3), both the FTD and WTD algorithms distribute computation uniformly between transmitter and receiver. For both, the histogram shaping, or mapping from the input to output distribution, occurs in real time. It is interesting to note that signals exhibiting the same properties as WTD signals, but generated by a different wavelet-based algorithm previously reported [277, 245], one that uses only a *single scale* of the wavelet transform, are asymmetric in their distribution of computation. For the latter, waveform generation amounts to one table lookup per PN chip but the demodulation computation is proportional to M, *without* division by the processing gain.

Featureless Characteristics. WTD waveforms have been subjected to tests that verify their featureless characteristics. A few results are shown in the following figures. A waveform sample (I and Q components) is shown in Figure 4.17. Comparison of such waveforms with Gaussian noise segments shows no obvious differences to the eye.

The left side of Figure 4.18 shows the histogram of one component of a WTD channel sequence in which the Gaussian character is apparent. The filters are Daubechies 30 in this case. We have seen other examples in which

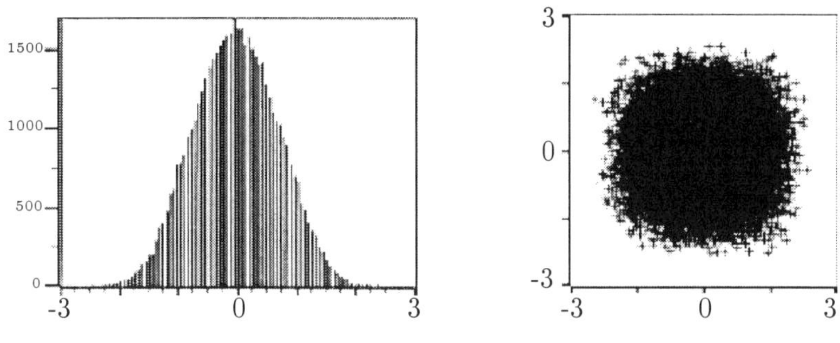

Figure 4.18 Component histogram and I/Q scatter plot of a WTD signal.

the histograms deviate noticeably from Gaussian, so one should not assume this result always occurs. Precise relationships between filter coefficients, filter length and number of stages are not known; filter lengths of 8 or 10 in a 3-stage system seem to give reasonably Gaussian histograms, and the approach to Gaussian does not necessarily converge rapidly or uniformly as the length increases. The I/Q scatter plot for the same signal is on the right side of Figure 4.18. It conveys the shape of the expected two-dimensional Gaussian pattern. For shorter filters, the scatter plot envelope is often more like a square. More complicated test results not shown here also confirm immunity to chip rate and carrier line detection via 2nd and 4th order cyclostationary processing.

4.3.5 Acquisition

This section presents a novel acquisition strategy for wavelet-based spread spectrum signals. Compared to conventional correlation or matched filter methods of PN code acquisition, the new approach promises considerable reduction of the total work performed in acquiring such codes, implying more favorable trade alternatives between acquisition time and processor throughput.

The new acquisition strategy is the result of very fundamental properties of the dyadic DWT, achieving capabilities apparently impossible when these attributes are absent. In this two-part algorithm, much of the correlation processing required in traditional techniques is replaced by highly efficient filter-bank processing in the first part; this considerably cuts the amount of correlation processing subsequently needed in the second part.

Using the new approach described in this paper, one can simultaneously search C time cells (code delay states) and D Doppler cells in time $O(\sqrt{CD})$ as opposed to $O(CD)$ and, despite the signals being Gaussian-like, perform all correlations using binary-valued, rather than complex-number, local references without loss in either communication performance or LPI/D.

Intuitive Development of the Algorithm. Consider the normal correlation or matched filter methods for PN code acquisition. For now we address delay uncertainty only, addressing frequency uncertainty later. For simplicity assume throughout that the search device is perfect, i.e. in examining the correct delay cell, it always acquires the code, in an incorrect cell, it never mistakenly thinks it has found something (the search processor is a $P_d = 1$, $P_{fa} = 0$ machine). Under these assumptions, a complete search of C cells, where T_l is the (fixed) dwell time per cell (subscript l denotes "linear") can take no more than time

$$T_{acq} = CT_l. \tag{4.10}$$

We use a worst-case acquisition time criterion throughout the chapter while recognizing the interest in also looking at statistical measures such as mean and variance.

The new test is a two-part algorithm. The key feature of the first part is that it eliminates *half* the remaining cells at each trial of the test (for which reason we call it the *geometric* test). This good news is tempered by the discovery that each trial takes twice as long as its predecessor. Clearly the efficiency drops rapidly with repetition. Thus, each test discards half as many possibilities as its predecessor, taking twice as long to do so; thus the effective search time per cell, which starts off quite small, increases by a factor of four per iteration, eventually becoming greater than the comparable time to search the cell by code correlation. Just before reaching this point of diminishing returns, the algorithm enters its second part, in which it applies some form of serial search over the remaining cells, which by this time are far fewer in number than the original C.

These two testing strategies define the whole algorithm. To make it quantitative, assume the first trial of the geometric test takes time T_g. The second time it takes $2T_g$, and the nth, $2^{n-1}T_g$. After N trials, the elapsed time is the time of the first part of the test, T_1:

$$T_1 = \sum_{n=1}^{N} 2^{n-1} T_g = (2^N - 1)T_g.$$

Assume that at this point the algorithm switches to serial search (the *linear* test). N trials of the geometric test leave $2^{-N}C$ cells in contention, which can be searched in time

$$T_2 = 2^{-N} C T_l.$$

The maximum acquisition time is just the sum of the two times,

$$T_{acq} = T_1 + T_2 = (2^N - 1)T_g + 2^{-N} C T_l.$$

Taking T_l and T_g to be fixed, we choose N to minimize T_{acq}. After the calculus we find the best time to switch is after the $\tilde{N}^{th}$ trial, where

$$N = \tilde{N} \equiv \frac{1}{2}\log_2(CT_l/T_g).$$

Of course this choice of N may not be an integer. If we can ignore this and proceed, the minimax acquisition time is found to be

$$\tilde{T}_{acq} \equiv T_{acq}|_{N=\tilde{N}} = 2\sqrt{CT_lT_g} - T_g. \tag{4.11}$$

It is easy to show that the value obtained by setting N equal to the integer nearest $\tilde{N}$ differs from (4.11) by only a factor close to 1. Letting $\hat{T}_{acq|g}$ denote the achievable minimum acquisition time, one can show that

$$2\sqrt{CT_lT_g} - T_g \leq \hat{T}_{acq|g} \leq 2\sqrt{(9/8)CT_lT_g} - T_g,$$

which is a rather tight bounding pair; the two sides are typically within 6% of one another.

Parallel Processing. If it is known that there are $P < C$ parallel correlators available, each capable of searching a state in time T_l, we can repeat the analysis in (4.5)–(4.7), replacing T_l with T_l/P, and finding

$$\tilde{N} = \frac{1}{2}\log_2(CT_l/PT_g)$$

and

$$\tilde{T}_{acq} = (2\sqrt{T_gT_l/P})\sqrt{C} - T_g. \tag{4.12}$$

Should the $O/\sqrt{P}$ behavior of (4.12) create an impression that parallelism has somehow been applied inefficiently, note that (4.12) is still an improvement on (4.3). The behavior is reasonable because only the second part of the test benefits from parallelism, precluding a full "Amdahl's rule" improvement in this case [123]. And of course, for large C both (4.11) and (4.12) are potentially much better than the fully linear search time, CT_l, of (4.10). In fact, one can show that the right-hand side of (4.11) cannot exceed the linear search time for any values of the linear and geometric test lengths.

Acquisition with Both Delay and Doppler Uncertainty. In a fully linear search, delay and Doppler cells must be tested simultaneously, since if the correlation reference is not a close match in both, the correlator will not accumulate a detectable value. Assume the frequency uncertainty encompasses D Doppler cells, thus to probe any one delay-Doppler pair requires the familiar dwell time, T_l. There are CD such cells, hence a maximum linear search time of

$$T_{acq|l} = CDT_l.$$

(Note the appendage of l to the subscript acq to denote the algorithm for which the acquisition time applies.)

If we could argue that the geometric search goes ahead unimpeded by the presence of Doppler offset, then N trials of the geometric test would reduce the number of time cells remaining to be searched to $2^N C$, while not diminishing the number of Doppler cells at all. Then the previous results would be adapted to the Doppler case simply by everywhere replacing C by CD, yielding

$$\tilde{N} = \frac{1}{2} \log_2 (CDT_l / PT_g)$$

and

$$\tilde{T}_{acq|l} \approx (2\sqrt{CDT_g T_l / P}) - T_g.$$

We claim that the requisite Doppler insensitivity of the geometric test is the norm for most practical scenarios. The demonstration of this is omitted, however.

Synchronization of the DWT. Synchronization of the DWT is prerequisite to code correlation. Recalling that the PN-modulated Walsh codeword sequence was encoded as the wavelet coefficients of the transmitted sequence, one can expect that, as a consequence of the lack of translational invariance in the wavelet transform, the receiver DWT extracts some *different* set of coefficients to represent its input signal if the timing relationship between the transmitter and receiver is incorrect. This different set of coefficients, when multiplexed into a code stream, cannot be expected to correlate well with the PN reference code under any relative shift.

The DWT filter bank is a linear, periodically time-varying system, a J-stage bank having 2^J states (period 2^J). Thus DWT synchronization can be accomplished by inserting a variable delay in front of the filter bank and stepping it through its 2^J states while performing a test to determine the correct state. It has been demonstrated that if the input to an orthogonal DWT filter bank is a white process — a reasonable model for a well-designed QPSK code, the DWT output will be a white process *independent of the synchronization state* [244]. Experimentally we have determined that the out-of-sync processes are quite Gaussian-like. However, there will be just one state, the correct one, for which the output will reproduce the QPSK input signal (plus noise, of course, the noise being just channel and receiver noise, nothing new introduced by the filter banks). To the extent that one can distinguish between binary and multilevel processes in the presence of the ambient noise environment, one has a statistical basis for a filter bank synchronization procedure.

At first, filter bank synchronization appears to be a nuisance, just another resource-consuming step whose value is not immediately apparent. As it turns out, it is the key to a fast algorithm. Because the filter bank is a periodic system, it can only distinguish among delay states modulo 2^J. When it rejects delay state n, it has also rejected all delays of the form $n + k2^J$ for integer k. This makes a filter bank synchronization test extremely powerful with respect to absolute time synchronization. Instead of accepting or rejecting one state per test (as in linear search), there is now the potential to eliminate a significant fraction (half or more) of the states with each trial of a test. To understand the specific algorithm requires a review of certain filter bank properties.

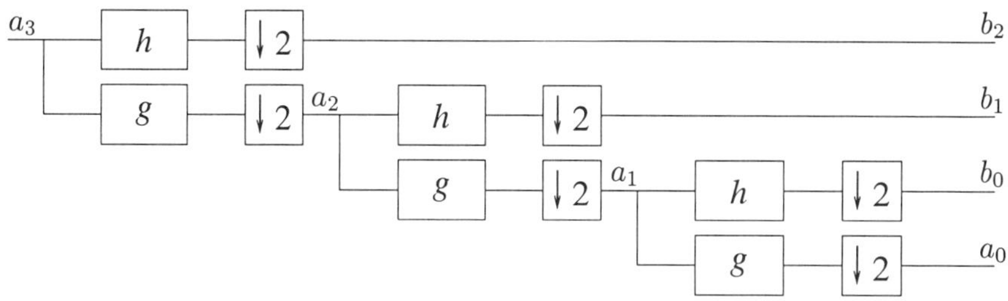

Figure 4.19 Receiver filter bank.

The Synchronization Algorithm. Figure 4.19 shows an example of a two-band dyadic receive filter bank. The high rate arm output, b_2, at the top of the filter bank has been decimated by two, and its outputs appear at half the chip rate. With respect to synchronization it is a period two device, i.e. it has only two synchronization states, even and odd. Therefore the output is only meaningful if the filter bank is synchronized modulo 2. The distinction between even- and odd-numbered states can be detected in this arm by observing the filtered sequence prior to the decimator and testing for whether the even or odd samples are more likely to contain a constant envelope signal. (The optimum decision rule for doing this has been derived but is omitted here.) When this determination is made correctly, the arm is synchronized and half the states are eliminated — *no code correlation is needed to do this.*

After the decision is made in the high-rate arm, the correct delay state is established and the decimated sequence a_2 is passed through the highpass filter h. Though output b_1 results from decimation by 4, there is no longer a 1 in 4 uncertainty as to the correct synchronization of this arm. Assuming success in the prior trial, the uncertainty is again 1 out of 2. Proper examination of the chips before the second decimator (identical to what was done for the first arm) will synchronize this arm, and now 3/4 of the states have been eliminated.

If the synchronization test is based on examining a fixed number of chips before deciding, the second trial will take exactly twice as long as the first because the symbol arrival rate is halved. Furthermore, the first trial eliminated 2^{J-1} states, the second eliminates 2^{J-2}. Per unit time, successive trials are less efficient in discarding states by a factor of 4.

Once the second arm is synchronized, we move to the next lowest rate arm and repeat. One could search the full filter bank in this manner until the uncertainty was 1 in 2^J. If the total initial uncertainty is less than 2^J states, the algorithm is finished. If not, there is more to do. Further resolution can be accomplished only by code correlation. Several states separated by delay 2^J may have to be searched in order to find the one correct state. For the moment, assume that the initial uncertainty does not exceed 2^J states.

Performed as above, the algorithm would be no more efficient than serial search, and is probably less so. It takes J trials to complete the test; the first is of length T_g, the second $2T_g$, et cetera. The total is $(2^J - 1)T_g$ identical to

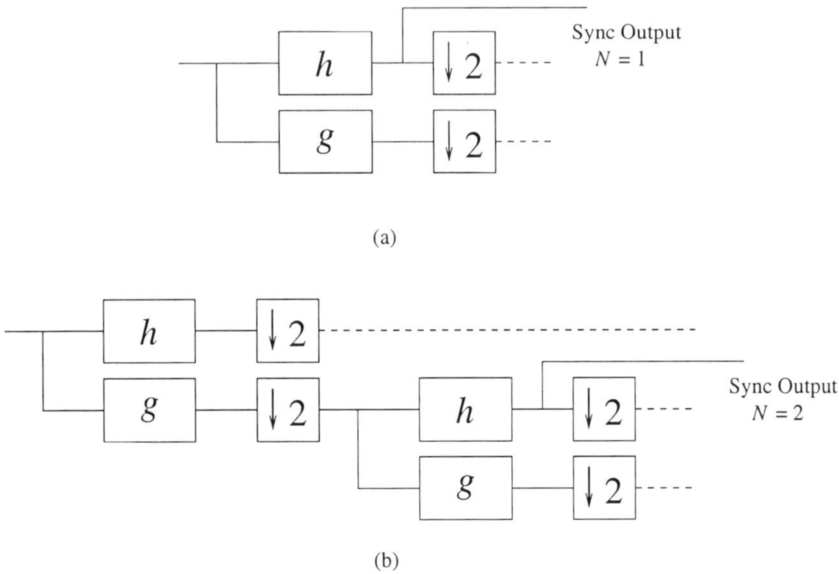

Figure 4.20 Processing for the geometric acquisition test. (a) First trial ($N = 1$) and (b) second trial ($N = 2$).

the serial search formula with the replacement $N \rightarrow J$, $T_l \rightarrow T_g$. The key to achieve speedup is in knowing when to stop the geometric test. This aspect is adequately covered by the analysis in Section 4.3.5.

Figure 4.20(a) shows the processing for the first trial ($N = 1$) of the geometric test. Only the highpass output of the bank is used and the lowpass filter is omitted. The acquisition output, however, is taken prior to the highpass decimator, requiring the full filter process to be executed. For a filter of L taps this requires L MACs per chip, the processing equivalent of L correlators.

In Figure 4.20(b) we diagram the second trial ($N = 2$) processing. The first stage highpass output is no longer needed, but the decimated first-stage lowpass output drives the highpass filter of the second stage, whose undecimated output provides the test statistics. The first stage filter performs $L/2$ MACs per chip, whereas the second requires L MACs per two chip intervals, again a net computation power of L MACs per chip. It is easy to show that this result holds for an arbitrary number of stages, yielding a constant acquisition processing load throughout the entire geometric test.

4.4 THE DETECTION OF HOPPED LPI SIGNALS USING TRANSFORM-BASED TECHNIQUES

In this section we discuss the detection of LPI signals in the presence of AWGN and present a method for detecting traditional frequency and time hopped signals (which may or may not also include direct sequence spreading) using a QMF bank tree. As will be demonstrated, this technique exploits a particular characteristic of hopped signals to improve detection performance when the interceptor's knowledge about the signal is limited.

As used here, "detection" refers to the process, by the interceptor, of determining whether or not LPI signals are present. It is a binary decision, either the interceptor decides no signals are present or it decides one or more signals are present. In the detection process, information is not necessarily obtained about how many signals are present or their characteristics. In this context, "feature extraction," refers to estimating bandwidths, hop rates, hop synchronization, and signal-to-noise ratios (SNRs) — features that may be used to characterize and distinguish among signals.

Furthermore, the term "waveform" is generally used to refer to whatever the interceptor receives, which is assumed to be WGN, possibly including an LPI signal. The term "signal," on the other hand, specifically refers to the LPI signal. We use "tile" to refer to a rectangular region of the time-frequency plane defined by the decomposition process of the LPI receiver. Note this is distinct from "cell" which denotes the LPI signal's hop cell.

4.4.1 Introduction — Why Use a Transform Based Technique

Much has been published on the detection of LPI signals [75, 83, 89, 328, 362]. Usually these references assume, as is done here, that the interceptor does not know the pseudo-random hopping code used by the transmitter and intended receiver. Sometimes they further assume the interceptor does not know the channelization, hop duration, or hop timing of the signal. In this case, a radiometer (energy detector) is often used.

The radiometer consists of a bandpass filter covering the LPI signal's entire band, W, a square law device, and an integrator to collect energy for time period T. When no signal is present and the only input to the radiometer is WGN, the output is a random variable (RV) with a χ-square probability distribution function (pdf). On the other hand, if a signal is present, the channel output is a RV with a non-central χ-squared pdf. Following the radiometer, a threshold detector determines signal presence.

By integrating under the χ-square and non-central χ-square curves from the threshold value to infinity, the theoretical probabilities of false alarm, and detection, respectively, can be determined. By varying the threshold, a family of values for each parameter can be found, and a receiver operating characteristic (ROC) curve can be plotted. This curve is helpful for comparing the performance of the radiometer against other intercept receiver architectures. Better receivers have a higher probability of detection for a given probability of false alarm.

The radiometer can be shown to be the optimal detector when nothing about the signal, other than its band, is known. On the other hand, an interceptor's performance can be greatly improved when he knows the channelization, hop duration, and timing, of the LPI signal, by using the filter bank combiner (FBC). The FBC consists of a bank of radiometers each tuned to a hop channel with integrators set to match the hop duration and synchronization. Each radiometer looks for an individual hop cell. The radiometer outputs are compared to thresholds and the individual decisions are combined as appropriate for the signal being detected.

Another way to look at the function of the FBC is to note it tiles the time-frequency plane, each tile corresponding to a possible hop cell. An individual detection decision is made for each tile with their total combination yielding the overall decision.

For our receiver, we assume we do not know the pseudo-random sequence used for hopping, nor do we know the channelization, hop duration, or synchronization. We do, however, know the signal is hopped, and therefore, that the signal's energy is concentrated at cell locations in the time-frequency plane. In this section, we exploit this fact to design a receiver that out-performs the radiometer and approaches FBC performance. As we demonstrate, our receiver achieves this by tiling the time-frequency plane and using the results to make a detection decision. In addition, the receiver can be used to look for multiple signals simultaneously and to make estimates of the channelization, hop duration and synchronization, and signal strength.

4.4.2 Mathematical Background

Before discussing the receiver's architecture, it is helpful to explore some of the general mathematics behind the decomposition of waveforms into orthogonal bases, and consider some of the statistical results when the waveform consists of WGN exclusively and when it consists of WGN plus a deterministic signal. Since our receiver samples and digitizes the received waveform before processing, we consider a digital waveform here.

Given the waveform, $f(n)$, we wish to decompose it as follows

$$f(n) = \sum_k a_k \Psi^*_{kn} \tag{4.13}$$

where Ψ_{kn} is the basis set with * denoting complex conjugation, n is the time index, and k is the function index ($n, k \in$ subset of integers). We next place the restriction of orthonormality on the basis functions,

$$\sum_k \Psi^*_{kn} \Psi_{mn} = \begin{cases} 1 & \text{for} \quad k = m \\ 0 & \text{for} \quad k \neq m. \end{cases}$$

By taking (4.13) and summing, over time, the product of $f(n)$ and one of the basis functions, we find a general formula for the coefficients

$$\sum_n f(n) \Psi_{mn} = \sum_n \sum_k a_k \Psi^*_{kn} \Psi_{mn} = \sum_n a_k \sum_k \Psi^*_{kn} \Psi_{mn} = a_m \tag{4.14}$$

where the last step follows from orthonormality. We are now in a position to see why this decomposition is so important. The energy of the waveform is the sum, over time, of the waveform squared

$$\sum_n |f(n)|^2 = \sum_n \left(\sum_k a_k \Psi^*_{kn} \right) \left(\sum_m a^*_m \Psi_{mn} \right) = \sum_k |a_k|^2$$

where, once again, the last step is due to orthonormality. This result, *Parseval's theorem*, is critical. Since we are looking for signal energy, we can apply the basis set to the waveform, as in (4.14), and then consider the values of the coefficients squared. A good basis set, for our purpose, concentrates the signal energy in as few of the coefficients as possible. An interceptor that knows what to look for can then disregard the other coefficients, reducing the probability of false alarm.

Since we are going to measure the values of coefficients of (4.14), we need to know their statistical properties. First, let's consider a waveform, $f(n)$, consisting entirely of zero-mean, bandlimited WGN with variance σ^2, i.e. $f(n) = \eta(n)$ where $\eta(n)$ is normally distributed, denoted as $N(0, \sigma^2)$. Also, we assume the value for each n is statistically independent of other values. The mean, or expected value of each of the coefficients is

$$E\left[a_k\right] = \sum_n E[\eta(n)]\Psi_{kn} = 0$$

and the variance is

$$\mathrm{Var}\left[a_k\right] = \sum_n \mathrm{Var}\left[\eta(n)\Psi_{kn}\right] = \sum_n \mathrm{Var}[\eta(n)]|\Psi_{kn}|^2 = \sigma^2 \tag{4.15}$$

since the operations to obtain a_k are linear and each a_k is distributed as $N(0, \sigma^2)$. Now, we need to consider the cross correlation between coefficients

$$\begin{aligned}
E[a_k a_m^*] &= E\left[\sum_n \eta(n)\Psi_{kn} \sum_p \eta(p)\Psi_{mp}^*\right] \\
&= E\left[\sum_n (\eta(n))^2 \Psi_{kn}\Psi_{mn}^* + \sum_n \eta(n)\Psi_{kn} \sum_{p\neq n} \eta(p)\Psi_{mp}^*\right] \\
&= \sum_n E[(\eta(n))^2]\Psi_{kn}\Psi_{mn}^* + \sum_n E[\eta(n)]\Psi_{kn} \sum_{p\neq n} E[\eta(p)]\Psi_{mp}^* \\
&= \sum_n \sigma^2 \Psi_{kn}\Psi_{mn}^* = 0 \text{ when } k \neq m.
\end{aligned}$$

Of course, when $k = m$, $E[(a_k)^2] = \sigma^2$. So, when the basis set is orthonormal, the coefficients are uncorrelated, and so, because we are dealing with a Gaussian waveform, are statistically independent.

We are also interested in the pdf for the square of the coefficients. This pdf can be found using transform techniques as

$$f_Y(y) = \sum_{i=1}^{k} \frac{f_X(x_i)}{|J'(x_i)|} \tag{4.16}$$

where $f_X(\cdot)$ is the pdf of the function to be transformed (Gaussian, in our case), $f_Y(\cdot)$ is the desired pdf, and $J(x)$ is the Jacobian of the transformation (here, $J(x) = x^2 \equiv y$). x_i are the values of the variate of the original function corresponding to y, y is the variate of the pdf we are trying to find, and k is the number of roots of $J(x)$. In our case, this leads to

$$f_{a_k^2}(y) = \frac{f_{a_k}(\sqrt{y})}{2\sqrt{y}} + \frac{f_{a_k}(\sqrt{y})}{2\sqrt{y}} = \frac{1}{\sqrt{2\pi y\sigma^2}} \exp\left[-y/(2\sigma^2)\right] \tag{4.17}$$

which, if the noise variance, σ^2, is normalized to 1, becomes the well-known χ-square distribution with one degree of freedom.

Often, we want to add several coefficients. The pdf of the sum of independent random variables can be found using the characteristic function associated with (4.17), resulting in

$$f_{a_k^2 a_{k+1}^2 \cdots a_{k+\gamma-1}^2}(y) = \frac{1}{2^{\gamma/2}\Gamma(\gamma/2)\sigma^\gamma} y^{\gamma/2-1} \exp\left[-y/(2\sigma^2)\right], \tag{4.18}$$

where $\Gamma(\cdot)$ is the gamma function. When σ^2 is normalized to 1, (4.18) becomes the χ-square function with γ degrees of freedom.

Now, let's examine the coefficients if the waveform consists of a deterministic signal plus bandlimited WGN, i.e.

$$f(n) = s(n) + \eta(n).$$

First we have

$$E[f(n)] = E[s(n)] + E[\eta(n)] = s(n)$$

$$\mathrm{Var}[f(n)] = \mathrm{Var}[s(n)] + \mathrm{Var}[\eta(n)] = \sigma^2.$$

Since the waveform variance is equal to the noise variance, we have from (4.15)

$$\mathrm{Var}[a_k] = \sigma^2.$$

From (4.14) we have

$$E[a_k] = \sum_n E[f(n)]\Psi_{kn} = \sum_n s(n)\Psi_{kn},$$

so

$$a_k \sim N\left(\sum_n s(n)\Psi_{kn}, \sigma^2\right).$$

We now want to find the pdf for the coefficient squared when the waveform consists of a deterministic signal plus bandlimited WGN. First, for notational convenience, let

$$m_k \equiv E[a_k] = \sum_n s(n)\Psi_{kn}.$$

The pdf for a_k is thus

$$f_{a_k}(x) = \frac{1}{\sqrt{2\pi\sigma^2}} \exp\left[-(x-m_k)^2/2\sigma^2\right].$$

As in (4.16), we have a Jacobian, $J(x) = x^2 = y$. As with (4.17), we have

$$f_{a_k^2}(y) = \frac{f_{a_k}(x_1)}{J'(x_1)} + \frac{f_{a_k}(x_2)}{J'(x_2)}$$

where $\sqrt{y} \equiv x_1 \equiv -x_2$. After some manipulation, we find

$$f_{a_k^2}(y) = \frac{1}{\sqrt{2\pi\sigma^2 y}} \exp\left[-(y+m_k)^2/2\sigma^2\right] \cosh\left[m_k\sqrt{y}/\sigma^2\right] \tag{4.19}$$

which, if σ^2 is normalized to 1, is the χ-square distribution with one degree of freedom and with a non-centrality parameter of m_k^2. Using the characteristic function corresponding to (4.19), we find the pdf for the sum of coefficients when both signal and noise energy are present to be

$$f_{a_k^2 a_{k+1}^2 \ldots a_{k+\gamma-1}^2}(y) = \frac{1}{2\sigma^2}\left(\frac{y}{\lambda}\right)^{\frac{(\gamma/2-1)}{2}} \exp\left[\left(-\frac{\lambda+y}{2\sigma^2}\right) I_{\gamma/2-1}\left(\frac{\sqrt{\lambda y}}{\sigma^2}\right)\right] \tag{4.20}$$

where $I_{\gamma/2-1}(\cdot)$ is the modified Bessel function of the first kind and order $\gamma/2-1$, and

$$\lambda = \sum_{i=k}^{k+\gamma-1} m_i^2$$

represents the signal energy contributed to the coefficients. When σ^2 is normalized to 1, (4.20) is the χ-square distribution with γ degrees of freedom and with a non-centrality parameter of λ.

By comparing the functions described by (4.18) and (4.20), we see how an interceptor can decide whether a signal is present. Figure 4.21 shows these functions for the particular case where we are adding two squared coefficients and σ^2 is equal to one. $f_n(Y)$ is the pdf when noise only is present, and $f_s(Y)$ is the pdf when noise and signal are present. As the figure indicates, we can set a threshold, T_h, and compare the sum of the squared coefficients against it, deciding a signal is present if the observed number exceeds the threshold. Given that a signal is present, the probability of detection is found by integrating under $f_s(Y)$ to the right of T_h (the shaded region labeled Q_d in the figure). Likewise, our probability of false alarm when no signal is present is computed similarly, and is indicated by Q_{fa}.

As previously indicated, to detect energy concentrations due to a signal we want to decompose the waveform into a basis set such that the signal energy is contained in as few terms as possible. Our ability to do this, of course, depends on both our knowledge of the signal and our ability to find an orthonormal basis set that spans the signal space. If no signal is present, the squared coefficients have a χ-square pdf. If one or more are present, the squared coefficients of functions containing signal energy have a χ-square pdf with a non-centrality parameter. Therefore, threshold detection is possible.

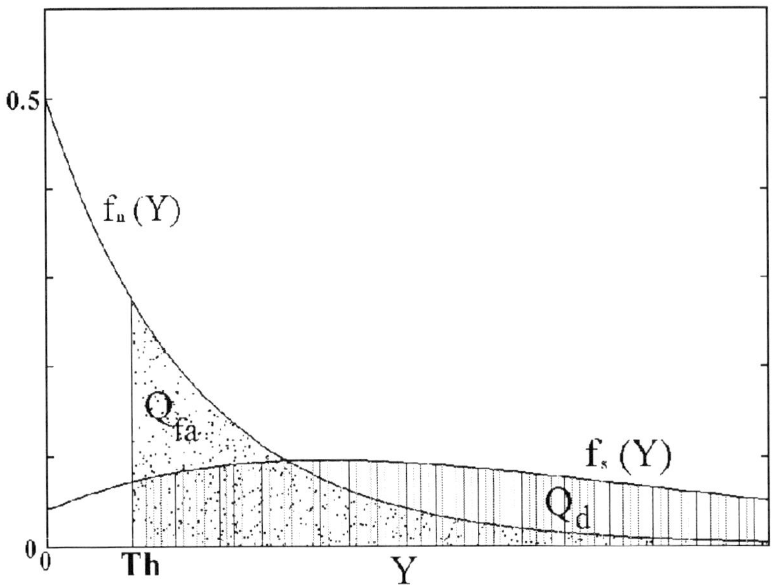

Figure 4.21 Chi-square probability distribution functions.

4.4.3 The Quadrature-Mirror Filter Bank Tree

We now demonstrate how we can use a QMF bank tree to accomplish the orthonormal decomposition described above. In [125] and [347] it is shown how orthogonal wavelets can be implemented using QMFs, filter pairs designed to divide the input signal energy into two orthonormal components based on frequency. As illustrated in Figure 4.22, we arrange the QMF pairs in a fully developed tree structure, connecting each of the outputs from every filter in one layer to a QMF pair in the next. (This is contrasted with the traditional wavelet decomposition, where only the lowpass filter in each layer is connected to the next layer.)

In [125] and [347] full trees are used for data compression and subband coding. There are several differences between their author's goal and ours. In particular, whereas they're primarily interested in the output from the last filter of each branch of the tree, we also use information from intermediate layers.

Referring to Figure 4.22, the function of the tree is easy to understand. Each QMF pair divides a digital input waveform into high frequency and low frequency components, with the filter transition centered at $\pi/2$. Here we assume a normalized digital input to the tree of one sample/second, with a bandwidth of π. Since each output has half the bandwidth, only half the samples are required to meet the Nyquist criteria (these sequences are decimated by two,

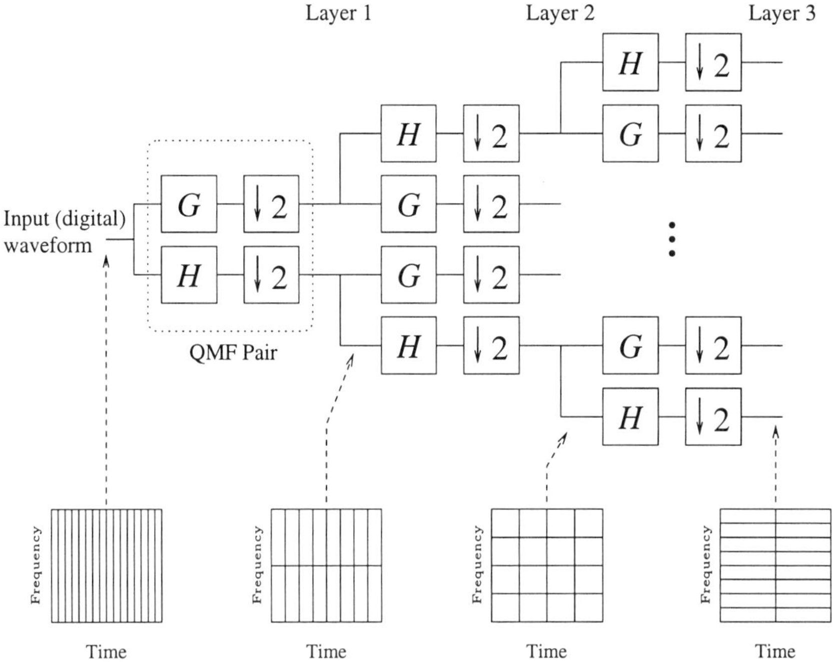

Figure 4.22 Quadrature-mirror filter bank tree.

giving us the same total number of samples output as were input). Each of the two resulting sequences are then fed into QMF pairs forming the next layer where the process is repeated on down the tree.

Because the transform performed by the QMF bank tree is linear, there is a fundamental limit on the minimum area of each of the tiles (0.5 Hz-sec.). However, we note the matrix of energy values output from each layer represent tiles that are twice as long in time and half as tall in frequency as the tiles in the previous layer. By properly comparing these matrices, it is possible to find concentrations of energy, and estimate their positions and sizes with high resolution in both time and frequency. Using this technique, we can decompose a waveform and estimate bandwidths, time widths, locations in the time-frequency plane, and the SNRs of individual FH and TH cells. This information, of course, can then be used by the interceptor to decide how many transmitters, and types thereof, are in operation.

4.4.4 Finite Impulse Response Filters with "Good" Energy Tiling Characteristics

Of course, it is not mathematically possible to find an finite impulse response (FIR) filter that perfectly divides the energy into tiles in both time and frequency. Any filter that is compact in time will have infinite extent in frequency, and vice versa. However, it is possible to approximate this desired condition with a good compromise; this is described more fully in [88] and only the results are presented here. Also, we generally consider the lowpass filter of the QMF pair here. It is well-known that once a suitable FIR filter, H, is found one obtains the highpass filter, G, by time reversing the coefficients and negating every other one. We now look at the properties of the desired equivalent filter.

In the time domain, we specify our requirement a bit differently from most of the literature. Conventionally, the total number of filter coefficients is usually minimized to reduce the amount of out-of-time energy collected. In reality, for any individual filter in the tree, only two of the coefficients collect signal from within the time dimension of the tile. The rest contribute to out-of-time energy. What we want then, is to declare two of the coefficients to be "main taps" and to make these as large as possible. All of the others, in turn, should be as small as possible. We also want the two main taps to be as nearly equal to one another as possible, so the energy of each of the two input samples falling in the tile's time span will be nearly equally represented. Any two adjacent taps can be designated the main taps. When constructing the complementary filter it is important to synchronize the outputs so that both filters cover the tile's beginning and ending at the same time. Fortunately, this is easy to accomplish by adding an even number of zero coefficients (pure delay) to the filters.

The Harr lowpass FIR filter has two coefficients, both with values of $1/\sqrt{2}$ [347]. This filter meets the wavelet requirements and also perfectly tiles the input energy in time. In this case, the energy in each output value is equal to the lowpass energy from two corresponding inputs. The pass region is also flat along the time dimension. Unfortunately, the Harr filter does not tile well in frequency.

In this respect, the opposite of the Harr filter is essentially one that has a "brick wall" frequency response. Except for one practical problem, an FIR filter with such a response can be constructed by selecting the coefficients as

$$h(n) = \frac{1}{\sqrt{2}} \operatorname{sinc}\left(\frac{n+0.5}{2}\right) \qquad n \in \text{integers} \tag{4.21}$$

where $\operatorname{sinc}(x) = \sin(\pi x)/\pi x$ when $x \neq 0$ and one when $x = 0$. Note that the main taps, $h(-1)$ and $h(0)$, are equally spaced around the center of the sinc function and will have significantly larger values than the other coefficients. The other coefficients, although necessary to achieve the proper tiling in frequency, do so at the expense of collecting out-of-time energy. Also note the two main taps are equal, so the two corresponding input values will contribute equally to the output tile's energy. The practical problem with (4.21) is in truncating the values for large $|n|$ — this leads to the Gibbs phenomenon, a "ringing" of the magnitude response near the transition region. One solution to this is to apply a Hamming window. Truncation and windowing affect the orthogonality of the filters, but this can be compensated by adjusting other values of (4.21)

Table 4.1 Coefficient values of 22-coefficient energy concentration filter.

n	$h(n)$	n	$h(n)$
0	-0.06937058780687	11	0.02100935003910
1	0.08787454185760	12	0.02296583051731
2	0.69303661557137	13	-0.02145780458290
3	0.69307535109821	14	-0.01348759528448
4	0.09491659263189	15	0.01318436272982
5	-0.09139339673362	16	0.01550256127783
6	-0.04245335524319	17	-0.01636308107201
7	0.04675991002879	18	-0.00627176286924
8	0.03613538459682	19	0.00993238693842
9	-0.03468231058910	20	-0.00105459770882
10	-0.02281230449607	21	-0.00083252852776

to give

$$h(n) = \sqrt{\frac{S}{2}}\text{sinc}\left(\frac{n+0.5}{C}\right) w(n)$$

where $w(n)$ denotes the Hamming window. For a 512-coefficient filter pair, $S = 1.006$, and $C = 1.994$, yields nearly orthogonal filters with a cross correlation of less than 0.001. We call this the "modified sinc filter."

A compromise between the Harr and modified sinc filters can be found by deriving two equations in terms of the filter coefficients, one each for the out-of-time and out-of-frequency energies for a particular layer and branch of the QMF bank tree. Summing these energies leads to the development of an objective function, which, since some energy will be both out-of-frequency and out-of-time, functions as an upper bound on the out-of-tile energy. Minimizing this function subject to the wavelet constraints gives us the filter we want, which we call the "energy concentration filter."

To remove the wavelet constraints, the objective function may be rewritten using the parametric mapping described in [382] where N filter coefficients are replaced with $(N/2) - 1$ parametric variables. This mapping is nice since any set of parametric variables yields a valid wavelet filter and all possible length N wavelet filters are mapped. Table 4.1 shows a 22-coefficient filter found using this technique to minimize the objective function for the lowpass filter in layer 1 of the QMF bank tree. In practice, we have found little difference between minimizing the objective function for layer 1 and minimizing for other layers. As we will see later, both the modified sinc filter and the energy concentration filter perform well in the QMF-based receiver.

4.4.5 A QMF-Based Receiver

One possible architecture for a QMF-based-receiver is shown in Figure 4.23. In accordance with this figure, the received waveform is bandpass filtered and

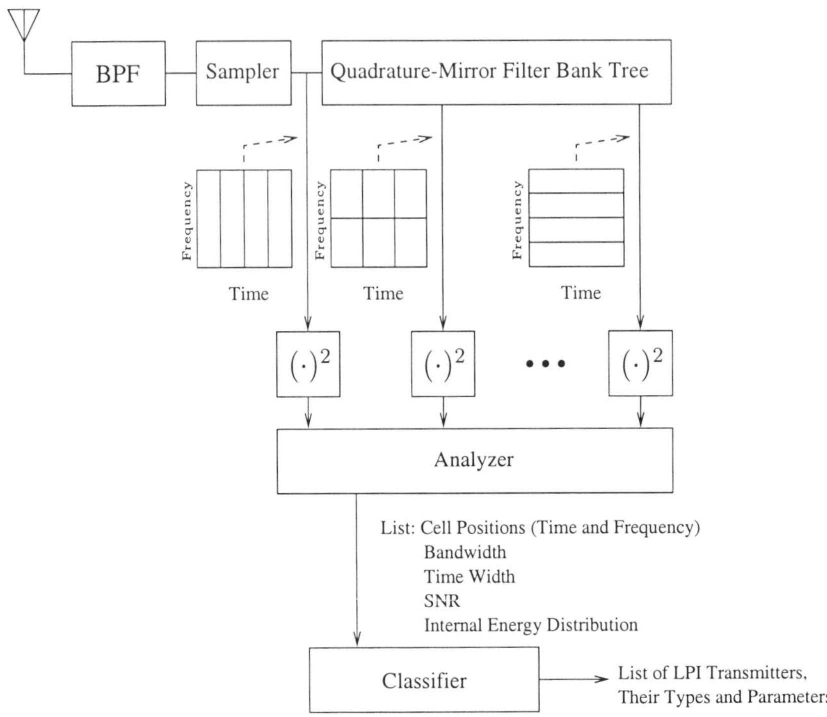

Figure 4.23 LPI receiver block diagram.

sampled at the Nyquist rate. The digital sequence is then fed to the QMF bank tree where it is decomposed. Matrices of values are output from each layer and are squared to produce numbers representing the energy in each tile. This information is then analyzed to determine locations, dimensions and SNRs of any hop cells. The output of the analyzer is a list of cells and these parameters. The classifier then takes the list from the analyzer, determines which cells belong to common transmitters, and outputs a list of transmitters and their features.

We now look at two algorithms that may be used in the analyzer portion of the receiver.

Analyzer Algorithm for Cells with Time-Bandwidth Products Greater than Unity. To develop this algorithm, let's assume there is a hop cell located within our receiver's passband and collection interval, and let's examine the output from a layer of a QMF bank tree whose tile dimensions are smaller than the cell dimensions in both time and frequency. Our best chance for detecting this cell in WGN is to add the energies from the tiles containing the cell's energy and to exclude all other tiles. The corresponding sum yields a RV with the pdf given in (4.20). If, on the other hand, the cell is not present, the pdf is given in (4.18). As we saw earlier, we can make a detection decision by comparing this RV against a threshold.

Our problem, of course, is that we do not know the size or location of the hop cell. If we did, we would know exactly which tiles to add, and (if we repeated this for every hop cell in the band and time interval collected) the receiver would, in essence, be a FBC. If, on the other hand, we added all of the tiles output from the layer, we would have a radiometer — and we would have spent a lot of useless effort decomposing the signal! Our algorithm works between these two extremes.

The algorithm begins by considering a single layer output from the QMF bank tree, and examining a rectangular "block" whose dimensions are the number of tiles in the time and frequency axes that just exceed the largest dimensions of the LPI cells we expect to detect. If we are looking for cells of widely varying dimensions, we may need to repeat the algorithm several times with different sized blocks. The algorithm exhaustively examines every block. Blocks will, of course, overlap with each other, and the goal is to find the ones that best cover each cell, and discard the rest.

The process for doing this is outlined in Figure 4.24. The algorithm computes the energies under every block, writing the results to the input list. It then sorts the list from largest energy to smallest. Then, on the assumption that the block that best covers a cell contains the most energy, the algorithm begins at the top of the list, saving largest energy blocks to the output list and discarding every block that overlaps. When the algorithm terminates, the output list contains the block energies and locations in the time-frequency plane. From this, the cells' SNRs and approximate locations are estimated. To estimate cell dimensions and to refine location estimates, marginal sums of energies are computed for each block along the time and frequency axes. Vectors of these two margins are then convolved with various sized rectangular windows with the largest results indicating those windows that best match the cell in size and location.

Many modifications of the last step are possible, in some cases improving the results. For example, resolution may be improved by using tiles from earlier layers of the QMF bank tree to compute the time marginal and later layers to compute the frequency marginal. The obvious limit here is that, as we go up or down layers, a tile's length in one dimension will eventually exceed that of the cell, and the individual tile's SNR will be too small. A modification of a different sort would be to replace the rectangular-shaped windows with ones more closely matched to the expected shapes of the cells' temporal and spectral envelopes.

Analyzer Algorithm for Cells with Unity Time-Bandwidth Products. Another algorithm may be used in the analyzer portion of the receiver for detecting cells with time-bandwidth products of unity. As we noted above, the time-bandwidth products of the tiles are always one-half, and so, if we happen to select a layer in which the tiles' duration in time is between one-half and one times the duration of a hop cell, the entire cell (with the exception of sidelobes)

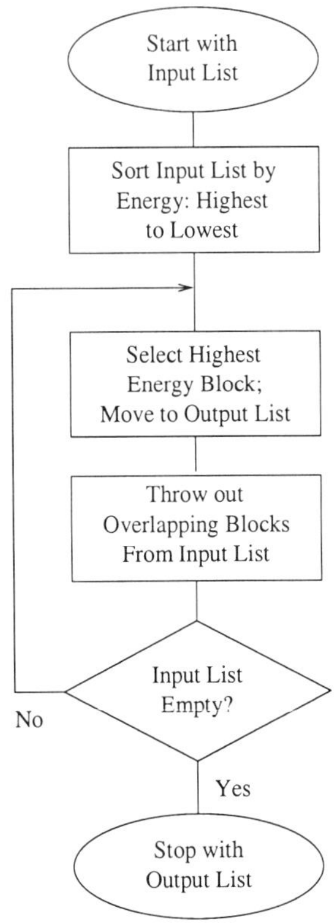

Figure 4.24 Process for selecting blocks. The "Input List" is an exhaustive list of block locations, dimensions, and energies. The "Output List" is a similar list of non-overlapping blocks.

will be contained within a 3×3 block of tiles. Therefore, when we know the hop rate of the cells we are trying to detect, within at least a factor of two, we can collect energies for all 3×3 blocks and then use the process shown in Figure 4.24 to locate cells. As above, the block locations and energy levels yield good estimates of the cells' SNRs and locations. In this case, our knowledge of the hop rate immediately gives us the cells' bandwidth.

When the hop rate is not known a priori, the algorithm may be extended by performing the process described in Figure 4.24 on the output of every layer of the QMF bank tree between one-half and one times the duration of the hop cells. This results in a list for each layer. These lists are then combined into one, and the process is performed one more time. This time the overlapping blocks come from different layers, but the block that best covers a particular cell should still contain the most energy. The result is a list of block energies, locations in the time-frequency plane, and block dimensions from the layer from which each block comes. From this information corresponding cell features may be estimated.

The Classifier. Blocks from regions of the time-frequency plane in which there are no cells are also included in the output lists generated by the algorithms. It is the job of the classifier, by threshold detection or some other criterion, to eliminate these from consideration.

In a practical receiver, further "filtering" may be done in the classifier, eliminating probable false alarms and signals that are not of interest to the interceptor. The classifier may even be adaptive, changing classification criteria based on present and past inputs and results. In the simulation described below we simply compare the threshold against energy from the highest energy block on the list from the analyzer.

Simulation. Unfortunately, a mathematical analysis of the performance of the two block algorithms described above appears to be intractable. Therefore, we perform numerical simulations. Representative simulation results obtained using the algorithm described for detecting cells with time-bandwidth products of unity are shown in Figure 4.25. In this case, a FH/TH signal, with (normalized) cell dimensions of 128 seconds by 1/128 Hz set to ten cells per observation interval, is intercepted. The two-sided noise energy density is normalized to 1 and the corresponding cell energy is set to 20. The interceptor, in this case, is assumed to know the hop rate to within a factor of two, so only a single layer of the QMF bank tree output is examined (layer 6, in this particular case).

The figure plots ROC curves, showing the probability of false alarm, P_{fa}, along the horizontal axis and the probability of detection, P_d, along the vertical axis. Each simulation curve is obtained by making 100 runs with waveforms containing noise only to obtain the probability of false alarm and then making 100 runs with both signal and noise to obtain the probability of detection. Rather than using a single threshold yielding a single point on the plot, a vector of thresholds is used to produce the entire curve. The figure shows three sets of runs, each for the 32-coefficient modified sinc filter and the 22-coefficient energy concentration filter (labeled "22-coefficient tile" in the figure). The theoretical performance of the radiometer is also shown for comparison. As is evident, our goal of out-performing the radiometer has been achieved.

4.5 ACKNOWLEDGEMENT

The authors would like to acknowledge Mr. Michael Lyall for his extensive analytical and simulation support in the development of WTD communications system concepts.

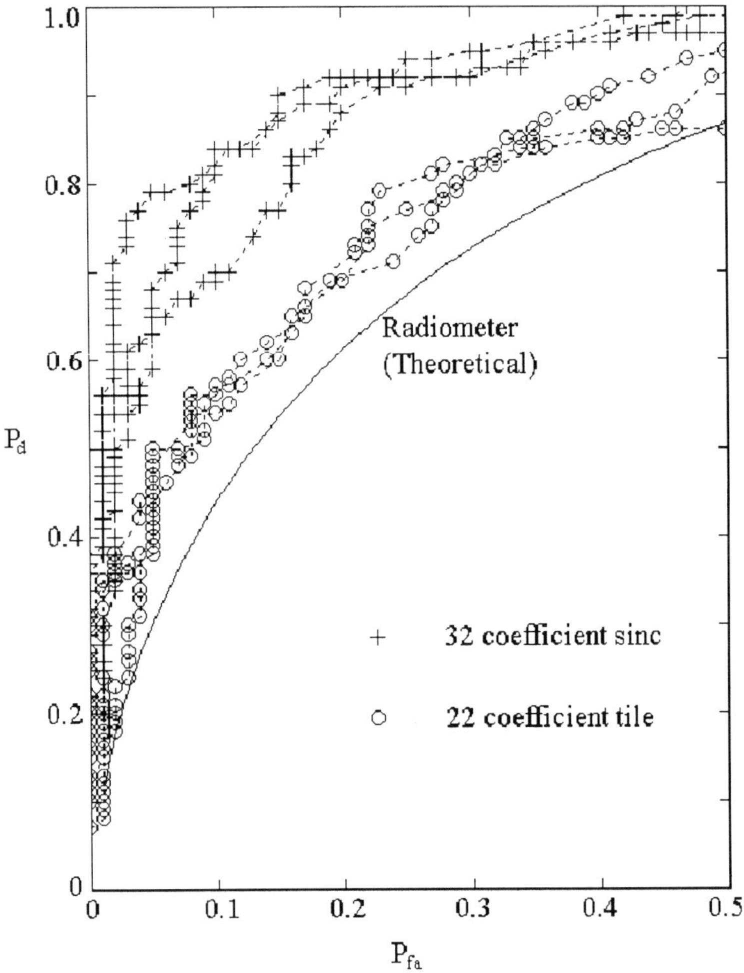

Figure 4.25 ROC for FH/TH signal detection.

5 DIGITAL SUBSCRIBER LINE COMMUNICATIONS SYSTEMS: ENVIRONMENT, PERFORMANCE, AND SPECTRAL COMPATIBILITY

Xueming Lin[1], Massimo Sorbara[1] and Ali N. Akansu[2]

[1]GlobeSpan Semiconductors Inc.,
Red Bank, NJ
[xlin,msorbara]@globespan.net

[2]New Jersey Institute of Technology
Newark, NJ
ali@hertz.njit.edu

5.1 INTRODUCTION

With digital subscriber line (DSL) technologies, the same cables that provide plain old telephone service (POTS) to customers are also used to provide high bit rate digital services to the end users. The bandwidths used in DSLs are significantly greater than those used in conventional voice service. These wideband signals crosstalk into other wire pairs in the same cable and may interfere with their corresponding signals. Hence, crosstalk is one of the key impairments that limit the performance of digital subscriber line systems.

In the design of DSL systems, spectral compatibility is important because any crosstalk that results from the deployment of any new DSL service should not degrade the performance of other services in the cable. Likewise, the existing services in the cable should not prevent the new DSL from meeting its performance objective.

In this chapter, we first present an overview of the spectral compatibility of the different DSL services deployed in the loop plant. Included are discussions on the loop plant environment, cable models, crosstalk models, descriptions of the various DSL spectra, and the compatibility of the various spectra with the

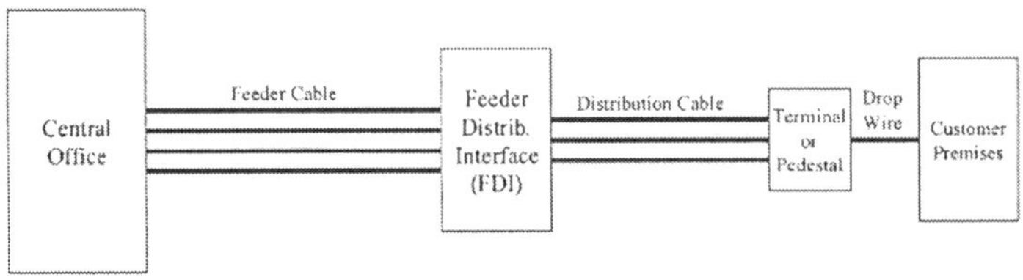

Figure 5.1 Architecture of the loop plant.

DSL services in the cable. Then, communications performance of DSL solutions are presented.

Although DSL systems have been designed for North American, European, and international markets, the focus of discussion in this chapter is primarily on the spectral compatibility of DSL in the North American network. The concepts presented in this chapter are applicable to the European and other international networks, but are outside the scope of this chapter.

5.2 THE LOOP PLANT ENVIRONMENT

POTS is provisioned to customers by routing twisted-wire pairs between the central office (CO) and the customer premises (CP) location. The twisted-wire pair that connects the CO to the CP is the subscriber loop, which may consist of sections of copper twisted-wire pairs of one or more different gauges.

Twisted-wire pairs are contained in cables that have large cross-sections near the CO. Within each cable, they are grouped into binders of 10, 25, or up to 50 wire pairs, and there could be up to 50 binder groups per cable.

Figure 5.1 shows the basic architecture of the loop plant, in which subscriber loops are provisioned [317]. Each subscriber loop can be divided into three portions of cable: feeder cable, distribution cable, and drop wire. Feeder cables provide links to and from the CO to a connection point, feeder distribution interface (FDI) or cross connect point in a concentrated customer area. Distribution cables provide links from the FDI to customer locations. The FDI provides connections of the wire pairs in the feeder cable with those in the distribution cable. The drop wires represent the portion of wire that extends from the terminal on a telephone pole to the CP or underground from the pedestal to the CP.

Because loop plant construction is completed prior to service request, distribution cables are made available to all existing and potential customer sites. It is common practice to connect a twisted-wire pair from a feeder cable to more than one wire pair in the distribution cable. Multiple connections from a feeder or distribution cable to more than one CP form bridged taps. Typically, only one customer is serviced at a time while the other bridged taps are left open circuited.

The loop plant was originally designed to provide customers with POTS. To insure proper quality of service, design rules were defined for subscriber loop provisioning. One set of rules that govern the distribution of twisted-wire pairs for voice service from the CO to the CP is the Resistance Design rules. Implemented in 1980-81 time frame, the design rules are summarized as follows:

1. Loop resistance is not to exceed 1500 Ohms.

2. Inductive loading is to be used whenever the sum of all cable lengths, including bridged taps, exceeds 15 kft.

3. For loaded cables, 88-mH loading coils are placed at 3 kft from the CO and thereafter at intervals of 6 kft.

4. For loaded cables, the total amount of cable, including bridged taps, in the section beyond the loading coil furthest from the CO should be between 3 kft and 12 kft.

5. There are to be no bridged taps between loading coils and no loaded bridged taps.

Revised Resistance Design (RRD) rules are defined in [78]. These rules require that the maximum loop resistance on an 18 kft twisted-wire pair between the CO to the CP must be less than 1300 Ohms and on loops between 18 kft and 24 kft, the maximum resistance is 1500 Ohms. Loading coils are applied to all loops greater than 18 kft or with loop resistance greater than 1300 Ohms.

Telephone cables are designed with different wire gauges ranging from 26 American Wire Gauge (AWG) (thin diameter resulting in higher resistance per unit length) to 19 AWG (thick diameter resulting in lower resistance per unit length). Since distances from the CO to each customer are different, distribution cables are equipped with different wire gauges to meet the resistance design guidelines and provide service to the maximum number of customers. On long loops, the distribution cables tend to use more coarse gauge wire in the regions closer to the subscriber location in order to minimize the total loop resistance. At the CO, the feeder cables tend to use fine diameter gauges (typically 26 AWG) to maximize the number of wire pairs being served.

Some customers may be located so far away from the CO that it may not be possible to meet the resistance design rules. If a subscriber loop is provisioned with a length greater than the maximum defined by the RRD rules, loading coils must be inserted in the loop to achieve proper voice quality. Note, however, that subscriber loops provisioned with loading coils are not suitable for support of wideband DSL services because loading coils introduce too much attenuation at frequencies outside the voice channel required by the DSL system. In short, loading coils must be removed from any subscriber loop that is to support a DSL-based service.

Another set of rules, called Carrier Serving Area (CSA) rules [314, 315, 355], define the distribution of twisted-wire pairs from digital loop carrier systems. The radius covered by CSA rules are up to 9 kft for 26-gauge wire and up to 12 kft for 24-gauge wire. The concept of CSA rules were originally developed for provisioning loops from digital loop carrier (DLC) systems in support of 56 kbps digital data service (DDS) and later slightly modified for the support of POTS from a DLC system. The CSA rules are defined as follows:

1. Non-loaded cable only.
2. Multi-gauge cable is restricted to two gauges.
3. Total bridged tap length may not exceed 2.5 kft and no single bridged tap may exceed 2 kft.
4. The amount of 26 AWG cable may not exceed a total length of 9 kft including bridged taps.
5. For single gauge or multi-gauge cables containing only 19, 22, or 24 AWG cable, the total cable length may not exceed 12 kft including bridged taps.
6. The total cable length including bridged taps of a multi-gauge cable that contains 26 gauge wire may not exceed $\left(12 - \frac{3(L_{26})}{9-L_{Btap}}\right)$ kft, where L_{26} is the total length of 26-gauge wire in the cable (excluding any 26-gauge bridged tap) and L_{Btap} is the total length of bridged tap in the cable. All lengths are in kft.

The above CSA guidelines do not include any wiring in the CO, any drop wiring or any wiring in the CP.

As the transport medium for wideband signals, the twisted-wire pairs introduce linear impairments to the signal. Specifically, the linear impairments are propagation loss, amplitude and delay distortion. The propagation loss in dB is directly proportional to the distance of the loop. Amplitude distortion results from the fact that signal components at higher frequencies experience more amplitude loss than components at low frequencies. To a first order, the amplitude response of the twisted-wire pair channel rolls off at roughly the square root of frequency. Finally, the delay is such that at low frequencies (less than approximately 10 kHz) there is very sharp variation in the group delay and at higher frequencies the delay response is relatively constant [355].

Test loops have been developed for AWG in T1E1.4 and metric cables in ETSI for Integrated Services Digital Network (ISDN) [7, 141], High-rate Digital Subscriber Line (HDSL) [314, 315, 142], and Asymmetric Digital Subscriber Line (ADSL) [6] systems. [355] provides a detailed description of cable modeling and [142] provides a comprehensive list of the test loops used for North American and European loop environments.

5.3 CROSSTALK MODELS: NEXT AND FEXT

Crosstalk generally refers to the interference that enters a communication channel, such as twisted-wire pairs, through some coupling path. The diagram in Figure 5.2 shows two examples of crosstalk generated in a multi-pair cable [355]. On the left-hand side of the figure, the signal source transmits a signal at full power on twisted-wire pair j. This signal, when propagating through the loop, generates two types of crosstalk in the other wire pairs in the cable. The crosstalk that appears on the left-hand side, in wire pair i, is called near-end crosstalk (NEXT) because it is at the same end of the cable as the cross-talking signal source. The crosstalk that appears on the right-hand side, in wire pair i, is called far-end crosstalk (FEXT) because the crosstalk appears on the end of the loop opposite to the reference signal source. In the loop plant, NEXT is generally far more damaging than FEXT because, unlike far-end crosstalk, near-end crosstalk directly disturbs the received signal transmitted from the far-end after it has experienced the propagation loss from traversing the distance from the far-end down the disturbed wire pair.

In a multi-pair cable, relative to one desired receive signal on one twisted-wire pair, all of the other wire pairs are sources of crosstalk. For DSL systems, the reference cable size for evaluating performance in the presence of crosstalk is a 50 pair cable. Reviewing the example shown in Figure 5.1, we see that relative to the received signal on wire pair i, the other 49 wire pairs are sources of crosstalk (both near-end and far-end).

5.3.1 Near-End Crosstalk Model

As described in the DSL standards and technical reports, for the reference 50 pair cable, NEXT coupling of signals in other wire pairs within the cable is modeled as

$$|H_{NEXT}(f)| = \chi_{49} f^{\frac{3}{2}} \left(\frac{N}{49}\right)^{0.6},$$

where $\chi_{49} = \frac{1}{1.134\times10^{13}}$ is the coupling coefficient for 49 NEXT disturbers, N is the number of disturbers in the cable, and f is the frequency in Hz. Note that the maximum number of disturbers (maximum value of N) in a 50 pair cable is 49. A signal source that outputs a signal with power spectral density (PSD), $PSD_{Signal}(f)$, injects a level of NEXT in a near-end receiver that is

$$PSD_{NEXT}(f) = |H_{NEXT}(f)|^2 PSD_{Signal}(f).$$

So, as illustrated in Figure 5.2, if there are signals in the cable with the same PSD, the PSD of the NEXT at the input to the near-end receiver on wire pair i is $PSD_{NEXT}(f)$.

Note from the above expressions that the crosstalk coupling is very low at the lower frequencies and the coupling increases 15 dB per decade with increasing frequency. For example, at 80 kHz, the coupling loss is 57 dB for 49 disturbers. The loss (in dB) for 49 disturbers at other frequencies may be computed using the following formula:

$$L_{49} = 57dB - 15\log_{10}\left(\frac{f}{80kHz}\right)$$

where L_{49} is the NEXT coupling loss in dB and f is the frequency in kHz.

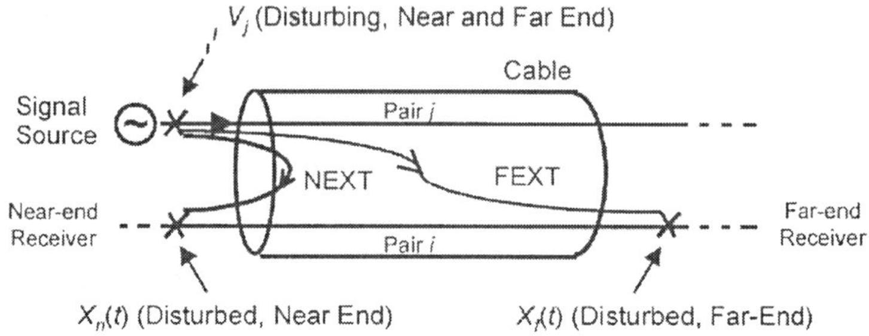

Figure 5.2 NEXT and FEXT in a multi-pair cable.

5.3.2 Far-End Crosstalk Model

In the same 50 pair cable, the FEXT coupling of signals in other wire pairs is modeled as

$$|H_{FEXT}(f)|^2 = kdf^2 \left(\frac{N}{49}\right)^{0.6} |H_{channel}(f)|^2,$$

where k is the coupling coefficient for 49 FEXT disturbers ($k = 8 \times 10^{-20}$), d is the coupling path distance in feet, f is the frequency in Hz, N is the number of disturbers and $H_{channel}(f)$ is the channel transfer function. Note that the coupling is small at low frequencies and higher at larger frequencies. The coupling slope increases 20 dB per decade with increasing frequency.

5.3.3 Comparison of NEXT with FEXT

If we compare the coupling coefficient for NEXT with that of FEXT, we see that the NEXT coupling is approximately six orders of magnitude greater than that of FEXT. Also notice that the coupling for NEXT increases 15 dB per decade with increasing frequency where as the coupling for FEXT increases 20 dB per decade with increasing frequency. A comprehensive study of loop plant cable characteristics, linear impairments and crosstalk can be found in [355].

5.3.4 Channel Capacities in the Presence of NEXT and FEXT

Let us assume that a 50 pair cable is filled completely with signals that have the same PSD. Figure 5.3 contains an example of NEXT and FEXT on a 9 kft 26-gauge loop. The figure plots the PSD (in dBm/Hz) of the transmit signal, the insertion loss of the 9 kft 26-gauge loop (in dB), and the PSDs of the received signal, the 49 NEXT disturbers, and the 49 FEXT disturbers. The bandwidth of the transmit signal is assumed to be approximately 700 kHz between its half-power points (-3 dB frequencies) with a PSD level of -40 dBm/Hz in the passband region. The PSD of the received signal is shaped by the insertion loss of the channel as shown in the figure. For example, at 500 kHz the insertion

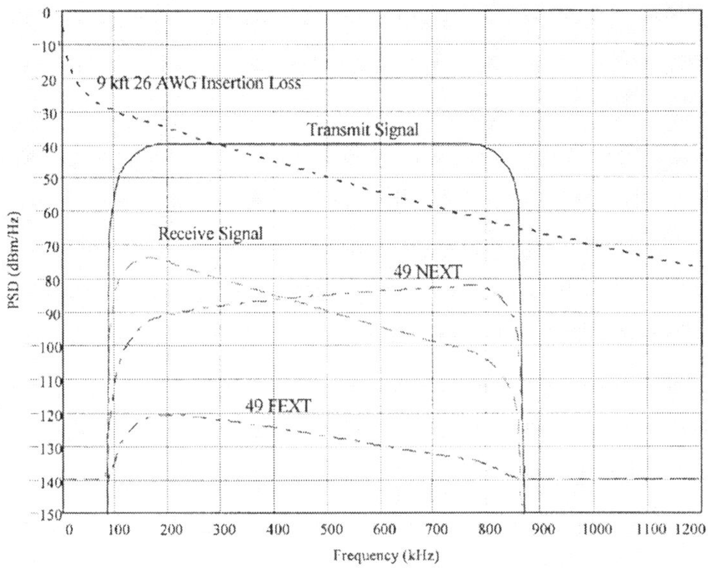

Figure 5.3 Comparison of NEXT and FEXT crosstalk levels.

loss of the channel is 50 dB; the resulting receive signal PSD is -90 dBm/Hz, which is 50 dB below its transmit level.

For the special case (as in the example in Figure 5.3) where the crosstalk comes from signals with the same PSD as that of the corresponding transmitter, NEXT is more specifically referred to as self-NEXT (or simply SNEXT) and FEXT is referred to as self-FEXT (or simply SFEXT).

In addition to the magnitude of the received signal, Figure 5.3 also shows the magnitudes of the NEXT and FEXT crosstalk PSDs. In each case, we assume 49 disturbers in the 50 pair cable. The signal-to-noise ratio (SNR) at the receiver input relative to NEXT is approximately 3.7 dB; the corresponding SNR relative to FEXT is approximately 40 dB. Note that the SNR relative to NEXT is proportional to the area between the received signal curve and the 49 NEXT curve. Similarly for FEXT, the SNR is proportional to the area between the received signal curve and the 49 FEXT curve.

To further quantify the effects of NEXT and FEXT, we compute the resulting Shannon (or channel) capacities. In general, the Shannon capacity for a given channel is

$$C = \int_0^\infty \log_2 \left[1 + SNR(f)\right] df,$$

where C is the channel capacity in bits per second, $SNR(f)$ is the signal-to-noise ratio density at the receiver input, and f is frequency in Hz. For the case of SNEXT, the channel capacity is 1.7 Mbps and for SFEXT, it is

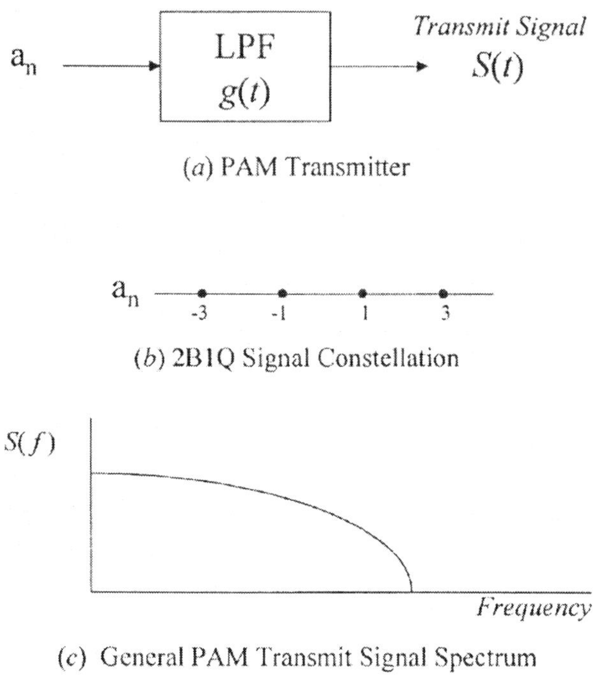

Figure 5.4 Baseband PAM transmitter model.

approximately 8.7 Mbps, a difference of approximately 7 Mbps. Clearly, as shown in the above example, NEXT interference strongly dominates over the effects of FEXT interference in the digital subscriber loop environment.

5.4 DSL MODULATION METHODS

This section provides a brief description of the modulation methods used in DSL systems. Included are descriptions of 2-binary 1-quaternary, discrete multitone, quadrature amplitude and carrierless amplitude and phase modulation.

5.4.1 2B1Q

2-binary 1-quaternary (2B1Q) is the modulation method used in ISDN and HDSL systems. This is also known as four level pulse amplitude modulation (PAM) or simply 4-PAM. Figure 5.4 provides a summary of a basic PAM (hence 2B1Q) transmitter, the 2B1Q signal constellation, and the resulting transmit signal spectrum. The transmitter, as shown in Figure 5.4(a), is simply a low-pass shaping filter. The impulse response g(t) of the lowpass filter defines the

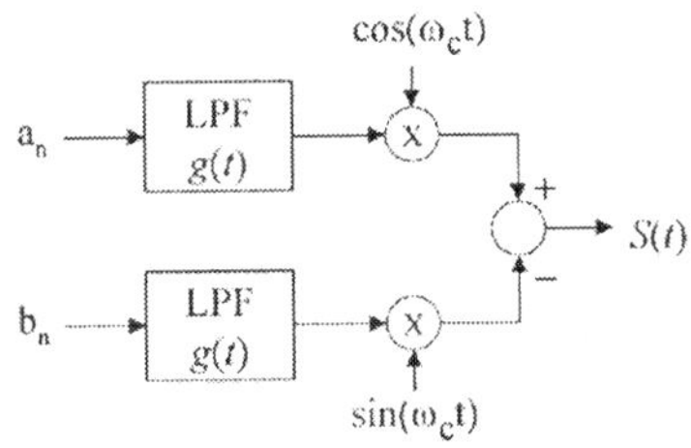

(a) QAM Transmitter

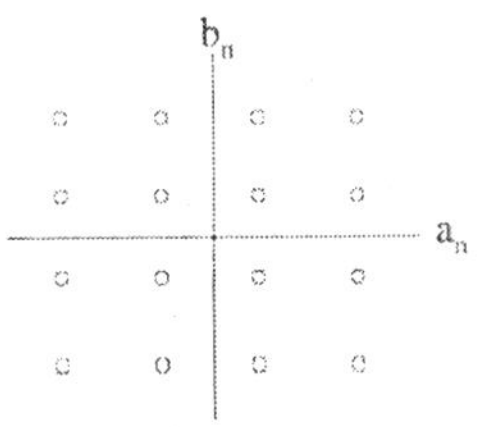

(b) 16-point Constellation

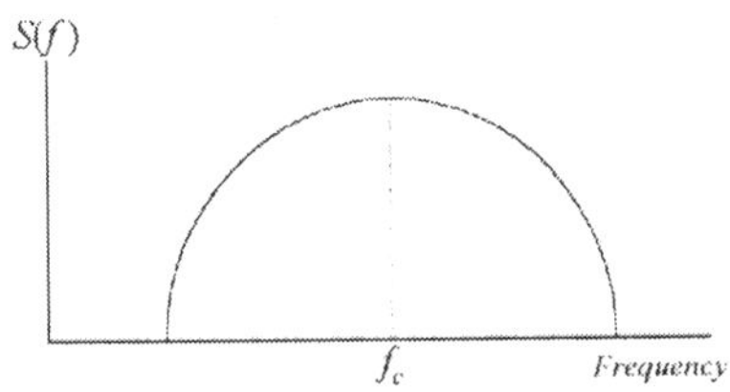

(c) Transmit Spectrum

Figure 5.5 Quadrature amplitude modulation.

spectral shape of the transmit signal, for which an example is shown in part (c) of the figure.

In 2B1Q, two bits of information are transmitted during each signaling interval. Hence, one of four voltage levels is transmitted for each symbol during the corresponding time. Figure 5.4(b) shows the constellation (or symbol alphabet) for 2B1Q. With the spectral shaping of the lowpass filter defined by impulse response g(t), the transmit signal is represented by

$$S(t) = \sum_{-\infty}^{\infty} a_n g(t - nT)$$

where a_n represents the transmit symbols, and T is the signaling interval. $S(t)$ is the signal transmitted on the transmission line.

5.4.2 *QAM and CAP Modulation*

Conventional bandwidth efficient modulation techniques, such as quadrature amplitude modulation (QAM), are widely used in digital communications systems. Carrierless amplitude and phase (CAP) signaling for DSL applications, proposed in [355, 167, 168], is a variant of QAM. Figures 5.5 and 5.6 illustrate the basic structures of QAM and CAP modulation schemes, respectively. A simplified structure of a CAP based transceiver is displayed in Figure 5.7. The

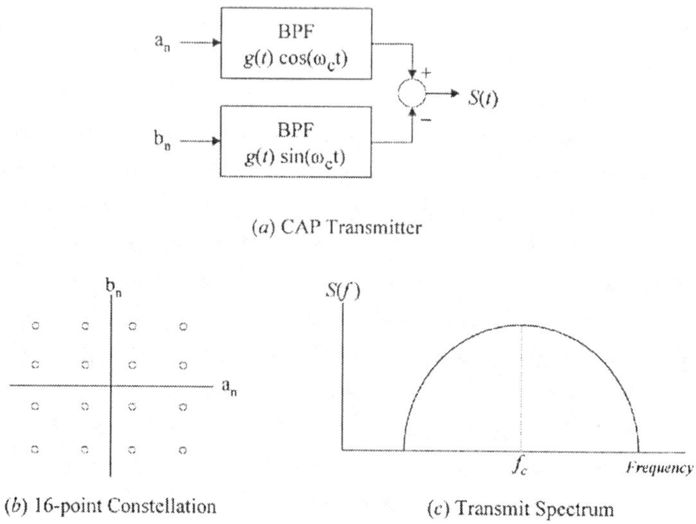

Figure 5.6 Carrierless amplitude and phase modulation.

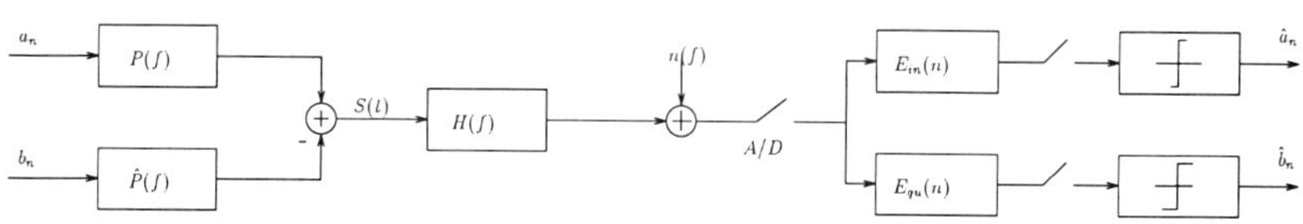

Figure 5.7 A simplified structure of a CAP modulation based transceiver system.

transmitted CAP signal is modeled as

$$S(t) = \sum_{-\infty}^{\infty} a_n p(t - nT) - \sum_{-\infty}^{\infty} b_n \tilde{p}(t - nT)$$

where a_n, b_n are the information bearing in-phase and quadrature symbols and $p(t)$ and $\tilde{p}(t)$ represent a Hilbert pair, both of which are transmitter passband spectral shaping filters — raised-cosine or square-root raised-cosine signals are often used in practice. Figure 5.8 displays the in-phase and quadrature CAP shaping filters using a raised cosine Nyquist filter [355].

In the implementation of a CAP based transceiver system, the CAP modulation signal is digitally generated by passing the bitstream to be transmitted through the $p(t)$ and $\tilde{p}(t)$. The incoming symbols are up-sampled (oversampled)

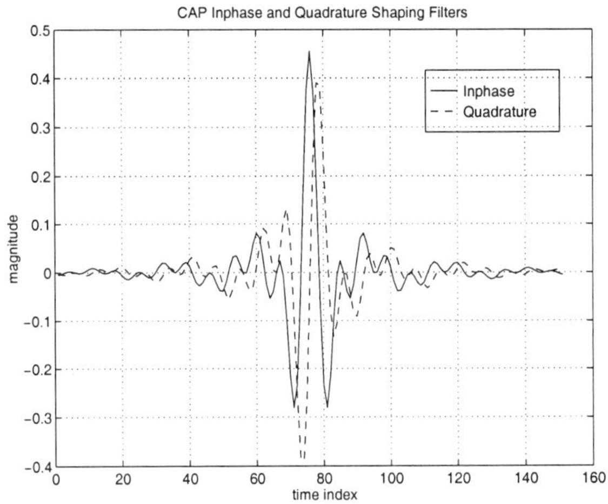

Figure 5.8 The coefficients of in-phase and quadrature shaping filters for a CAP based system.

by a factor of K such that the CAP shaping filters are $\frac{T}{K}$ fractionally spaced. Here, T is the symbol period. K is selected as 4 or 5 in general. The in-phase and quadrature shaping filters can have time spans of 10–20 periods. For the case of K=5 with shaping filters with a time span of 10 periods, $p(t)$ and $\tilde{p}(t)$ have 50 coefficients. At the transmitter side, the discrete output samples of the shaping filter are converted to an analog signal using a D/A converter. A lowpass analog filter is used before it is put through a wire splitter.

Let us design a CAP-based ADSL transceiver with 6.72 Mbps downstream and 250 kbps upstream. The downstream signal constellation is of size 256 and the upstream signal constellation size is 64. Figure 5.9 shows the 64 CAP constellation. As we know, significant signal distortion, namely, intersymbol interference is introduced by this transmission channel. At the receiver, an adaptive fractionally spaced linear feed-forward filter and symbol spaced decision feedback equalizer (DFE) transceiver structure effectively cancels the intersymbol interference. In a CAP-based system, noise predictive (NP), hybrid or conventional DFE structures can be used [356]. Figure 5.10 illustrates the structure of NP DFE used in a CAP based system. The equalized symbol is sliced to make a decision.

In order to prevent error propagation due to the nature of the DFE structure, the Tomlinson precoder needs to be implemented in a CAP based system. For different loop plants under different field noise conditions, different transmission rates can be realized. The CAP based Rate-Adaptive ADSL (RADSL) transceiver has been used in a field trial and commercially deployed around the world.

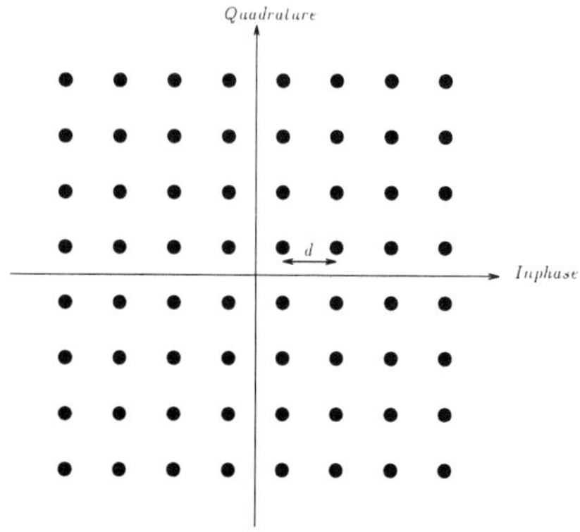

Figure 5.9 Constellation of 64 CAP signaling.

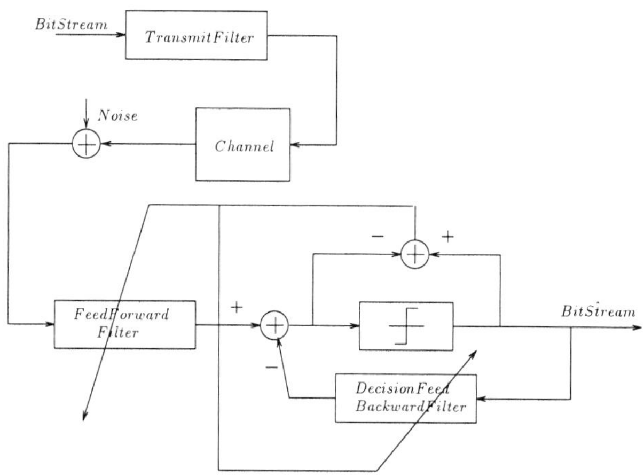

Figure 5.10 Noise predictive DFE structure implemented in CAP transceiver.

5.5 MULTICARRIER MODULATION

Multicarrier modulation techniques utilize a set of modulation line codes for digital communications. Ideally, these modulation line codes divide the transmission channel into a number of independent subchannels. The transmultiplexed digital signals can be transmitted independently in different subchannels. We can model this channel into many parallel brick wall subchannels. As the number of subchannels increases, the brick wall model becomes more

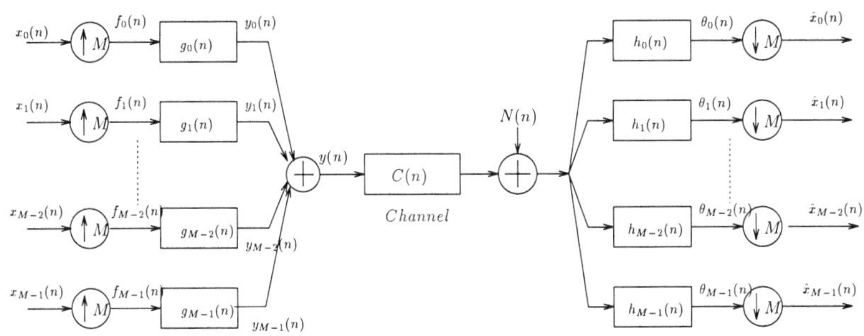

Figure 5.11 Basic structure of a multicarrier modulation based digital transceiver.

accurate. There is no need to consider any intersymbol interference in any subchannel. A one-tap complex equalizer can be implemented in subchannels to recover the transmitted (multiplexed) signal. Naturally, multicarrier modulation represents a frequency division type multiplexer.

The basic structure of a multicarrier modulation based transceiver is displayed in Figure 5.11. It is shown as a multirate M-band synthesis/analysis filter bank transmultiplexer. In a multicarrier modulation system, a set of M frequency-selective orthogonal functions $\{g_i(n)\}$ are used. The subsymbols $\{x_0(n), x_1(n), \ldots x_{M-1}(n)\}$ are formed by grouping blocks of an incoming bitstream via a constellation scheme like QAM or PAM. The dynamic parsing of the incoming bitstream to the subsymbol is determined by the achievable SNR in all subchannels. The subchannels that suffer less attenuation or less interference carry more bits of information [39].

The modulation filters of the transceiver form an orthogonal function set. The design methodologies of synthesis/analysis filter banks are applicable for this application [2, 337].

5.5.1 Discrete Multitone Transceiver Techniques for DSL Communications

Discrete multitone (DMT) uses the discrete Fourier transform (DFT) as its modulation/demodulation basis function set. Each subcarrier modulation function is indeed one of the DFT basis functions. Since there are fast Fourier transform algorithms which make the computational complexity very low, DMT is very attractive to industry. Many companies are working on DMT based techniques for DSL applications. As a matter of fact, ANSI has adopted DMT as its standard in the ADSL [6].

Current DMT-based ADSL communication systems utilize 512-size DFT basis functions as their orthogonal subcarriers in a synthesis/analysis (transmultiplexer) transform configuration. These subcarriers spectrally overlap although they are orthogonal. The orthogonality of DMT comes from implementation of the guard interval (cyclic prefix) in the transmitter. The idea of the cyclic

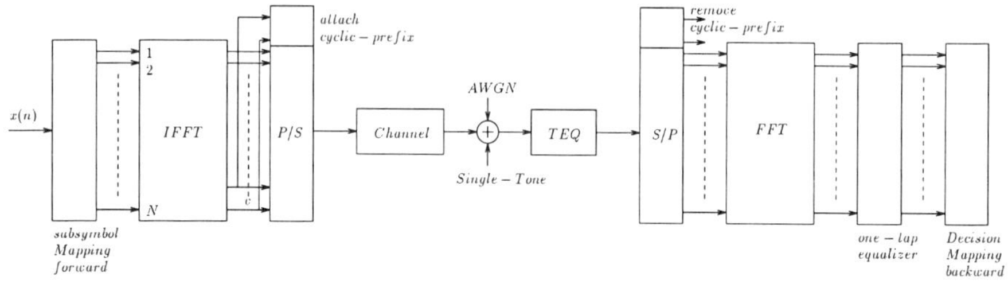

Figure 5.12 An implementation of DFT-based DMT transceiver.

prefix first proposed by Peled and Ruiz [249] is further detailed in [1]. Figure 5.12 displays an implementation of a DFT-based DMT transceiver system wherein additive white Gaussian noise (AWGN) and single-tone interference are injected; NEXT, FEXT or other types of noise may be injected in alternative channel environments.

The transmission channel has a high dynamic range due to the transmission loss at the higher frequency range. Therefore, it generates severe distortion at the transmitted signal. In order to eliminate the intersymbol interference as well as the interchannel interference, a cyclic prefix is used. Its length, ν, should be larger than the duration of the transmission channel impulse response.

At the transmitter, the transmitted subsymbols can use QAM constellation in each subchannel. The size of the QAM constellation depends on the noise condition of each subchannel. This turns into an optimal bit allocation problem (water pouring). The parsed subsymbols are arranged in symmetric Hermitian pairs such that the inverse DFT (IDFT) transformed sequence is real-valued. The cyclic prefix is attached to the IDFT modulated super-symbol (block) sequence. Assume that the length of cyclic prefix is ν, we copy the last ν samples of IDFT transformed data samples to the beginning of the super-symbol. Then, the composite super-symbol is of size $512+\nu$. The composite super-symbol is converted from parallel to serial and passed through a D/A converter. The modulated signal is then transmitted to the splitter of DSL loop plant. The cyclic prefix makes the linear convolution the same as circular convolution. The addition of cyclic prefix operation will naturally reduce the effective transmission rate by $\frac{N}{N+\nu}$. Here, N is the size of IDFT/DFT transform. The reduction of transmission by $\frac{N}{N+\nu}$ is the penalty paid to utilize an efficient modulation and demodulation algorithm. Therefore, ν should not be too big compared to N, the size of IDFT/DFT. In ANSI ADSL standard, N is 512 and the length of cyclic prefix ν is chosen as 32.

As it is displayed in Figure 5.13, the practical loop impulse response has a much longer duration compared to ν when the sampling rate is high. If we

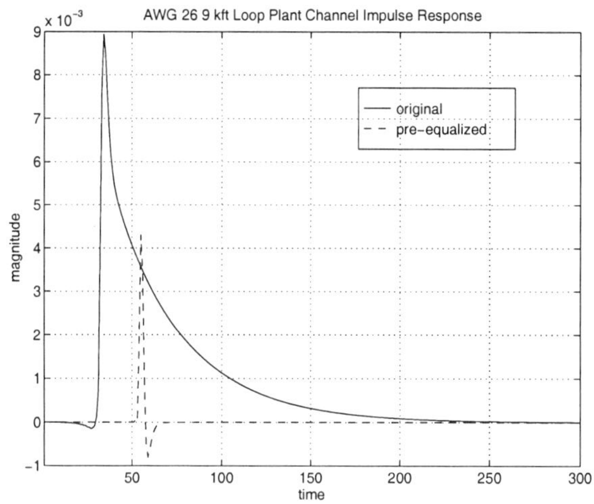

Figure 5.13 Original channel impulse response and TEQ pre-equalized channel impulse response in a DMT transceiver system.

keep the cyclic prefix length longer than the channel impulse response, the throughput is reduced. In order to increase the throughput, a pre-equalizer, called a time domain equalizer (TEQ), is designed. It reduces the duration of the composite channel impulse response. Figure 5.13 shows the original impulse response of AWG 26-gauge wire with length of 9 kft and sampling frequency of 2.048 MHz. Obviously, the duration of the channel is long compared to N=512. Using a TEQ, the channel impulse response is reduced to less than ν, where ν is set as 32 by the ANSI ADSL standard [316]. The TEQ pre-equalized channel impulse response is also displayed in Figure 5.13.

At the receiver, the received signal is sampled by an A/D converter with the same sampling frequency. The cyclic prefix part of received signal is discarded first. Then the size-N DFT demodulation is performed. Due to the presence of the cyclic prefix, all the subchannels are independent. In order to recover a transmitted subsymbol, a one-tap complex equalizer is used at the DFT output. The one-tap complex equalizer for each subchannel is obtained from the inverse of the DFT transform for the composite channel impulse response. A decision is made after the one-tap complex equalizer and the sliced subsymbols are mapped back to form the composite data stream.

5.6 DSL SIGNAL SPECTRA

This section describes the PSD of each DSL transmit signal deployed in the network. The DSLs include ISDN, HDSL, Symmetric Digital Subscriber Line (SDSL), ADSL and RADSL. The following sections describe the signal power spectra of each DSL.

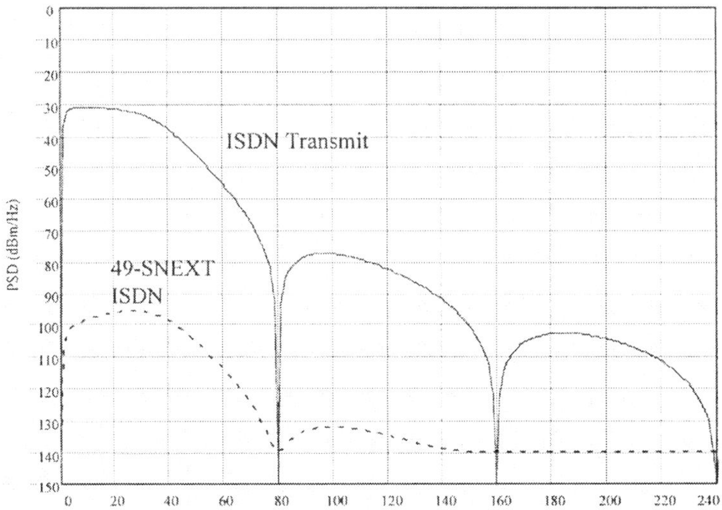

Figure 5.14 ISDN transmit signal and 49 NEXT spectra.

5.6.1 *ISDN*

ISDN Basic Rate provides symmetrical transport of 160 kbps on the subscriber line. The line code is 2B1Q and the corresponding transmit signal PSD is expressed as

$$PSD_{ISDN}(f) = \frac{5}{9}\frac{V_p^2}{R}\frac{\left[\sin\left(\frac{\pi f}{f_0}\right)\right]^2}{\left(\frac{\pi f}{f_0}\right)^2}\frac{1}{1+\left(\frac{f}{f_{3dB}}\right)^4}, \quad 0 \leq f < \infty$$

where f_{3dB}=80 kHz, f_0=80 kHz, V_p=2.5 Volts and R=135 Ω. The transmit power of ISDN is nominally 13.5 dBm. In Figure 5.14, the solid line plots the PSD of the ISDN transmit signal. Also shown in the figure (dotted plot) is the PSD of 49 NEXT disturbers from ISDN.

Both the upstream and downstream signals occupy the same frequency band, so echo-cancellation is used to separate the two directions of transmission on the subscriber line. In such a system, ISDN is subject to SNEXT. The deployment objective for ISDN is to operate on non-loaded loops that range up to 18 kft in the presence of SNEXT. These include all loops that meet RRD rules.

5.6.2 *HDSL*

In North America, HDSL is a service that provides the transport of T1 (1.544 Mbps) signals between the CO and the CP. This service is deployed using two subscriber lines, where the bit rate is 784 kbps on each wire pair and half of the

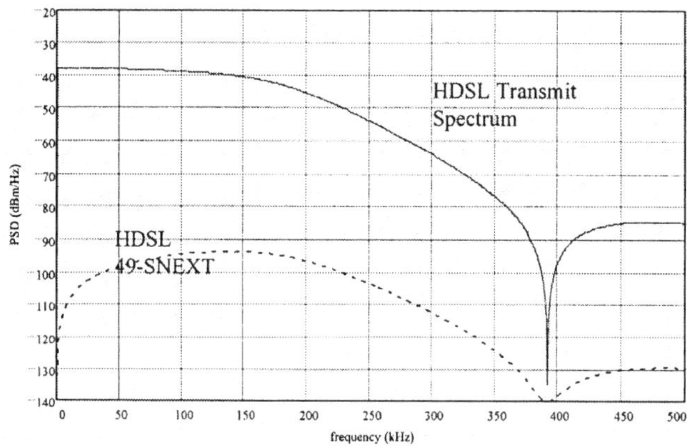

Figure 5.15 HDSL transmit signal and 49 NEXT spectra.

T1 payload is carried on each pair. The line code is 2B1Q and the corresponding transmit signal PSD is expressed as

$$PSD_{HDSL}(f) = \frac{5}{9}\frac{V_p^2}{R}\frac{\left[\sin\left(\frac{\pi f}{f_0}\right)\right]^2}{\left(\frac{\pi f}{f_0}\right)^2}\frac{1}{1+\left(\frac{f}{f_{3dB}}\right)^8}, \quad 0 \le f < \infty$$

where f_{3dB}=196 kHz, f_0=392 kHz, V_p=2.7 V and R=135 Ω. The transmit power of the HDSL signal is 13.5 dBm. In Figure 5.15, the solid line plots the PSD of the HDSL transmit signal. Also shown in the figure (dotted plot) is the PSD of 49 NEXT disturbers from HDSL.

Both the upstream and downstream signals occupy the same frequency band, so echo-cancellation is used to separate the two directions of transmission on the subscriber line. In such a system, HDSL is subject to SNEXT. The deployment objectives for HDSL is operation on loops that meet carrier serving area requirements while operating in the presence of SNEXT.

5.6.3 SDSL

SDSL, or MSDSL, defines the transport of multirate symmetric DSL services on a single twisted-wire pair. SDSL solutions deployed today are echo-cancellation based and are implemented using CAP and 2B1Q technologies.

The SDSL bit rates considered here are 160 kbps (ISDN), 384 kbps, 784 kbps (on a pair of an HDSL system) and 1560 kbps. Both CAP and 2B1Q solutions are considered for these rates. Figure 5.16 shows the transmit signal PSD plots for the SDSL line signals using CAP The spectral shaping of each transmit signal is square-root raised cosine with a roll-off factor of 15%. For each bit rate, the spectrum is scaled and shaped for a transmit power of 13.5 dBm. Figure 5.17 shows the 2B1Q SDSL PSD plots. Each system uses NRZ (non-

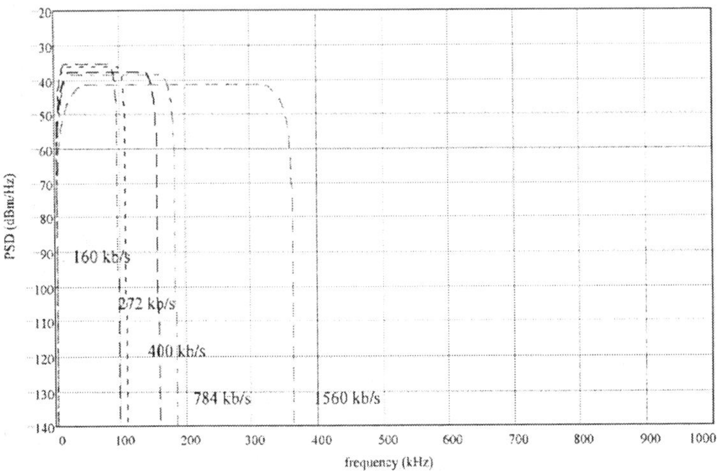

Figure 5.16 CAP SDSL transmit signal spectra.

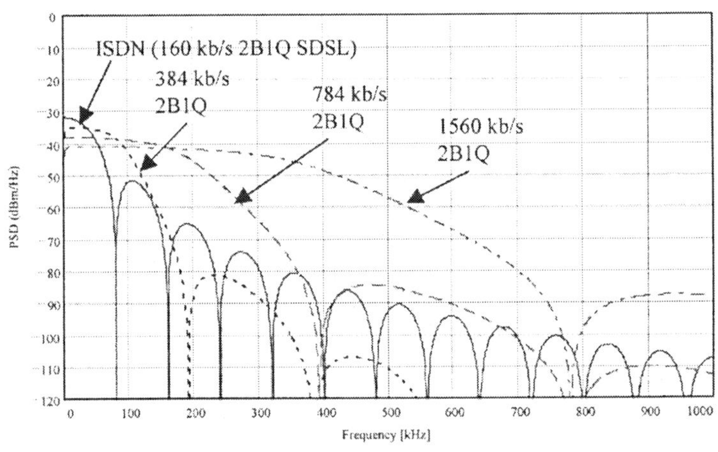

Figure 5.17 2B1Q SDSL transmit signal spectra.

return to zero) pulses followed by an Nth order Butterworth filter. The 160 kbps spectrum is the same as that defined for ISDN in T1.601, which defines a 2nd order Butterworth filter for out-of-band energy attenuation. The remaining signals use a 4th order Butterworth filter for out-of-band energy attenuation. The first null in each spectrum defines the signal bandwidth. Note that for 784 kbps and 1560 kbps systems, the bandwidth of the CAP systems is roughly half that of the 2B1Q systems.

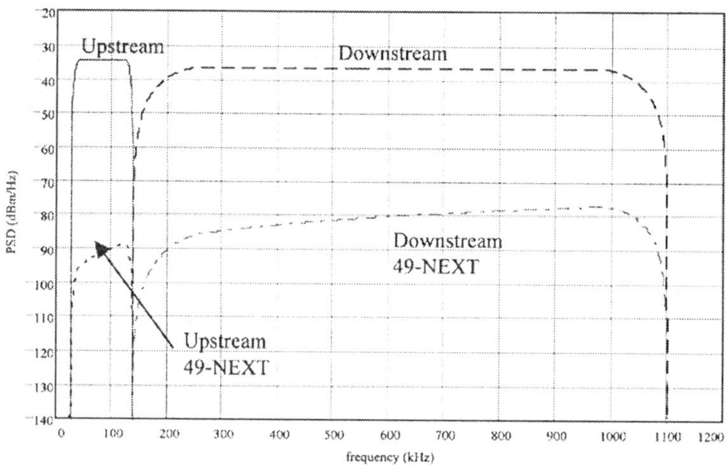

Figure 5.18 DMT ADSL FDM transmit signal spectra.

The 2B1Q systems (SDSL, ISDN, and HDSL) do not use any coding or forward error correction; the CAP systems use a 2-dimension 8-state trellis code that provides a 4 dB asymptotic coding gain. For SDSL, both the CAP and 2B1Q systems are affected by SNEXT. If the cable is deployed with multiple services, consideration must be given to the bandwidth of the SDSL signals because these signals induce NEXT in other services.

5.6.4 ADSL

Figure 5.18 shows the transmit spectra and NEXT spectral plots of the upstream and downstream DMT (or RADSL) channels. DMT is a variable bit rate system and the actual bandwidths of the upstream and downstream channels may vary depending on the bit rate and noise environment. The plots in Figure 5.18 are idealized in that they display the maximum possible useful bandwidth for the upstream and downstream channels. Not shown in the figure are the out-of-band energy levels. Note that the PSD levels shown in the figure are peak values as opposed to root-mean-squared values. No guard band is specified between the upstream and downstream channel. Details of the DMT PSD masks are given in [316].

The spectra shown in Figure 5.18 use frequency division multiplexing (FDM) to separate the upstream and downstream channel. If the cable has only FDM based ADSL systems deployed, there is no SNEXT; system performance would be limited by SFEXT. T1.413 also defines an echo-cancelled version of ADSL where the downstream channel completely overlaps the upstream channel. In this case, NEXT is injected between the upstream and downstream channel. The channel most impacted is the upstream channel, where the NEXT from the downstream channel completely covers the upstream band.

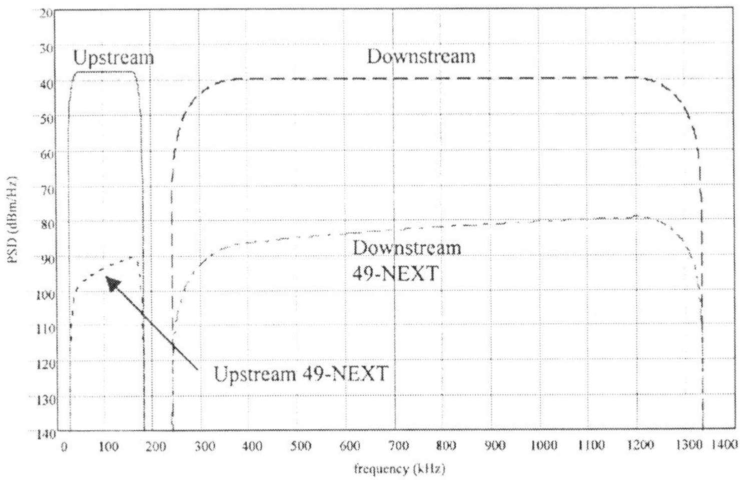

Figure 5.19 RADSL signal and crosstalk spectra.

5.6.5 CAP RADSL

Figure 5.19 shows the PSD plots of the CAP RADSL upstream and downstream channels along with their 49 disturber NEXT spectra. The upstream channel PSD mask is -38 dBm/Hz (rms) and that for the downstream channel is -40 dBm/Hz. The upstream channel uses frequencies from 25 kHz to 181 kHz (worst-case assuming 15% raised-cosine shaping) and the downstream channel uses frequencies from 240 kHz up to approximately 1.1 MHz. The complete definition of the CAP RADSL PSD masks are given in [318]. RADSL systems have only been deployed using FDM for separation of the upstream and downstream channels. Hence, with RADSL, there is no NEXT generated between the upstream and downstream channels.

5.6.6 T1 Alternate Mark Inversion

The PSD of the T1 line signal is assumed to be a 50% duty-cycle random alternate mark inversion (AMI) code at 1.544 Mbps. The single-sided PSD is represented by the following expression:

$$PSD_{T1} = \frac{V_p^2}{R_L}\frac{2}{f_0}\left[\frac{\sin\left(\frac{\pi f}{f_0}\right)}{\frac{\pi f}{f_0}}\right]^2 \sin^2\left(\frac{\pi f}{2f_0}\right)\frac{1}{1+\left(\frac{f}{f_{3dB-Sh}}\right)^6}\frac{f^2}{f^2+f_{3dB-Xf}^2},$$

where $0 \leq f < \infty$, f_0=1.544 MHz, V_p=3.6 V and R_L=100 Ω. In this case, f_{3dB-Sh}=3.0 MHz is the 3 dB frequency of a third order Butterworth lowpass shaping filter, and f_{3dB-Xf}=40 kHz is the highpass transformer coupling cutoff

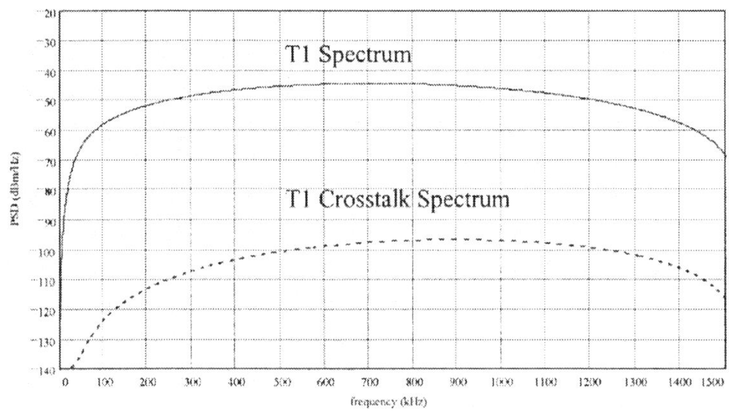

Figure 5.20 T1 AMI signal and crosstalk spectra.

frequency. Figure 5.20 shows a plot of the T1 AMI transmit signal PSD along with the 49 disturber NEXT PSD.

5.6.7 Computation of Performance

To compute the resulting SNR margin, we first need to compute the resulting output signal-to-noise ratio (SNR_{out}) and then take the difference from the reference SNR value (SNR_{ref}) that corresponds to a bit-error-rate (BER) of 10^{-7}. The first two columns in Table 5.1 lists the SNR_{ref} values for the uncoded signal constellations considered in this study. To include the gain effects from the trellis code, the coding gain is subtracted from the uncoded signal's SNR_{ref}.

Table 5.1 List of reference SNRs of various line codes.

Line code (without trellis coding)	$SNR_{ref-uncoded}$ (dB)
4-CAP/2-PAM	14.5
8-CAP	18.0
16-CAP/4-PAM	21.5
32-CAP	24.5
64-CAP/8-PAM	27.7
128-CAP	30.6
256-CAP/16-PAM	33.8

The output SNR of the DFE in the CAP/QAM receiver is computed in [356] as

$$SNR_{out} = \exp\left[\frac{1}{f_B}\int_0^{f_B} \ln\left(1+\sum_{n=0}^{\infty}\frac{S_R(f+nf_B)}{N_X(f+nf_B)+N_0}\right)df\right].$$

In the PAM receiver it is computed as

$$SNR_{out} = \exp\left[\frac{1}{f_B}\int_{-\frac{f_B}{2}}^{\frac{f_B}{2}} \ln\left(1+\sum_{n=0}^{\infty}\frac{S_R(f+nf_B)}{N_X(f+nf_B)+N_0}\right)df\right].$$

The SNR margin (M) is computed as

$$M = SNR_{out} + G_{TC} - SNR_{ref,coded}$$

where G_{TC} is the coding gain of the trellis code and $SNR_{ref,uncoded}$ is the reference SNR of the uncoded line code (second column in Table 5.1). Alternatively, this equation may be expressed as

$$M = SNR_{out} - SNR_{ref,coded}$$

where $SNR_{ref,coded} = SNR_{ref,uncoded} - G_{TC}$.

5.7 SPECTRAL COMPATIBILITY OF DSL SYSTEMS

This section describes the spectral compatibility of the echo-canceled (EC) DSL systems relative to the DSLs deployed in the loop plant. The echo-cancelled DSLs included are ISDN, HDSL, and SDSL. For all spectral compatibility studies done in this paper, only 50 pair cables with 26-gauge wire are assumed in each case.

5.7.1 ISDN

The transmit spectrum of ISDN is shown in Figure 5.14. Since ISDN is an echo-canceled symmetric transport system, we need to consider the effects of SNEXT. Since the SNEXT crosstalk spectrum completely overlaps the ISDN transmit spectra, we expect this disturber to dominate over other disturbers whose spectra only partially overlap.

In the evaluation of ISDN transceiver performance, echo-canceler performance is considered for the ISDN transceiver. 70 dB of echo-cancellation has been achieved in practical ISDN transceivers. If there is no crosstalk in the cable, then the performance of the ISDN transceiver is limited by the performance of the echo-canceler. Specifically, consider the scenario where we have a 50 pair cable of 26-gauge wire. If in this cable we have only a single ISDN transmission system deployed and the remaining 49 wire pairs are not used, then the reach of an ISDN transceiver operating at a BER of 10^{-7} with 6 dB of margin is 20.5 kft on 26-gauge wire. When we add one additional ISDN system in the cable, the added single SNEXT disturber reduces the reach to 20 kft.

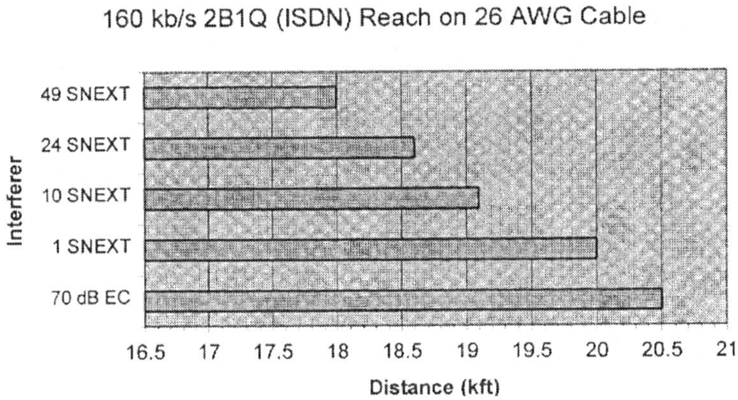

Figure 5.21 ISDN reach as a function of SNEXT.

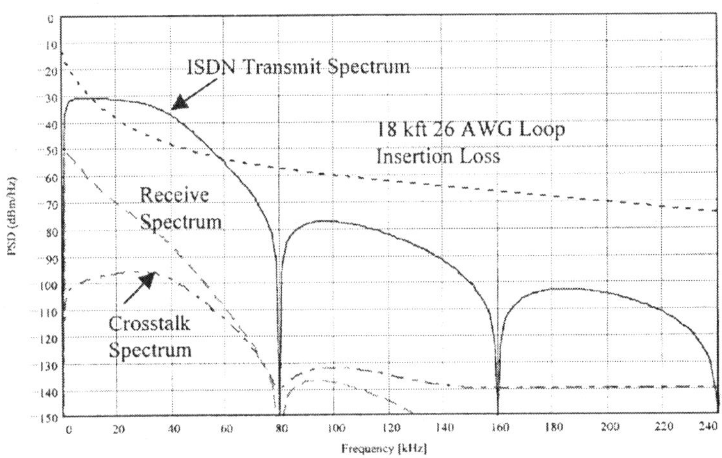

Figure 5.22 Spectral plots of ISDN reach with 49 SNEXT.

With 10 SNEXT disturbers, the reach is 19.1 kft and with 25 disturbers the reach is 18.6 kft. Finally, if the whole 50 pair cable is filled only with ISDN systems, the maximum achievable reach of an ISDN system would be 18 kft, limited by SNEXT. Figure 5.21 shows a summary of the ISDN reach in the environment of SNEXT.

Figure 5.22 shows the spectral plots of the transmit and receive signals of an ISDN system operating on an 18 kft 26-gauge loop. Also shown in the figure are the insertion loss of the 18 kft loop and the 49 SNEXT plus residual echo spectrum. The area between the received signal and crosstalk curves defines the received signal to noise ratio.

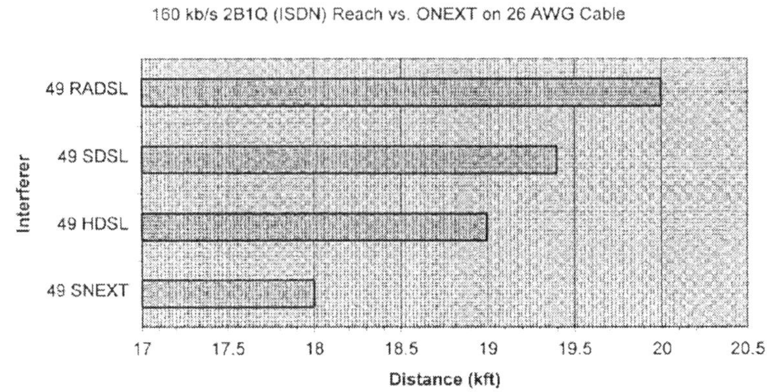

Figure 5.23 ISDN reach as a function of other NEXT disturbers.

We now consider the case when the cable deploys a mixture of DSL services. Other DSLs considered are HDSL, SDSL, ADSL (RADSL) upstream, and ADSL (RADSL) downstream. For each case we consider the worst-case scenario measuring the reach of ISDN in the presence of 49 disturbers from the "other" DSL in question. Figure 5.23 shows a comparison of the ISDN reach on 26-gauge wire in the presence of 49 disturbers from each of the "other" DSLs. Because of the total spectral overlap, SNEXT is the worst disturber to ISDN as compared to the "other" DSLs , since they only have a partial overlap of their spectra with that of ISDN. Deploying "other" services in the same cable with ISDN has less impact on the performance of ISDN than if only ISDN is deployed in the cable.

5.7.2 HDSL

The transmit spectrum of HDSL is shown in Figure 5.15. As with ISDN, HDSL is also an echo-canceled system. In the evaluation of HDSL transceiver performance, echo-canceler performance is considered for the HDSL transceiver. 70 dB of echo-cancellation has been achieved in practical HDSL transceivers.

If there is no crosstalk in the cable, then the performance of the HDSL transceiver is limited by the performance of the echo-canceler. Specifically, consider the scenario where we have a 50 pair cable of 26-gauge wire. If in this cable we have only a single HDSL transmission system deployed and the remaining 49 wire pairs are not used, then the reach of an HDSL transceiver operating at a BER of 10^{-7} with 6 dB of margin is 13.7 kft on 26-gauge wire. When we add one additional HDSL system in the cable, the added single SNEXT disturber reduces the reach by 1.7 kft to 12 kft. With 10 SNEXT disturbers, the reach is 10.6 kft and with 25 disturbers the reach is 10.1 kft. Finally, if the whole 50 pair cable is filled only with HDSL systems, the maximum achievable

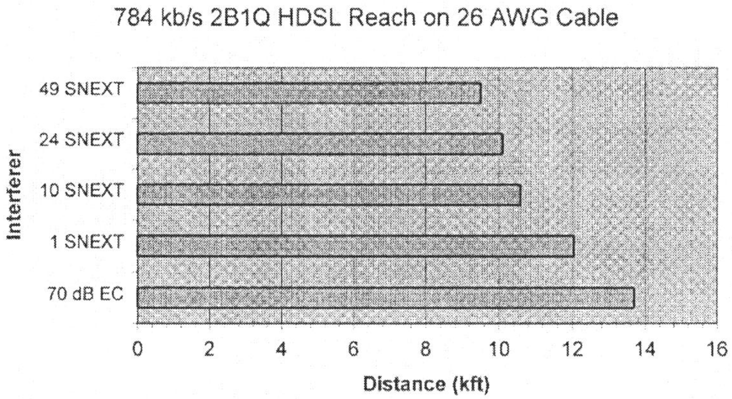

Figure 5.24 HDSL reach as a function of SNEXT.

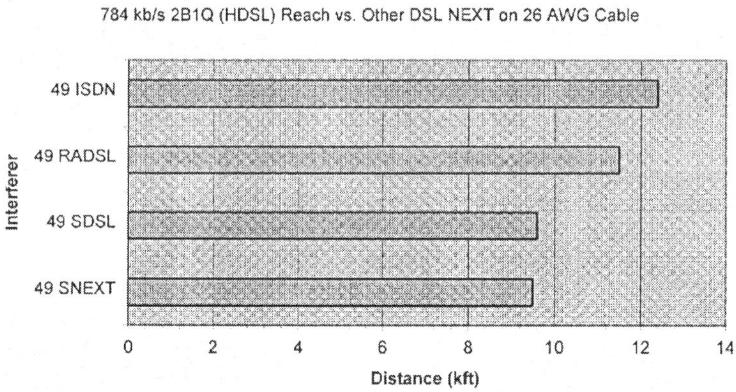

Figure 5.25 HDSL reach as a function of NEXT from other services.

reach of an HDSL system would be 9.5 kft, limited by 49 SNEXT. Figure 5.24 summarizes HDSL reach as a function of SNEXT.

Consider now the cases where HDSL is deployed in the cable with a mixture of DSL services. Relative to HDSL, the "other" DSLs considered are ISDN, SDSL, CAP RADSL upstream, and CAP RADSL downstream. For each case we consider the worst-case scenario measuring the reach of HDSL in the presence of 49 disturbers from the "other" DSLs in question. Figure 5.25 shows a comparison of the HDSL reach on 26-gauge wire in the presence of 49 disturbers from each of the "other" DSLs. Because of the total spectral overlap, SNEXT is the worst disturber to HDSL as compared to the "other" DSLs, since the other DSLs only have a partial overlap of their spectra with that of HDSL. Since the other spectra in the cable have NEXT spectra that do not fully overlap with

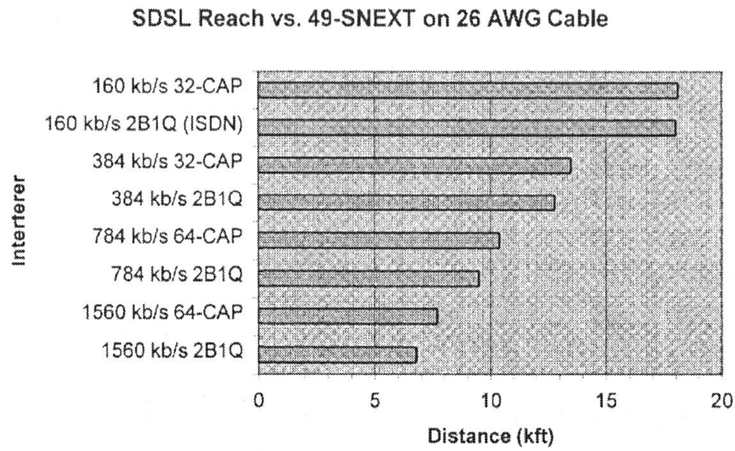

Figure 5.26 SDSL reach versus 49 NEXT.

the HDSL transmit spectrum, the overall interference is less than the NEXT from other similar HDSL signals.

5.7.3 SDSL

As mentioned earlier, all SDSL systems are echo-cancelled systems. Therefore, all SDSL systems are subject to SNEXT. In fact, SNEXT is the dominating disturber to SDSL. The wider the signal bandwidth, the greater the level of SNEXT injected in the signal. The maximum bit rate of the signal is directly proportional to the signal bandwidth. Given these relations, the reach of SDSL systems decreases with increasing bandwidth and, hence, increasing bit rate. Figure 5.26 shows the reach of SDSL systems relative to SNEXT.

Since SDSL performance is limited by SNEXT, crosstalk from other DSL systems (ONEXT) does not have as much impact on SDSL reach, assuming that the PSDs of other DSLs have comparable PSD mask levels. However, depending on the signal bandwidth, SDSL systems may impact the performance of other DSL systems such as DMT ADSL or CAP RADSL. The spectral compatibility of SDSL in other systems is discussed in the subsequent sections. In summary, SNEXT is the dominant disturber in the SDSL.

5.7.4 (R)ADSL

The signal spectra for the CAP RADSL upstream and downstream channels are shown in Figure 5.19. Note that RADSL is a variable bit rate and symbol rate system; hence, the bandwidths of the upstream and downstream channels vary depending on channel conditions. Shown in the figure are the maximum bandwidths for both channels.

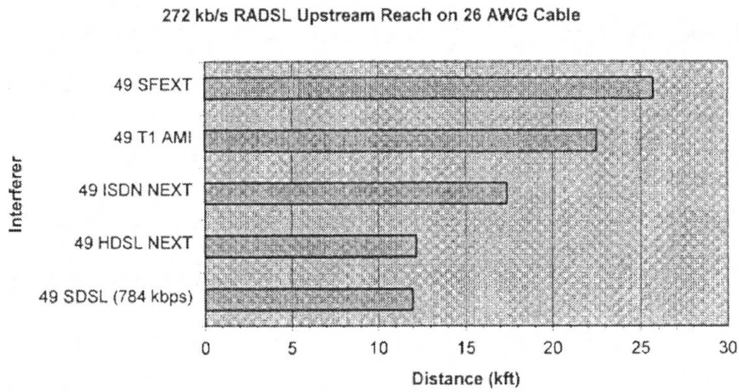

Figure 5.27 272 kbps upstream RADSL reach versus other NEXT.

Since RADSL is a frequency division multiplexed (FDM) system, there is no SNEXT associated with it in the cable. With an FDM system, there is SFEXT associated with both the upstream and downstream channels; but as we have shown earlier, the magnitude of SFEXT is orders of magnitude lower than that of NEXT. However, if the cable is mixed with DSLs having signal spectra traveling in opposite directions from those of RADSL, the upstream and downstream channels of RADSL will be subject to NEXT for the "other" signals.

Figure 5.27 shows the performance of a 272 kbps RADSL upstream signal in the presence of SFEXT and NEXT from "other" DSL services, which include ISDN, HDSL, 784 kbps SDSL, and T1 AMI. Since their spectra fully overlap, NEXT from HDSL and SDSL limit the reach of the RADSL upstream channel. Near-end crosstalk from T1 AMI has little affect on the RADSL upstream channel because the bulk of the T1 AMI energy is in the frequency neighborhood of 772 kHz and the T1 crosstalk energy in the upstream frequency band is relatively low. The presence of NEXT from HDSL and SDSL degrades the maximum possible reach of upstream RADSL by nearly 12 kft when compared to upstream channel SFEXT and by nearly 8 kft when compared to the 680 kbps downstream channel reach in the presence of downstream SFEXT (Figure 5.28).

Figure 5.28 shows the performance of 680 kbps downstream RADSL in the presence SFEXT and NEXT from "other" services. Note that the downstream channel in the presence of SFEXT has a shorter reach than the upstream channel because the downstream channel has greater loss at the higher frequencies. As shown in Figure 5.28, the downstream channel reach in the presence of SFEXT is approximately 17 kft. The dominant disturber to the RADSL downstream channel is NEXT from T1 AMI, since its maximum energy is at 772 kHz. The best case performance is against SFEXT and NEXT from 784 kbps SDSL (no spectral overlap). The frequency band of HDSL has less of an over-

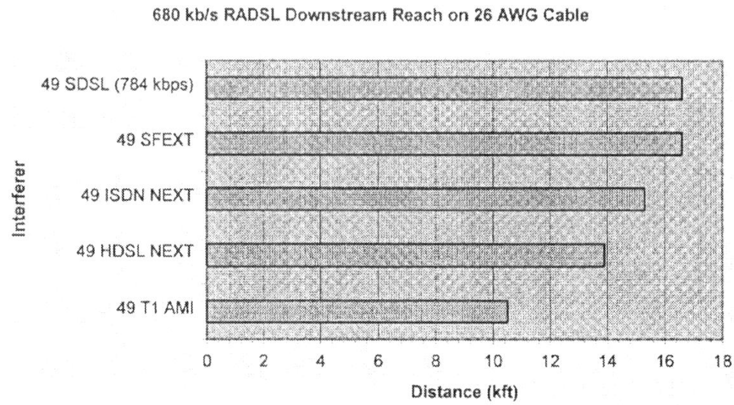

Figure 5.28 680 kbps downstream RADSL reach versus other DSL NEXT.

lap with the downstream channel than it does with the upstream channel, so its impact on the downstream channel is significantly less. This is similar for interference from ISDN. In summary, the best case scenario for deployment of an FDM based system such as RADSL is to fill up the cable completely with RADSL and include no echo-cancelled services in the cable. Since the upstream and downstream and downstream channels in an FDM system occupy the same frequencies, there is no NEXT; instead, there is FEXT which has orders of magnitude less interference.

5.7.5 DMT ADSL

In this section, we consider the spectral compatibility of FDM based DMT ADSL. Figure 5.18 shows the transmit and NEXT spectral plots of the upstream and downstream DMT ADSL channels. The spectral compatibility of DMT ADSL and RADSL are similar in that neither system has SNEXT to deal with. They both have SFEXT and they must deal with NEXT from other services in the same cable. As with RADSL, DMT ADSL is a variable bit rate system and the actual bandwidths of the upstream and downstream channels vary depending on the bit rate and noise environment. Shown in Figure 5.18 is the maximum possible useful bandwidth for the upstream and downstream channels. To evaluate the spectral compatibility of the DMT upstream channel with other services, we examine the reach of 192 kbps DMT upstream channel in the presence of crosstalk from other DSL services.

Figure 5.29 shows a comparison of the reach of a 192 kbps upstream DMT system in the presence of NEXT from HDSL, T1 AMI, ISDN, 784 kbps SDSL, and SFEXT. Clearly, SFEXT is the best noise environment providing the least amount of interference. T1 AMI also provides low interference in the upstream channel because the AMI signal energy is very low in the DMT upstream channel band. The dominant disturbers in the upstream channel are HDSL and

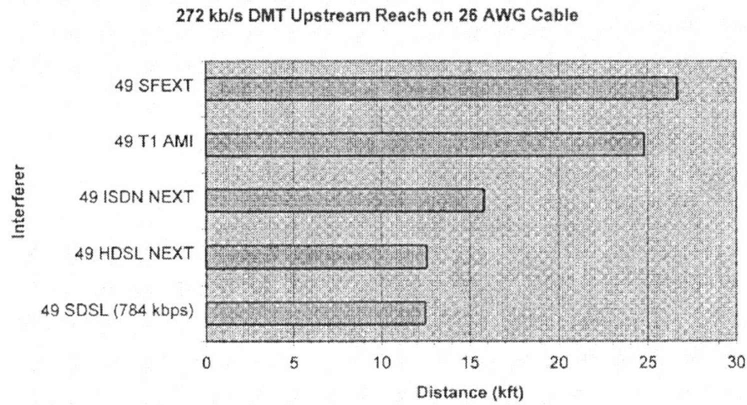

Figure 5.29 Upstream DMT spectral compatibility with other DSLs.

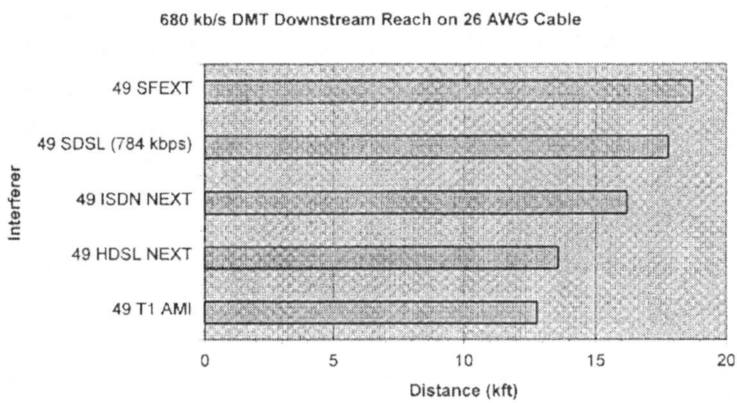

Figure 5.30 Downstream DMT spectral compatibility with other DSLs.

SDSL because the NEXT from these services provides full bandwidth overlap with the DMT upstream channel. The ISDN spectrum has partial overlap with the DMT upstream channel, so the upstream channel reach is greater than the cases for HDSL and SDSL.

To evaluate the spectral compatibility of the DMT downstream channel with other services, we examine the reach of a 680 kbps DMT downstream channel in the presence of crosstalk from other DSL services. Figure 5.30 shows a comparison of the reach of a 680 kbps downstream DMT system in the presence of NEXT from HDSL, T1 AMI, ISDN, 784 kbps SDSL, and SFEXT. As with the upstream channel, SFEXT is the best noise environment providing the least amount of interference; however, its reach is lower than the upstream channel because the loop has higher loss in the frequencies of the downstream channel.

Contrary to the upstream, T1 AMI provides the dominant level of interference in the downstream channel because the AMI signal energy is highest in the DMT downstream channel band. HDSL, because of the significant bandwidth overlap with the downstream channel, is the next dominant disturber in line. ISDN and SDSL have the least impact of NEXT in the DMT downstream channel. The degradation in reach from T1 AMI versus the best case of SFEXT is approximately 6 kft; the corresponding reach from HDSL is approximately 5 kft.

In summary, HDSL and SDSL are dominant disturbers in the upstream channel of ADSL. T1 AMI is the dominant disturber in the ADSL downstream channel. The best case for deployment of FDM ADSL services is to fill the cable completely with ADSL and eliminate all NEXT. If the cable contains a mixture of DSLs, then NEXT from HDSL and SDSL are the dominant disturbers in the upstream channel and T1 AMI and HDSL are the dominant disturbers in the downstream channel.

5.8 SPECTRAL COMPATIBILITY OF RADSL WITH DMT ADSL

When we compare the spectra of the RADSL upstream and downstream channels with those of DMT, we see that there is some overlap between the RADSL upstream and DMT downstream channel. A result of this overlap is that when the systems are deployed in the same cable they inject NEXT into each other. However, as seen in Figures 5.18 and 5.19, this overlap is minimal.

Figure 5.31 shows a composite plot of 49 NEXT from each of RADSL upstream, DMT upstream, HDSL, and ISDN signals. Clearly, the level of spectral compatibility of each of these signals with the DMT downstream channel is a function of the amount of overlap with the downstream channel. The upstream DMT crosstalk spectrum is that defined in T1.413 [6]. The RADSL upstream crosstalk spectrum is that of 136 kBaud square-root raised-cosine spectrum with 15% excess bandwidth and the spectrum assumes 50 dB out-of-band attenuation. To demonstrate the spectral compatibility of these signals in the DMT downstream, we present the margin computed in the DMT downstream channel in the presence of crosstalk from each of the above DSL disturbers.

Table 5.2 shows the margin of an FDM based 6.784 Mbps DMT receiver in the presence of various disturbers on a 9 kft 26-gauge test loop. The margin of the downstream channel with crosstalk from 20 FDM DMT upstream disturbers and that from 20 RADSL upstream disturbers are both about 5.5 dB. The margin from 20 HDSL disturbers is about 5 to 6 dB worse, i.e. 4.9 dB. NEXT from HDSL dominates disturbance in the 6.784 Mbps DMT downstream.

Table 5.3 shows the margin of an FDM based 1.72 Mbps DMT receiver in the presence of various disturbers on a 13.5 kft 26-gauge test loop. The margins of the downstream channel with crosstalk from 24 FDM DMT upstream disturbers and 24 RADSL upstream disturbers are 7.0 dB and 7.4 dB, respectively. The margin from 24 ISDN disturbers is about 3 to 3.4 dB worse, i.e. 4.0 dB. Disturbance from HDSL was not considered because HDSL is not deployed on

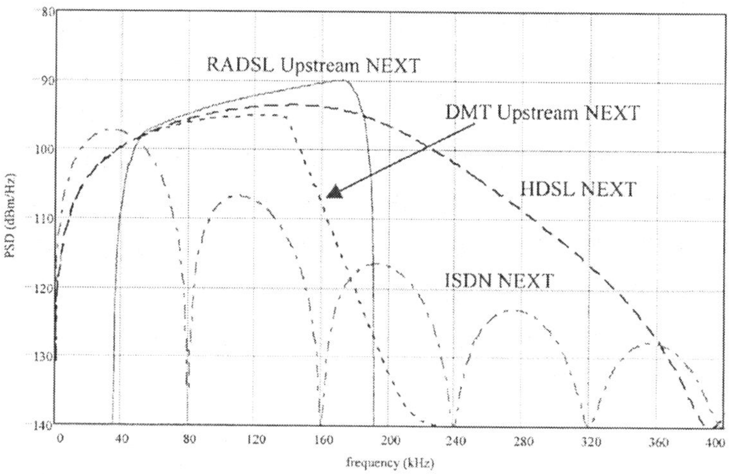

Figure 5.31 RADSL upstream, DMT upstream, HDSL and ISDN crosstalk spectra.

Table 5.2 Spectral compatibility in the 6.784 Mbps DMT downstream.

DSL interferer	Downstream 6.784 Mbps FDM DMT margin
20 HDSL	4.9 dB
20 FDM DMT upstream	5.4 dB
20 RADSL upstream	5.5 dB

Table 5.3 Spectral compatibility in 1.72 Mbps DMT downstream.

DSL interferer	Downstream 1.72 Mbps FDM DMT margin
24 ISDN	4.0 dB
24 FDM DMT upstream	7.0 dB
24 RADSL upstream	7.4 dB

loops greater than the CSA range. In summary, NEXT from ISDN dominates disturbance in the 1.72 Mbps DMT downstream.

In summary, although the RADSL upstream channel has a slightly greater bandwidth than the DMT upstream, its out-of-band efficiency is greater that the upstream DMT channel defined in T1.413. With 50 dB out-of-band attenuation of the RADSL upstream spectrum, the spectral compatibility in the DMT downstream channel is the same as that from the DMT upstream chan-

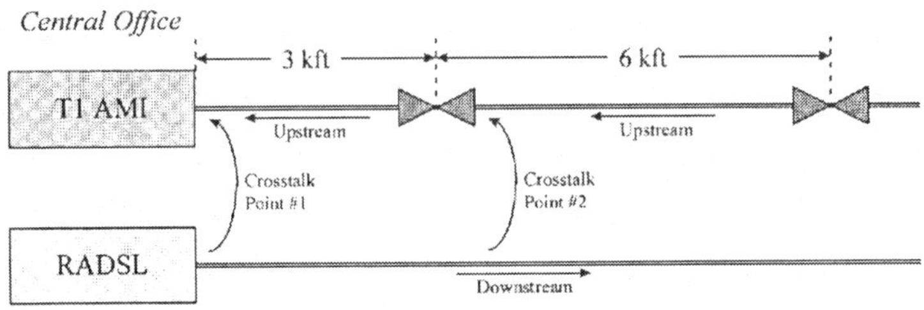

Figure 5.32 Crosstalk scenarios for DSLs in T1 AMI.

nel. In either case, HDSL and ISDN are greater disturbers in the downstream channel than either of the DMT or RADSL upstream channels. So RADSL is spectrally compatible with T1.413 ADSL.

5.8.1 T1 AMI

Figure 5.32 shows the system model for determining the spectral compatibility of the downstream RADSL channel in the T1 AMI. Since T1 AMI is a repeated link, there are numerous points to consider observing the crosstalk and they are labeled Crosstalk Points 1 and 2 in the figure.

In the conventional provisioning of T1 links, the first repeater is placed at a maximum of 3 kft from an end-point and a maximum of 6 kft between repeaters. Note, however, that T1 lines were originally designed as trunk lines to interconnect CO and the wire gauges used were 22 AWG or 19 AWG. Since the distribution plant usually uses 26 AWG wire (thinner than 22 or 19 AWG) directly out of the CO, the provisioning rules used in the distribution plant are not ubiquitously known. However, in this study, we assume a worst-case scenario of 26-gauge wire using the same repeater spacing rules for the trunk plant.

The RADSL downstream transmit signal is strongest at the CO transmitter output. So at the CO (Crosstalk Point 1 in Figure 5.32) the RADSL downstream signal introduces the strongest level of crosstalk in the T1 AMI receiver. The T1 AMI signals have maximum energy at the transmitter output of both the end units and repeaters. In the first repeater span, the loop segment length is 3 kft and so the received AMI signal would not be attenuated as much as it would be in a mid-span repeater spacing of 6 kft. At Crosstalk Point 2, the downstream RADSL signal is attenuated by 3 kft of cable and so the crosstalk level in the first repeater is attenuated by that amount.

To estimate the performance of the AMI signal, we compute the SNR at the AMI receiver input and crosstalk points 1 and 2. In each case, the number of downstream RADSL disturbers assumed is 24. The SNR is measured in two ways: (1) at the T1 AMI center frequency of 772 kHz, and (2) averaged throughout the entire T1 AMI band to the first null, i.e. 0 to 1.544 MHz. In the

Table 5.4 Spectral compatibility computation results with 24 RADSL disturbers.

Crosstalk point	Center frequency (722 kHz)		Averaged SNR (0–1544 kHz)	
	SNR (dB)	Margin (dB)	SNR (dB)	Margin (dB)
#1	18.7	1.2	18.8	1.1
#2	18.8	1.3	25.5	8.0

AMI receiver, we assume that the receiver provides automatic gain adjustment, no equalization, and ideal time sampling. To achieve a BER of 10^{-7}, we assume a 17.5 dB SNR is required for the three level signal at the input to the AMI receiver. The margin achieved is the difference in SNR at the receiver input and the 17.5 dB reference value.

Table 5.4 shows the spectral compatibility computation results with 24 RADSL downstream channel disturbers. The third column shows the SNR margin seen at the AMI receiver inputs measured at the AMI center frequency of 772 kHz. For both crosstalk points, the SNR margin is roughly 1 dB so the T1 AMI system should still provide service with better than 10^{-7} BER. The last column shows input SNR averaged over the entire T1 AMI band. At crosstalk point 1, the averaged SNR is roughly the same as the SNR at the center frequency. For crosstalk point 2, however, the averaged SNR is greater than the center frequency SNR because of the greater attenuation suffered by the AMI signal on the 6 kft loop.

In summary, an estimation of spectral compatibility of RADSL with T1 AMI has been provided using very pessimistic assumptions. Specifically, the same provisioning rules of repeater spacings on 22- and 19-gauge wire are applied to 26-gauge wire; the losses seen on 26-gauge wire are significantly greater. In all cases of 24 RADSL disturbers in a T1 AMI receiver, the input SNR is at least 1 dB greater than that required for achieving 10^{-7} BER performance. If the repeater spacings are shorter than those assumed here, then the margins improve. Based on this data, we expect RADSL to not degrade T1 AMI service in the distribution plant.

5.9 SUMMARY OF SPECTRAL COMPATIBILITY

In the telco loop plant there are numerous types of digital subscriber lines deployed. The DSLs deployed can be categorized in two different types: (1) symmetric EC DSLs and (2) asymmetric (FDM/EC) DSLs. The first class of EC DSLs includes ISDN, HDSL, and SDSL, the second class include ADSL and RADSL. The modulation technologies for the DSLs include 2B1Q for ISDN and HDSL, CAP for HDSL and SDSL, DMT for ADSL and CAP for RADSL. The echo-cancelled systems of ISDN, HDSL, and SDSL use the same spectra for the transmission of upstream and downstream data. Assuming the same transmit power for each of these systems, the worst-case crosstalk performance occurs

Table 5.5 Summary of EC DSL theoretical performance.

EC DSL type	Reach with 49 SNEXT, 10^{-7} BER and 6 dB margin on 26 AWG
160 kbps 32-CAP	18.1 kft
160 kbps 2B1Q (ISDN)	18.0 kft
384 kbps 32-CAP	13.5 kft
384 kbps 2B1Q (ISDN)	12.8 kft
784 kbps 64-CAP	10.4 kft
784 kbps 2B1Q (ISDN)	9.5 kft
1560 kbps 64-CAP	7.7 kft
1560 kbps 2B1Q (ISDN)	6.8 kft

when the cable is filled with the same type of EC systems. For a 50-pair cable, the worst-case crosstalk is 49 SNEXT. Table 5.5 contains a summary of the theoretical performance of the various EC DSL systems in the presence of 49 SNEXT.

Each system in Table 5.5 assumes a transmit-signal power of 13.5 dBm. In addition to SNEXT, 70 dB of echo-cancellation is assumed in the receiver. The reach values computed in Table 5 are for a 10^{-7} BER and 6 dB of margin. Note that the CAP systems contain a trellis code and the performance values in the table include 4 dB of coding gain. The 2B1Q transceivers do not contain any trellis coding. As shown in Table 5.5, the reach of the echo-cancelled DSL systems is inversely proportional to the bit rate of the DSL. The ISDN system uses 160 kbps 2B1Q. HDSL is a dual-duplex system transporting a T1 (1.544 Mbps) payload on two twisted-wire pairs running at 784 kbps on each pair. The 2B1Q HDSL system in the above table corresponds to the 784 kbps 2B1Q entry, while the CAP HDSL system corresponds to the 784 kbps 64-CAP entry.

The second type of DSL system is asymmetric FDM/EC, which includes DMT ADSL and CAP RADSL. Both the DMT ADSL and CAP RADSL systems use frequency division multiplexing to separate the upstream and downstream channels. If a cable is filled with only FDM systems, SFEXT limits the performance. Since SFEXT is orders of magnitude less than SNEXT, the reach of FDM can be significantly greater than that where NEXT is present. So contrary to that of EC systems, the best case crosstalk environment occurs when the whole cable is filled with the same FDM system.

Another version of ADSL is an echo-canceled version where the wideband downstream channel also utilizes the narrowband frequencies of the upstream channel. In this case, the NEXT of the downstream channel severely limits the reach of the upstream channel because the downstream channel bandwidth completely covers that of the upstream.

Since the cable simultaneously contains EC and FDM type systems, performance of the DSLs in the presence of NEXT from "other" systems must be

Table 5.6 Other DSL NEXT in the RADSL.

"Other" NEXT disturber	272 kbps RADSL Upstream-Reach	680 kbps RADSL Downstream-Reach
49 SFEXT	25.7 kft	16.6 kft
49 SDSL (784 kbps)	12 kft	16.6 kft
49 ISDN NEXT	17.4 kft	15.3 kft
49 T1 AMI	22.5 kft	10.5 kft
49 HDSL NEXT	12.2 kft	13.9 kft

Table 5.7 Other DSL NEXT in the FDM-based DMT ADSL.

"Other" NEXT disturber	272 kbps RADSL Upstream-Reach	680 kbps RADSL Downstream-Reach
49 SFEXT	26.7 kft	18.7 kft
49 SDSL (784 kbps)	12.5 kft	17.8 kft
49 ISDN NEXT	15.8 kft	16.2 kft
49 T1 AMI	24.8 kft	12.8 kft
49 HDSL NEXT	12.6 kft	13.6 kft

considered. The DSLs discussed in this paper contain comparable PSD mask values. Therefore, the crosstalk in EC systems is dominated by SNEXT. However, the varying bandwidths of the EC systems introduce different levels of NEXT in the ADSL and RADSL systems.

Table 5.6 shows the theoretical reach of the RADSL channels in the presence of crosstalk from other systems. Both the upstream and downstream RADSL receivers assume 4 dB of coding gain. HDSL and 784 kbps SDSL have the greatest impact on the performance of the RADSL upstream channel because both spectra completely cover the RADSL upstream band. On the downstream channel, T1 AMI has the greatest impact because the downstream channel frequencies experience greater loop loss and the T1 AMI crosstalk energy is at its maximum.

Table 5.7 shows the theoretical reach of the FDM based DMT ADSL channels in the presence of crosstalk from other systems. Both the upstream and downstream RADSL receivers assume 4 dB of coding gain. Because of the different spectral placement of the upstream and downstream channels, the theoretical performance of the DMT systems differ slightly. As with RADSL, the downstream channel is affected most by NEXT from T1 AMI; the upstream channel is affected most by NEXT from HDSL and 784 kbps SDSL. Because of the lower start frequency, the DMT upstream channel is affected more from ISDN NEXT.

Although the RADSL upstream channel has a slightly greater bandwidth than the DMT upstream, its out-of-band efficiency is greater that the upstream DMT channel defined in T1.413. With 50 dB out-of-band attenuation in the RADSL upstream spectrum, the spectral compatibility in the DMT downstream channel is the same as that from the DMT upstream channel. In either case, HDSL and ISDN are greater disturbers in the downstream channel than the DMT or RADSL upstream channels. So RADSL is spectrally compatible with T1.413 ADSL.

5.10 PERFORMANCE OF ADSL SYSTEMS

In the field of ADSL, there are two competing transmission methods under consideration: single-carrier systems such as QAM or its variant, CAP modulation, and orthogonal frequency division multiplexing (OFDM) techniques such as DMT. In industry, DMT has been adopted as the ANSI standard for ADSL application while CAP is widely used as the de facto standard. There has been much interest in performance comparison of single-carrier DFE based systems, such as CAP or QAM, with DMT. The performance parameters investigated for the CAP and DMT systems are the achievable transmission capacity and SNR margin. This chapter provides a performance comparison of CAP and DMT transmission methods under numerous DSL scenarios. We evaluate the performance of these technologies on selected CSA loops. We emphasize that the implementation issues of any technique are beyond the focus of this study. Therefore, the results represent ideal cases for both schemes.

From a signal processing point of view, DMT divides the channel into N subchannels while CAP uses only one channel. As shown later, these two signaling techniques yield the same theoretical performance bounds.

5.10.1 A Capacity Bound for a DFE-Based Single Carrier System

The probability of symbol error for single-carrier broadband QAM or CAP minimum mean-squared error decision feedback equalization (MMSE-DFE) based transceiver system can be found as [167, 43]

$$P_e = N_e Q\left(\sqrt{\frac{d_{min}^2}{4\sigma_{DFE}^2}}\right)$$

where σ_{DFE}^2 is the variance of the DFE mean-squared error and d_{min} is the minimum distance between any two points in a constellation. For individual dimension of M-ary QAM modulation technique, we assume that the transmitted signal power is S_x where

$$S_x = \frac{d_{min}^2}{12}\left(M^2 - 1\right) = \frac{d_{min}^2}{12}\left(2^{R_{bitrate}T_{symbol}} - 1\right)$$

where $R_{bitrate}$ is the transmission bit rate and T_{symbol} is the symbol duration. The SNR of this MMSE-DFE based transceiver system is given as $SNR_{DFE} = \frac{S_x}{\sigma_{DFE}^2}$.

The maximum achievable channel rate, $R_{MMSE-DFE}$, is [167, 43]

$$R_{MMSE-DFE} = \frac{1}{T}\log_2\left(1 + \frac{SNR_{MMSE-DFE}}{\Gamma}\right)$$

where

$$SNR_{MMSE-DFE} = \exp\left[\frac{1}{W}\int_0^W \ln\left(1 + \sum \frac{S_{psd}(f)|H(f)|^2}{N_0}\right) df\right] - 1.$$

The last term, "-1," in this expression comes from the unbiased version of the MMSE-DFE derived from [43]. This SNR is defined for the additive Gaussian noise environment.

In general, the symbol rate is approximately the same as the bandwidth ($\frac{1}{T} \simeq W$) and the maximum achievable bit rate is expressed as

$$R_{MMSE-DFE} = W\log_2\left(1 + \frac{SNR_{MMSE-DFE}}{\Gamma}\right).$$

If there is some crosstalk such as NEXT and FEXT in the bandwidth efficient single-carrier broadband transceiver system, the channel capacity bound is calculated as

$$C = \int_0^W \log_2\left(1 + \frac{S_{psd}(f)|H(f)|^2}{[N_0 + |X(f)|^2 S_{psd}(f)]\,\Gamma}\right) df. \tag{5.1}$$

All of these calculations are based on the well-known Shannon capacity theorem. We assume that any real number QAM for each subchannel is realizable and the crosstalk noise is of a Gaussian type. The background AWGN at the receiver input is assumed to have a spectral density N_0 of -140 dBm/Hz. The PSD of the transmitted signal is -40 dBm/Hz within the transmission band. A square-root raised-cosine shaping filter can be used. We also assume that the targeted BER is 10^{-7} for the transceiver system. Using the CAP signal PSD function, $S_{psd}(f)$, in (5.1), we obtain the channel capacity bound by numerical integration.

5.10.2 A Capacity Bound for the DMT Transceiver

The bandlimited channel has the frequency transfer function $H(f)$. Figure 5.33 displays a typical DSL loop plant frequency response using AWG 24 with 12 kft length.

DMT consists of a large number of independent subcarriers. Suppose that the well-known QAM modulation technique is used for each of these subchannels in DMT transceivers. For demonstration purpose, we assume that N independent subchannels are divided from the practical loop plant. If we assume that each subchannel is an ideal brick wall-like channel and uses QAM modulation, the multicarrier modulation based transceiver system can be modeled as N parallel subchannels transmitting bitstreams in each independent subchannel. After transmission, the bitstream can be multiplexed into one high bit rate

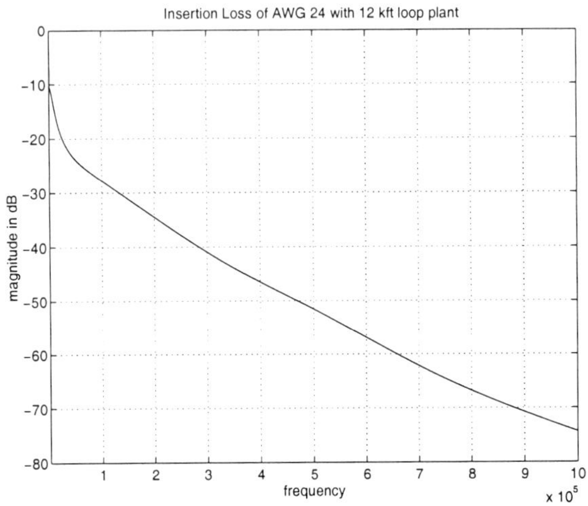

Figure 5.33 Frequency transfer function of a typical DSL loop: AWG 24 loop plant with 12 kft length.

data stream. The overall achievable bit rate is the summation of bit rate in all the subchannels.

For the ith subchannel, we have the corresponding subchannel magnitude transfer function $|H_i(f)|$. The PSD of additive white Gaussian noise is $\frac{N_0}{2}$. We assume that the probability of symbol error, P_e, is the same for all subchannels of the DMT transceivers. Assuming the transmitted signal power in the ith subchannel is P_i and the bandwidth of the ith subchannel is W_i, and using two-dimensional symbol QAM modulation, the number of bits assigned to the ith subchannel can be expressed as [168]

$$n_i = \log_2 \left(1 + \frac{3P_i}{N_0 W_i \left[Q^{-1} \left(\frac{P_e}{N_e} \right) \right]^2} \right) \tag{5.2}$$

where N_e is the number of adjacent constellations; $N_e = 4\left(1 - \frac{1}{\sqrt{2^{n_i}}}\right) \simeq 4$ for large n_i.

SNR gap is defined as $\Gamma = \frac{\gamma_m}{3\gamma_{code}} \left[Q^{-1} \left(\frac{P_e}{N_e} \right) \right]^2$ where γ_m is the system design target SNR margin and γ_{code} is the coding gain [39]. For uncoded QAM in each subchannel, $\gamma_{code} = 1$ (0 dB). We also assume the target SNR margin $\gamma_m = 1$ (0 dB) for demonstration purposes. One can rewrite (5.2) using the SNR gap definition as

$$n_i = \log_2 \left(1 + \frac{P_i}{N_0 W_i \Gamma} \right).$$

If we increase the number of subcarriers, N, in the modulation, eventually there are N subchannels in this bandlimited loop. Whenever N approaches infinity, we can model the subchannels using index of frequency f. Let's denote $S_{psd}(f)$ as the PSD function of the transmitted signal in subchannel index of f. The received signal PSD at the receiver input is $S_{psd}(f)|H(f)|^2$, then the number of bits assigned for the subchannel at tone f is obtained as

$$n(f) = \log_2\left(1 + \frac{S_{psd}(f)|H(f)|^2}{N_0\Gamma}\right).$$

The total achievable bit rate is therefore obtained as [39, 168]

$$R_{DMT} = \int_0^W \log_2\left(1 + \frac{S_{psd}(f)|H(f)|^2}{N_0\Gamma}\right) df.$$

If there is some crosstalk interference in the bandlimited channel, then the total achievable bit rate is similarly calculated as

$$R_{DMT} = \int_0^W \log_2\left(1 + \frac{S_{psd}(f)|H(f)|^2}{[N_0 + |X(f)|^2 S_{xt}(f))\,\Gamma}\right) df$$

where $S_{xt}(f)$ is the crosstalk interference signal power spectrum and $|X(f)|$ is the crosstalk coupling filter.

5.10.3 Achievable Bit Rates and Comparisons

It can be shown that when the complexity is not a concern and SNR is high in a Gaussian noise environment, the performance of multicarrier modulation based system and MMSE-DFE based single-carrier modulation based system are the same [1]. It is assumed that the roll-off factor of single-carrier broadband transmission shaping filter, α, is zero for a fair comparison. In this case, we use the unfolded spectrum in the derivation. The MMSE-DFE based CAP system has the maximum achievable bit rate as

$$\begin{aligned} R_{CAP} &= R_{symbol}\log_2\left(1 + \frac{SNR_{DFE-MMSE}}{\Gamma}\right) \\ &= W_{CAP}\log_2\left(1 + \frac{\exp\left[\frac{1}{W_{CAP}}\int_0^{W_{CAP}} \ln\left(1 + \frac{S(f)|H(f)|^2}{N_0}\right) df\right] - 1}{\Gamma}\right). \end{aligned}$$

Whenever $\frac{SNR_{DFE-MMSE}}{\Gamma} \gg 1$ we can rewrite the above formula as

$$\begin{aligned} R_{CAP} &\simeq W_{CAP}\log_2\left(\frac{\exp\left[\frac{1}{W_{CAP}}\int_0^{W_{CAP}} \ln\left(\frac{S(f)|H(f)|^2}{N_0}\right) df\right]}{\Gamma}\right) \\ &= \int_0^{W_{CAP}} \log_2\left(\frac{S(f)|H(f)|^2}{N_0\Gamma}\right) df. \end{aligned}$$

Replacing W_{CAP} with W_{DMT}, where W_{CAP} and W_{DMT} are the bandwidths of the transmitted signals, in this expression yields the achievable bit rate for DMT [167]. Clearly, if $W_{CAP} = W_{DMT}$ then $R_{DMT} = R_{CAP}$ for the high SNR case.

As reported in [167], the maximum achievable bit rate is the same in the NEXT environment [167]. Under certain conditions, this claim is correct. In [168], the condition for the equivalence is mentioned as the integrand of (5.1) is larger than 1. In other words, the number of bits for each subchannel should be larger than one. In a practical implementation, this condition always holds for DMT. It is necessary to note that the frequency bandwidth or operation symbol rate of a DFE-based single-carrier transceiver system should be chosen properly. If the symbol rate is not properly selected, the achievable bit rate may be lower than DMT's. If the optimum bandwidth is chosen, the achievable bit rate is nearly the same for DMT as MMSE-DFE based QAM or CAP system.

In the SNEXT only scenario, the DFE based CAP system has the achievable bit rate

$$\begin{aligned} R_{CAP} &= R_{symbol} \log_2 \left(1 + \frac{SNR_{DFE-MMSE}}{\Gamma}\right) \\ &= W_{CAP} \log_2 \left(1 + \frac{\exp\left[\frac{1}{W_{CAP}} \int_0^{W_{CAP}} \ln\left(1 + \frac{|H_c(f)|^2}{|H_x(f)|^2}\right) df\right] - 1}{\Gamma}\right). \end{aligned}$$

Whenever $\frac{SNR_{DFE}}{\Gamma} \gg 1$, we can rewrite the above equation as

$$\begin{aligned} R_{CAP} &\simeq W_{CAP} \log_2 \left(\frac{\exp\left[\frac{1}{W_{CAP}} \int_0^{W_{CAP}} \ln\left(1 + \frac{|H_c(f)|^2}{|H_x(f)|^2}\right) df\right]}{\Gamma}\right) \\ &\simeq \int_0^{W_{CAP}} \log_2 \left(\frac{|H(f)|^2}{|H_x(f)|^2 \Gamma}\right) df. \end{aligned}$$

This approximation holds if $\frac{|H(f)|^2}{|H_x(f)|^2} \gg 1$. Similar to the AWGN only scenario, R_{DMT} for the SNEXT case can be written as

$$R_{DMT} \simeq \int_0^{W_{DMT}} \log_2 \left(\frac{|H_c(f)|^2}{|H_x(f)|^2 \Gamma}\right) df.$$

This approximation also holds if $\frac{|H(f)|^2}{|H_x(f)|^2 \Gamma} \gg 1$. $R_{MMSE-DFE} = R_{DMT}$ under this condition; this conclusion is valid for any Gaussian noise environment.

5.10.4 SNR Margin Performance Comparison

SNR margin is the measure used to evaluate the performance of HDSL and ADSL systems. In ANSI T1E1 technical subcommittee discussions, a variety of SNR margin results were forwarded as contributions. But they have different calculations in many cases. Fair comparison should be performed under the same test conditions.

Table 5.8 Required SNR to achieve 10^{-7} error rate using MMSE-DFE technique for different constellation size in single-carrier modulation schemes with and without 4 dB coding gain.

Constellation	4	8	16	32	64	128	256
SNR without coding	14.5	18.0	21.5	24.5	27.7	30.6	33.8
SNR with coding	–	14.0	17.5	20.5	23.7	26.6	29.8

SNR margin is defined as the difference of achievable SNR with the SNR required for the specified target transmission bit rate and the transmission error rate. In ADSL applications, the transmission error rate is set to be less than 10^{-7}. From system design point of view, one must design the transceiver system SNR as high as possible, approaching the theoretical bound. But we should evaluate the attainable SNR to make sure that certain level of SNR margin can be achieved. There is 6 dB SNR margin requirement for ADSL applications.

In single-carrier modulation techniques based transceiver system, let's say a CAP based transceiver system is implemented. Assume a MMSE-DFE technique is utilized. The required SNRs to achieve the transmission error rate of 10^{-7} using different constellation size modulation are listed in Table 5.8. Considering the coding gain of 4 dB, we need to increase the constellation size by 1 bit. The required SNR are also listed in Table 5.8.

For the multicarrier modulation based transceiver system such as DMT technique, the theoretical SNR margin is calculated as the difference of achievable average SNR of all subcarrier and the required SNR for target transmission error rate. It is defined average SNR as [39]

$$\overline{SNR} = \Gamma\left\{\left[\prod_{i}^{\overline{N}}\left(1+\frac{SNR_i}{\Gamma_i}\right)^{\frac{1}{\overline{N}}}\right]-1\right\},$$

where Γ is the overall SNR gap and Γ_i implies the SNR gap of the ith subchannel. It is assumed that $\Gamma = \Gamma_i$. $\overline{N}$ is the total number of subchannels used in a DMT transceiver. The subchannels in which $b_i > 1$ bit are defined as usable subchannels. The total number of bits transmitted per symbol (superblock), b, is $b = \sum_{i=1}^{\overline{N}} b_i$. In a high SNR environment, we can ignore the ± 1s. Accordingly,

$$\overline{SNR} = \left[\prod_{i}^{\overline{N}} SNR_i\right]^{\frac{1}{\overline{N}}} = \left[\prod_{i}^{\overline{N}} SNR(f_i)\right]^{\frac{1}{\overline{N}}}.$$

Now, the SNR margin is defined as [39]

$$SNR_{margin} = \frac{10}{W_{DMT}} \int_{f_{start}}^{f_{end}} \log_{10}(SNR(f))\, df - \frac{10b}{\overline{N}} \log_{10} 2 + \gamma_{code} - \Gamma_{gap}, \tag{5.3}$$

where W_{DMT} is the bandwidth used for DMT system. Assume that b_{DMT} is the transmission bit rate of DMT and T_{symbol} is the duration of DMT supersymbol. Then, b must be equal to $b_{DMT}T_{symbol}$, yielding

$$\frac{b}{\overline{N}} = \frac{b_{DMT}T_{symbol}}{\overline{N}} = \frac{b_{DMT}}{\frac{\overline{N}}{T_{symbol}}} = \frac{b_{DMT}}{W_{DMT}}.$$

One can then rewrite (5.3) as

$$\begin{aligned} &SNR_{margin}^{DMT} \\ &= \frac{10}{W_{DMT}} \int_{f_{start}}^{f_{end}} \log_{10}(SNR(f))\, df - 10\frac{b_{DMT}}{W_{DMT}} \log_{10} 2 + \gamma_{code} - \Gamma_{gap}. \end{aligned}$$

In MMSE DFE-based CAP system, we calculate $SNR_{ref-cap}$ for each CAP signaling corresponding to symbol error rate $P_e = 10^{-7}$. Assume b_{CAP} is the transmission bit rate and m is the number of bits per symbol for CAP signaling. For 256 CAP signaling, $SNR_{ref-cap}$ is about 30.8 dB. Then, we can write

$$\begin{aligned} SNR_{ref-cap} &= 10\log_{10}\left(\frac{Q^{-1}(\frac{P_e}{2})}{3}\right) + 10\log_{10}(2^m - 1) \\ &= \Gamma_{gap} + 10\frac{b_{CAP}}{W_{CAP}} \log_{10} 2 \end{aligned} \tag{5.4}$$

and

$$SNR_{DFE} = 10\log_{10}\left[\exp\left(\frac{1}{W_{CAP}} \int_{f_{start}}^{f_{end}} \ln\left[1 + SNR(f)\right] df\right) - 1\right].$$

Therefore,

$$\begin{aligned} &SNR_{margin}^{CAP} \\ &= SNR_{DFE} - SNR_{ref-cap} + \gamma_{code} \\ &= \frac{10}{W_{CAP}} \int_{f_{start}}^{f_{end}} \log_{10}\left[1 + SNR(f)\right] df - 10\frac{b_{CAP}}{W_{CAP}} \log_{10} 2 - \Gamma_{gap} + \gamma_{code}. \end{aligned} \tag{5.5}$$

From (5.4) and (5.5), we see that if both CAP and DMT systems transmit the same digital bit rate, and use the same transmitted signal PSD and the same frequency range with the same noise condition, SNR_{margin} performance bounds of both techniques are the same. In the derivation, ± 1s are ignored. It means that high SNR case is assumed. In ADSL and HDSL practical application scenarios, $SNR(f)$ is much larger than 1.

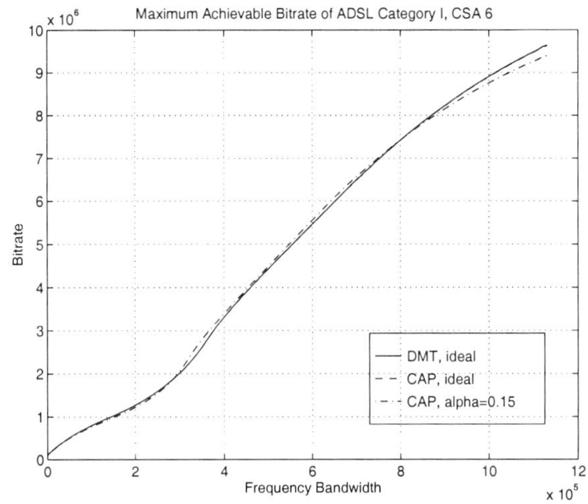

Figure 5.34 Maximum achievable bit rate in ADSL, Category I, CSA 6.

5.10.5 Performance Simulation Results and Discussions

We evaluated the theoretical bounds for the achievable bit rate as well as the SNR margin for HDSL and ADSL applications. It is shown that these performance bounds are the same for both DMT and CAP modulation line codes under the same test conditions. Numerical performance results are in accord with the theoretical analysis.

In the simulation, ADSL Category I Test Platform is used according to ADSL standard [315]. The FDM types of downstream and upstream signaling are implemented. The starting point of upstream signal is 20 kHz such that the POTS is available in the low 4 kHz band. The downstream signal power spectrum starts at 165 kHz. There is some separate gap between upstream and downstream signal. In maximum achievable bit rate calculation, the SNR margin and coding gain are assumed to be 0 dB. In SNR margin calculation for both DMT and CAP-DFE scenario, 5 dB coding gain is assumed.

Figures 5.34 and 5.35 display the maximum achievable bit rate for DMT and CAP on CSA 6 and CSA 7 under the ADSL Category I noise scenario, respectively. SNR margins are listed in Table 5.9. It is found from these results that the single-carrier and multicarrier modulation schemes provide the same performance upper bounds for the communication channels considered.

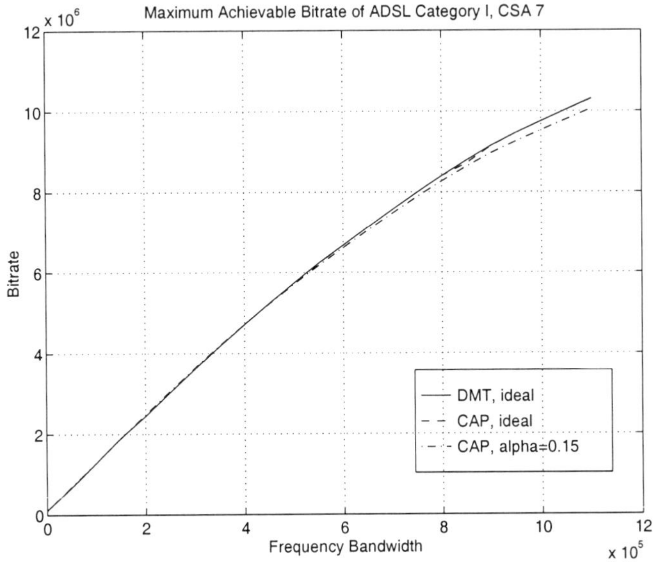

Figure 5.35 Maximum achievable bit rate in ADSL, Category I, CSA 7.

Table 5.9 ADSL standard, Category I, SNR margin in dB for DMT and CAP DFE: 256 coded CAP 6.72 Mbps downstream, 64 coded CAP 250 kbps upstream.

Loop Number	SNR margin Downstream, DMT	SNR margin Downstream, CAP	SNR margin Upstream, DMT	SNR margin Upstream, CAP
T1.601(7)	8.8	8.9	8.7	8.7
T1.601(13)	12.1	12.1	9.3	9.4
CSA(4)	10.7	10.7	22.6	22.7
CSA(6)	10.6	10.6	13.1	13.1
CSA(7)	10.8	10.8	25.8	25.7
mid-CSA	7.6	7.5	59.8	59.9

6 MULTISCALE DETECTION

Nurgün Erdöl

Florida Atlantic University
Boca Raton, FL

erdol@acc.fau.edu

Detection of signals in noise has been studied independently of wavelets for nearly half a century. Its success has resulted in significant advances in the broad areas of digital communications, radar/sonar, and pattern recognition. It is also a general area that, since the introduction of wavelets and multiresolution analysis into signal processing, has attracted research attention for detection of short duration signals embedded in correlated noise, detection in nonstationary environments, and detection and estimation in environments with low signal-to-noise ratios. The reason that makes multiscale analysis seem like the right tool is the natural agreement between the much celebrated and well known theory of matched filtering and the availability of not one, but a family of scalable basis functions called wavelets. It prompts the question if there can be wavelets that match signals or if signals transmitted can be synthesized from wavelets. Such matching between signals and bases have also suggested the possibility that wavelet bases can have properties that approximate those of the Karhunen-Loève (KL) basis of certain covariance functions.

The use of wavelets and related bases for detection has been previously considered in the literature. Whereas the Gabor transform has been applied to the detection of transients in [97], the first examples of using wavelets as matched filters are given in [331] and [98]. The KL approach to detection with wavelets proposed in [84] is presented here in greater analytical depth.

In this chapter we analyze the properties of wavelets to represent random processes efficiently. We explore the wavelet defined multiresolution analysis (MRA) subspaces as a setting for optimal detection strategies. This chapter is organized in three sections. The introduction, Section 6.1, is a review of classical theory for the detection of a known waveform in Gaussian noise. It establishes the basic components of optimal detection as whitening and matched filtering, both of which require knowledge of eigenfunctions and eigenvalues of the noise process. As eigenvalues and functions, or KL bases, depend on the statistics of the random process, and are usually difficult to compute, we look to wavelets and wavelet defined MRA subspaces to provide the proper environment for their approximation and/or computation. In Section 6.2, we develop KL transforms on the Hilbert space L^2 of square integrable functions, and also on wavelet MRAs. In Section 6.2.1, we derive the relationship between eigenfunctions (~KL bases) of a random process and the KL bases that represent their restrictions to MRA subspaces. We comment on its implication in the finite dimensional case in Section 6.2.2. In Section 6.2.3, we characterize random processes that accept wavelets as their KL bases and follow with an analysis of stationarity and white noise over MRA subspaces. In Section 6.3, we propose an optimal detection strategy for each subspace and show that it can have overall optimal performance. We close with a conclusion of the results.

6.1 INTRODUCTION

A binary detection problem [340, 122][1] is described by two hypotheses

$$\begin{aligned} H_1 &: x(t) = s(t) + \eta(t), \\ H_0 &: x(t) = \eta(t) \qquad t \in T_o \end{aligned} \tag{6.1}$$

where $s(t)$ is the message signal of known shape and energy, $\eta(t)$ is zero-mean, second-order Gaussian random noise with a continuous, square-integrable covariance function

$$E\left[\eta(t)\eta(u)\right] = r_\eta(t,u), \tag{6.2}$$

and T_o is the observation interval. It is assumed that the noise energy in the observation interval $\int_{T_o} r_\eta(t,t)dt$ is finite.

The fundamental components of detection are decision making and signal processing.

Decision Making

The two established principles of decision making are the Bayes and the Neyman-Pearson criteria. Bayes statistics leads to the likelihood ratio (LR) test which is derived to minimize the risk of making a wrong decision. In terms of transition probabilities,

$$LR\left(\mathbf{X}\right) = \frac{P\left(\mathbf{X}|H_1\right)}{P\left(\mathbf{X}|H_0\right)}$$

where $P\left(\mathbf{X}|H_\bullet\right)$ represents the conditional probability that the signal $x(t)$ is received given the hypothesis $H_\bullet$ is true. In an ideal system, $LR\left(\mathbf{X}\right) = \infty$ if $x(t)$ contains the message signal, and $LR\left(\mathbf{X}\right) = 0$ if the received signal is just noise; in this case, detection is possible with zero probability of error. In general, a decision is made by comparing $LR\left(\mathbf{X}\right)$ against a threshold. Naturally, if

[1] Most of the material in this section can be found in detail in these references.

$LR(\mathbf{X})$ clusters around two points that are separated by a sufficient distance, the threshold can be chosen with a higher degree of confidence. The purpose of an ideal receiver can thus be expressed as one that, in some sense, comes close to the ideal values for the LR. Expressions for the LR are more useful in terms of sufficient statistics.

A general expression for the LR uses the eigenfunctions, $\theta_k(t)$, and the eigenvalues, λ_k, of the noise covariance function $r_\eta(t,u)$. These quantities satisfy the eigenvalue equation

$$\int_{T_o} r_\eta(t,u)\theta_k(u)\,du = \lambda_k\theta_k(t) \qquad t \in T_o,\ k = 1, 2, \ldots$$

Equivalently, the coefficients of $\eta(t)$ in the series[2]

$$\eta(t) = \sum_k \eta_k\theta_k(t), \tag{6.3}$$

are uncorrelated, i.e. $E[\eta_j\eta_k] = \lambda_k\delta_{jk}$, where δ_{jk} is the Kronecker delta. Using Mercer's formula, its covariance can be written as

$$r_\eta(t,u) = \sum_k \lambda_k\theta_k(t)\theta_k(u)\,.$$

The series in (6.3) is called the KL expansion of the noise signal. The input and the message signals also admit expansion in terms of the noise KL basis functions where the expansion coefficients are

$$x_k = \int_{T_o} x(t)\theta_k(t)\,dt \quad \text{and} \quad s_k = \int_{T_o} s(t)\theta_k(t)\,dt. \tag{6.4}$$

The LR can be expressed as

$$LR = \exp\left[\sum_k \frac{x_k s_k}{\lambda_k} - \frac{1}{2}\sum_k \frac{s_k^2}{\lambda_k}\right]. \tag{6.5}$$

Equivalently, we can write

$$LR = \exp\left[\sum_k \frac{\eta_k s_k}{\lambda_k} \pm \frac{1}{2}\sum_k \frac{s_k^2}{\lambda_k}\right]$$

with + if $H_1$is true and − if H_0 is true. Both LR expressions are valid only if the quantity

$$d^2 = \sum_k \frac{s_k^2}{\lambda_k} \tag{6.6}$$

[2]All sums are assumed to be from $k = 1$ to ∞ unless stated otherwise.

is finite. Perfect detection, i.e., detection with zero probability of error, is achieved if d^2 diverges and LR becomes infinite for H_1 and 0 for H_0. Also called *singular* detection, this condition can be achieved in two ways:

I. If all the eigenvalues are nonzero, then the message signal coefficients, s_k^2, can be chosen to be proportional to λ_k. If $s_k^2 = c_k^2 \lambda_k$, then $d^2 = \sum_k c_k^2$ will diverge if the rate of decay of c_k^2 is slow enough. Two examples are (i) $c_k^2 = c^2$ for all k. The signal energy is proportional to $\sum_k \lambda_k$ which is finite since it is the energy of the noise in the observation interval; (ii) $c_k^2 = c^2/k$ causes $d^2 = \lim_{N \to \infty} c^2 \sum_{k=1}^{N} 1/k$ to diverge while keeping signal energy $\sum_k \lambda_k / k < \sum_k \lambda_k$ finite.

II. If the noise covariance has zero eigenvalues, then singular detection can be achieved if at least one signal component, s_j, corresponding to the zero eigenvalue, λ_j, is non-zero.

In all cases of practical interest, detection is non-singular. If the noise is white, then $d^2 = \lambda^{-1} \sum_k s_k^2$. This implies that detectability in white noise is proportional only to the signal-to-noise ratio (SNR) in $x(t)$ where $\lambda = E\left[\eta_k^2\right]$ is the variance or expected value of the energy of each coefficient, η_k, in (6.3). For non-white noise, d^2 defines a generalized concept of distance which is the inner product of two positive valued (possibly infinite length) vectors $\{s_k^2\}$ and $\{1/\lambda_k\}$. In all implementations, infinite series expansions are truncated. Detectability is given by a finite sum

$$d_K^2 = \sum_{k \in \mathcal{K}} \frac{s_k^2}{\lambda_k} \leq d^2, \tag{6.7}$$

where $\mathcal{K}$ denotes the $K-$ element subset $\{k_1, k_2, \ldots k_K\}$ of the set of positive integers. For a given signal energy, $\xi_s = \sum_{k \in \mathcal{K}} s_k^2$, maximum detectability may be found by minimizing d_K^2 subject to the condition of total energy being ξ_s. Applying standard optimization theory results, we set the objective function

$$\mathcal{O}(\mathbf{s}, \Upsilon) = d_K^2 + \Upsilon(\xi_s - \sum_{k \in \mathcal{K}} s_k^2)$$

and differentiate it with respect to the elements, s_k, of $\mathbf{s} = (s_{k_1}, s_{k_2} \ldots s_{k_K})$ and the Lagrange multiplier Υ. d_K^2 is maximized if all the signal energy is concentrated in one coefficient corresponding to the minimum eigenvalue λ_{k_1}, where it is assumed that the eigenvalues in (6.7) are distinct and ordered such that $\lambda_{k_1} < \lambda_{k_2} < \ldots < \lambda_{k_K}$. Accordingly, we have

$$s_{k_1} = \sqrt{\xi_s}, \qquad s_{k_j} = 0 \quad \text{if } j \neq 1, k_j \in \mathcal{K} \tag{6.8}$$

and the Lagrange multiplier $\Upsilon = 1/\lambda_{k_1}$. The optimal value of detectability becomes $d_K^2 = \xi_s / \lambda_{k_1}$. Increasing detectability also increases the probability of detection in the Neyman-Pearson sense [122]. Decision is made by comparison of the sufficient statistic, $\sum_k x_k s_k / \lambda_k$, to a threshold determined by the probability of false detection and d^2.

Signal Processing

As shown in (6.5), decision making requires the computation of a sufficient statistic given by the first term of the log of LR

$$g = \sum_k \frac{x_k s_k}{\lambda_k}, \tag{6.9}$$

where x_k and s_k are given in (6.4). The sufficient statistic, g, can be computed by the use of the linear filter

$$\check{h}(t,u) = \sum_k (\lambda_k)^{-1/2}\, \theta_k(t)\, \theta_k(u). \tag{6.10}$$

The output due to $s(t)$ is given by

$$\check{s}(t) = \int \check{h}(t,u)s(u)\, du = \sum_k s_k\, (\lambda_k)^{-1/2}\, \theta_k(t), \tag{6.11}$$

and, the outputs, $\check{x}(t)$ and $\check{\eta}(t)$, due respectively to $x(t)$ and $\eta(t)$, are similar. Using the series form of $\check{s}(t)$ and $\check{\eta}(t)$, g can be shown to be

$$g = \int \check{s}(t)\check{x}(t)\, dt. \tag{6.12}$$

The following remarks are of importance:

- The noise output, $\check{\eta}(t)$, of $\check{h}\,(t,u)$ is white and of unit variance. $\check{h}\,(t,u)$ is described as a whitening filter.
- The energy of the signal component of g under H_1 is $\left[\sum_k s_k^2/\lambda_k\right]^2$.
- Signal-to-noise power ratio (SNR_g) in g under H_1 is $d^2 = \sum_k s_k^2/\lambda_k$.
- g as given in (6.12) is the sampled output of filter $\check{s}(-t)$ to input $\check{x}(t)$ which consists of signal $\check{s}(t)$ plus white noise $\check{\eta}(t)$. The filter $\check{s}(-t)$ is called a matched filter [340, 122] and outputs the highest signal-to-noise ratio attainable with a linear filter.

The process described above to compute g can be summarized as the output of two functional units in cascade: a whitener, as defined in (6.10), and a matched filter, as in (6.11). Alternatively, it can be given by

$$g = \int_{T_o} x(t)q(t)\, dt$$

where

$$\begin{aligned} q(t) &= \int_u s(u) \int_v \check{h}(t,v)\check{h}(u,v)\, dv\, du \\ &= \sum_k \left(\frac{s_k}{\lambda_k}\right) \theta_k(t). \end{aligned} \tag{6.13}$$

$q(t)$, as expressed in (6.13), also satisfies the *Fredholm* equation [340],

$$\int_{T_o} r_\eta(t,u)q(u)\,du = s(t). \tag{6.14}$$

Substituting (6.13) into (6.14), one sees that the effective operation to obtain $s(t)$ is to filter the noise covariance by $\int_v \check{h}(t,v)\check{h}(u,v)dv$ so that its output is the covariance of white noise, $\delta(t,u)$.

With the use of g, the received signal $x(t) = s(t) + \eta(t)$ of (6.1) can be expressed as

$$x(t) = g\ s(t) + y(t)$$

where g and $y(t)$ are statistically independent under both hypotheses and the statistics of $y(t)$ are invariant under either hypothesis. When the noise is white, $q(t)$ is proportional to $s(t)$ and matches the message $s(t)$. (6.14) becomes

$$\int_{T_o} r_\eta(t,u)s(u)\,du = \lambda s(t)$$

making $s(t)$ an eigenfunction of $r_\eta(t,u)$ corresponding to the eigenvalue λ. Since any complete orthonormal (ON) basis satisfies the eigenvalue equation of a white-noise covariance function, this is equivalent to the choice of one of the basis functions to be $s(t)$. This result is in agreement with the choice of s_k for optimal detectability which was given earlier in (6.8).

We have observed that optimal detection methods are based on some form of noise whitening. Whitening requires knowledge of the eigenvalues and eigenfunctions, which is often difficult and sometimes impossible to find. Expanding a signal in an arbitrary basis yields coefficients, e_k, that are random variables [246]. Letting $\mu_{jk} = E\left[e_j e_k\right]$,

- If $E\left[e_j e_k\right] = \mu(j-k)$, then the coefficients are stationary. Ergodicity can be used to obtain statistics using "time" averages. The corresponding covariance matrix is Toeplitz, which is equidiagonal and circulant. If the coefficients are also decorrelated so that $\mu(j-k) = \mu(0)\delta_{jk}$ then the random sequence e_k is white with variance $\mu(0)$.

- If $E\left[e_j e_k\right] = \mu_{jk}$ is not a function of $|j-k|$, then e_k is a nonstationary sequence. If the coefficients are decorrelated, then $\mu_{jk} = \mu_{jj}\delta_{jk}$ and the sequence $e_j/\sqrt{\mu_{jj}}$ is white with unit variance.

Approximations and simplifications, even if they are valid under special conditions, are sought for in different bases. In the next section, spaces and subspaces defined by wavelets are analyzed for their properties to represent noise efficiently. We first define the characteristics of KL bases on MRA subspaces. We show how wavelet coefficients can be decorrelated. Then we derive the properties of covariance functions that have wavelets as their KL basis. Finally we address the question of stationarity.

6.2 KL TRANSFORMS ON MRA SUBSPACES

The dyadic wavelet transform [205] analyzes the space of square integrable (finite energy) functions, L^2, into embedded subspaces that form a multiresolution analysis. Its filter bank structure leads to efficient implementations, thus making it a commonly utilized form of the wavelet transform. It has been shown [93, 368, 369, 324], that the orthogonal wavelet transform provides a natural setting in which to analyze and synthesize fractional Brownian motion (FBM) and other $1/f$ processes. They show the extent to which wavelets approximate the KL basis functions of $1/f$ process. The inverse question explores the structure of processes for which wavelets are eigenfunctions. The question is motivated by the wish to characterize such processes so that we may predict the behavior of already existing models, such as FBMs, under the wavelet transform, develop new models and use them to synthesize innovations for signaling in communication systems and develop detection strategies.

In this section, we seek answers to the following questions:

- Can we find the KL basis of a random process by finding the KL bases of its projection onto the MRA subspaces? We show that the answer is no and go on to show how these different bases are related. (Section 6.2.1)
- We repeat the previous question for the finite dimensional case. (Section 6.2.2)
- If wavelets are the eigenfunctions of a process, what are the characteristics of its covariance function and eigenvalues? (Section 6.2.3)
- Can wavelets represent stationary processes? Are there processes that have wavelet coefficients that form a stationary sequence? (Section 6.2.4)

The answers to these questions together with any *a priori* knowledge about the noise environment can be used to find the eigenfunctions and eigenvalues needed for optimal detection.

For notational clarity, we present the following well known relationships about wavelets. We use $\mathbf{R}$ and $\mathbf{Z}$ to represent the set of real numbers and integers, respectively. Dyadic wavelets [61] are formed by the dilations and translations

$$\psi_{jk}(t) = 2^{j/2}\psi(2^j t - k)$$

of the mother wavelet $\psi(t)$. Its associated scaling function is represented by $\varphi_{jk}(t)$ for j and $k \in \mathbf{Z}$. If $\{\psi_{jk}(t)\}_{jk}$ are ON, then for any $f(t) \in L^2(\mathbf{R})$, we can write the following expansions.

$$f(t) = \sum_{jk} d_{jk}\psi_{jk}(t) \tag{6.15}$$

$$f(t) = \sum_{j=J}^{\infty}\sum_{k} d_{jk}\psi_{jk}(t) + \sum_{k} c_{Jk}\varphi_{Jk}(t)$$

where[3]

$$d_{jk} = \langle f(t), \psi_{jk}(t) \rangle$$

and

$$c_{jk} = \langle f(t), \varphi_{jk}(t) \rangle .$$

The coefficients above satisfy

$$\begin{aligned} c_{jk} &= \sum_n h^0_{n-2k} c_{j+1,n} \\ d_{jk} &= \sum_n h^1_{n-2k} c_{j+1,n} \\ c_{j+1,k} &= \sum_n \left[c_{jn} h^0_{2n-k} + d_{jn} h^1_{2n-k} \right] \end{aligned}$$

where the filter coefficients h^0_n and h^1_n satisfy the two-scale equations

$$\begin{aligned} \varphi_{jk}(t) &= \sum_n h^0_{2k-n} \varphi_{j+1,n}(t), \\ \psi_{jk}(t) &= \sum_n h^1_{2k-n} \varphi_{j+1,n}(t). \end{aligned}$$

Orthonormal wavelets form a MRA by partitioning $L^2(\mathbf{R})$ into a family of embedded subspaces, $\mathbf{V}_j$, spanned by $\varphi_{jk}(t)$, so that for all integers j and k, the following hold:

$$\mathbf{V}_j \subset \mathbf{V}_{j+1}$$

$$\bigcup_{j=-\infty}^{\infty} \mathbf{V}_j = L^2(\mathbf{R}) \quad \text{and} \quad \bigcap_{j=-\infty}^{\infty} \mathbf{V}_j = \{0\}.$$

The disjoint subspaces, $\mathbf{W}_j$, form the orthogonal complement of $\mathbf{V}_j$ in $\mathbf{V}_{j+1}$, i.e., $\mathbf{V}_{j+1} = \mathbf{W}_j \oplus \mathbf{V}_j$. $\mathbf{W}_j$ are spanned by $\{\psi_{jk}(t)\}_k$. We define $f_j(t) = Q_j f(t)$ and $f^j(t) = P_j f(t)$ as the projection of $f(t)$ onto the subspace $\mathbf{W}_j$ and $\mathbf{V}_j$ respectively. We have the relationships

$$f_j(t) = \sum_k d_{jk} \psi_{jk}(t) \qquad \text{and} \qquad f^j(t) = \sum_k c_{jk} \varphi_{jk}(t) \tag{6.16}$$

with the respective autocovariance functions

$$r_j(t,u) = E\left[f_j(t) f_j(u)\right] \qquad \text{and} \qquad r^j(t,u) = E\left[f^j(t) f^j(u)\right].$$

The next section searches for KL bases over MRA subspaces.

[3]The inner product $\langle f, g \rangle = \int fg$.

6.2.1 *KL bases over MRA subspaces*

When analyzing signals in the multiresolution subspaces described in Section 6.2, it is useful to know how KL bases defined over subspaces $\mathbf{W}_j$, $\mathbf{V}_j$ and $L^2(\mathbf{R})$ are related. The key question is whether or not the decorrelation of coefficients on two disjoint subspaces such as $\mathbf{V}_j$ and $\mathbf{W}_j$ implies the same for coefficients on $\mathbf{V}_{j+1} = \mathbf{V}_j \oplus \mathbf{W}_j$. Another relevant example of disjoint subspaces are $\mathbf{W}_j$ and $\mathbf{W}_{j+1}$. It is not difficult to see that KL bases found for the constituent subspaces do not comprise a KL basis for their sum. This means that if we are interested in finding a basis that decorrelates all the coefficients $\{d_{jk}\}$, for all k and $0 \le j \le J$, then we need to be working in $\bigcup_{j=0}^{J} \mathbf{W}_j$. We will formulate these concepts for the sake of rigor as well as to gain insight into the problem.

We will work in a general setting of subspaces, $\mathbf{A} = \mathbf{A}_0 \oplus \mathbf{A}_1$ in $L^2(\mathbf{R})$, where $\mathbf{A}_0$ and $\mathbf{A}_1$ are disjoint. For simplicity of notation, we will use such variables as $f(t)$, $y(t)$ and $z(t)$ exclusive of their previously defined meanings. We will assume $f(t) = y(t) + z(t) \in \mathbf{A}$ has orthogonal projections $y(t) \in \mathbf{A}_0$ and $z(t) \in \mathbf{A}$. The corresponding covariance functions $r_\bullet(t,u)$ are assumed to have the eigenvalues, $\lambda_k^\bullet$,and orthonormal eigenfunctions $\theta_k^\bullet(t)$, where $\bullet$ means f, y or z. Defining $(\bullet)_k = \langle \bullet, \theta_k^\bullet \rangle$, we write the series expansions

$$f(t) = \sum_{k=1}^{\infty} f_k \theta_k^f(t),$$

$$y(t) = \sum_k y_k \theta_k^y(t) \quad \text{and} \quad z(t) = \sum_k z_k \theta_k^z(t).$$

Mercer's formula gives

$$r_f(t,u) = \sum_k \lambda_k^f \theta_k^f(t) \theta_k^f(u)$$

with similar expressions for $r_\bullet(t,u)$ for y and z. The expansion coefficients are uncorrelated: $E[f_k f_l] = \lambda_k^f \delta_{kl}$, $E[y_k y_l] = \lambda_k^y \delta_{kl}$ and $E[z_k z_l] = \lambda_k^z \delta_{kl}$. Clearly, $\{\theta_k^y(t)\}_k$ and $\{\theta_k^z(t)\}_k$ form complete orthonormal bases, respectively, of $\mathbf{A}_0$ and $\mathbf{A}_1$, and $\{\theta_k^y(t), \theta_k^z(t)\}_k$ and $\left\{\theta_k^f(t)\right\}_k$ are two different orthonormal bases spanning $\mathbf{A}$. The covariance function above can be expressed in terms of these bases as

$$\begin{aligned} r_f(t,u) &= \sum_k \lambda_k^y \theta_k^y(t) \theta_k^y(u) + \sum_k \lambda_k^z \theta_k^z(t) \theta_k^z(u) \\ &\quad + \sum_{kl} E[y_k z_l] \left[\theta_k^y(t) \theta_l^z(u) + \theta_k^y(u) \theta_l^z(t)\right]. \end{aligned}$$

As long as the signals $y(t)$ and $z(t)$ in the subspaces are correlated, the set $\{\theta_k^y(t), \theta_k^z(t)\}_k$ does not form a KL basis for $f(t)$ on $\mathbf{A}$. Since $\mathbf{A}_0, \mathbf{A}_1 \subset \mathbf{A}$, the following expansions are valid:

$$\theta_k^y(t) = \sum_n q_{fy}(n,k)\theta_n^f(t) \tag{6.17}$$

$$\theta_k^z(t) = \sum_n q_{fz}(n,k)\theta_n^f(t) \tag{6.18}$$

$$\theta_k^f(t) = \sum_n \left[q_{fy}(k,n)\theta_n^y(t) + q_{fz}(k,n)\theta_n^z(t)\right]$$

where $q_{fy}(n,k) = \left\langle \theta_n^f(t), \theta_k^y(t) \right\rangle$ and $q_{fz}(n,k) = \left\langle \theta_n^f(t), \theta_k^z(t) \right\rangle$. Orthonormality of the bases $\left\{\theta_k^f(t)\right\}_k$, $\{\theta_k^y(t)\}_k$ and $\{\theta_k^z(t)\}_k$ and the orthogonality of $\theta_k^y(t)$ and $\theta_k^z(t)$ imply

$$\begin{aligned} \sum_n q_{fy}(n,k)q_{fy}(n,l) &= \delta_{kl}, \\ \sum_n q_{fz}(n,k)q_{fz}(n,l) &= \delta_{kl}, \\ \sum_k \left[q_{fy}(n,k)q_{fy}(m,k) + q_{fz}(n,k)q_{fz}(m,k)\right] &= \delta_{nm}, \\ \sum_n q_{fy}(n,k)q_{fz}(n,l) &= 0. \end{aligned} \tag{6.19}$$

Analogous KL expansions can be defined for the discrete coefficients $\{y_k\}_k$. $\{z_k\}_k$ and $\{f_k\}_k$ in terms of eigenvectors of subspaces of $l_2(R)$, the space of square summable or finite energy sequences. We first note that the discretization that generates the coefficients is not linear, as the expansions are given in terms of different bases. Hence f_k is not the sum of y_k and z_k, even though $f(t)$ is the sum of $y(t)$ and $z(t)$. Since $y(t)$ and $z(t)$ are orthogonal complements of $f(t)$, then

$$y_k = \langle f(t), \theta_k^y(t)\rangle \quad \text{and} \quad z_k = \langle f(t), \theta_k^z(t)\rangle\,. \tag{6.20}$$

The resolution of identity, or Parseval's rule, dictates conservation of energy:

$$\sum_k f_k^2 = \sum_k y_k^2 + \sum_k z_k^2$$

or

$$\sum_k \lambda_k^f = \sum_k \lambda_k^y + \sum_k \lambda_k^z. \tag{6.21}$$

Using (6.17) and (6.18) in (6.20), we have the decomposition relationships

$$y_k = \sum_n f_n q_{fy}(n,k), \tag{6.22}$$

$$z_k = \sum_n f_n q_{fz}(n,k), \tag{6.23}$$

and the reconstruction relationship

$$f_k = \sum_n y_n q_{fy}(k,n) + \sum_n z_n q_{fz}(k,n). \tag{6.24}$$

Since the elements of each of the sequences y_k and z_k are uncorrelated, using (6.22) and (6.23) in $E[y_k y_l]$ and $E[z_k z_l]$ yields

$$\lambda_k^y = \sum_n \lambda_n^f q_{fy}(n,k) q_{fy}(n,k), \tag{6.25}$$

$$\lambda_k^z = \sum_n \lambda_n^f q_{fz}(n,k) q_{fz}(n,k). \tag{6.26}$$

The correlation between y_k and z_l can be expressed by

$$E[y_k z_l] = \sum_n \lambda_n^f q_{fy}(n,k) q_{fz}(n,l). \tag{6.27}$$

Using (6.24) in $E[f_k f_l]$, we also obtain

$$\begin{aligned}\lambda_k^f \delta_{kl} &= \sum_n \lambda_n^y q_{fy}(k,n) q_{fy}(l,n) + \sum_n \lambda_n^z q_{fz}(k,n) q_{fz}(l,n) \\ &\quad + \sum_{nm} E[y_n z_m]\left[q_{fy}(k,n) q_{fz}(l,m) + q_{fy}(l,n) q_{fz}(k,m)\right].\end{aligned}$$

These results can be expressed using infinite dimensional matrices. We define vector $\mathbf{u} = [y_1 y_2 \ldots y_n, \ldots z_1 z_2 \ldots z_n, \ldots]^T$ and $\mathbf{f} = [f_1 f_2 \ldots f_n \ldots]^T$ .The respective correlation matrices are $\Lambda_f = E[\mathbf{ff}^T] = \mathrm{diag}(\lambda_1^f \lambda_2^f \ldots \lambda_n^f)$ and $\Lambda_u = E[\mathbf{uu}^T]$

$$\mathbf{R}_u = \begin{bmatrix} \Lambda_y & \mathbf{R} \\ \mathbf{R}^T & \Lambda_z \end{bmatrix} \tag{6.28}$$

where $\Lambda_y = \mathrm{diag}(\lambda_1^y \lambda_2^y \ldots \lambda_{n_y}^y \ldots)$, $\Lambda_z = \mathrm{diag}(\lambda_1^z \lambda_2^z \ldots \lambda_n^z \ldots)$ and $[\mathbf{R}]_{mn} = E[y_m z_n]$ for $m, n \in \{1, 2, \ldots\}$. From (6.25)–(6.27), we have

$$\mathbf{R}_u = \mathbf{Q} \Lambda_f \mathbf{Q}^T$$

where

$$\mathbf{Q} = \begin{bmatrix} q_{fy}(1,1) & q_{fy}(2,1) & \ldots & q_{fy}(n,1)\ldots \\ \vdots & \vdots & \ddots & \vdots \\ q_{fy}(1,n) & q_{fy}(2,n) & \ldots & q_{fy}(n,n)\ldots \\ \vdots & \vdots & \ddots & \vdots \\ q_{fz}(1,1) & q_{fz}(2,1) & \ldots & q_{fz}(n,1)\ldots \\ \vdots & \vdots & \ddots & \vdots \\ q_{fz}(1,n) & q_{fz}(2,n) & \ldots & q_{fz}(n,n)\ldots \\ \vdots & \vdots & & \vdots \end{bmatrix}.$$

From (6.19), $\mathbf{Q}^{-1}=\mathbf{Q}^T$, i.e. $\mathbf{Q}$ represents a unitary operation, thus we also have

$$\Lambda_f = \mathbf{Q}^T\mathbf{R}_\mathbf{u}\mathbf{Q}.$$

Clearly, $\lambda_\bullet^f$ are the eigenvalues and the columns of $\mathbf{Q}$ are the eigenvectors of $\mathbf{R}_\mathbf{u}$. The transformation $\mathbf{f} = \mathbf{Q}^T\mathbf{u}$ is the KL transform of $\mathbf{u}$. Analogous results are obtained in the finite dimensional case.

6.2.2 *Finite dimensional KL bases*

For the finite dimensional case, we define the $M = N + K$ length vector $\mathbf{u}_M=[y_1y_2\ldots y_N, z_1z_2\ldots z_K]^T$. The $M \times M$ covariance matrix, $\mathbf{R}_{uM} = E\left[\mathbf{u}_M\mathbf{u}_M^T\right]$, is given by

$$\mathbf{R}_{uM} = \begin{bmatrix} \Lambda_{yN} & \mathbf{R}_{NK} \\ \mathbf{R}_{NK}^T & \Lambda_{zK} \end{bmatrix}$$

where $\Lambda_\bullet$ are the finite size versions of the diagonal matrices used in (6.28) in the previous section. Unitary transformation, $\mathbf{Q}_M$, operating on $\mathbf{u}_M$ results in $\mathbf{f}_M=[f_{1M}f_{2M}\ldots f_{M,M}]^T$ so that $\Lambda_{fM} = E\left[\mathbf{f}_M\mathbf{f}_M^T\right]= \text{diag}(\lambda_{1M}^f\lambda_{2M}^f\ldots\lambda_{MM}^f)$ is given by

$$\Lambda_{fM} = \mathbf{Q}_M\mathbf{R}_{uM}\mathbf{Q}_M^T.$$

Since the trace of matrix Λ_{fM} is equal to the trace of $\mathbf{R}_{uM}$, we have energy conservation and (6.21) holds for the finite case. Clearly, as N and K get large, $\mathbf{Q}_M \to \mathbf{Q}$ and $\Lambda_{fM} \to \Lambda_f$. We can follow the migration of eigenvalues $\{\lambda_{nN}^y\}$ and $\{\lambda_{kK}^z\}$ to $\left\{\lambda_{mM}^f\right\}$ by considering the case $M = N + K$ as K is increased from 0 to N one step at a time. Each additional element in $\mathbf{u}_M$ inserts one row and one column into $\mathbf{R}_{uM}$, thus changing it to $\mathbf{R}_{uM+1}$. When $K = 0$, $M = N$ and $\{\lambda_{nN}^y\} = \left\{\lambda_{nN}^f\right\}$. When $K = 1$, the eigenvalues of $\mathbf{R}_{uN+1}$are$\left\{\lambda_{nN+1}^f\right\}$ and they separate $\left\{\lambda_{nN}^f\right\}$, the eigenvalues of $\mathbf{R}_{uN}$. Assuming that the eigenvalues are numbered in decreasing order and distinct, we have [297]

$$\lambda_{1N+1}^f > \lambda_{1N}^f > \lambda_{2N+1}^f > \ldots > \lambda_{N,N}^f > \lambda_{N+1,N+1}^f.$$

One concludes then, that the condition number, i.e. the ratio of the largest to the smallest eigenvalue, increases as the matrix gets larger. A similar argument can be made when the size of the covariance matrix is increased by keeping $K = 0$ and increasing N. It is well known that a covariance matrix becomes ill conditioned as the condition number increases. A well known remedy to the ill conditioning phenomenon is to resize the covariance matrix by choosing only the largest eigenvalues and their corresponding eigenvectors.

Finding eigenvectors and eigenvalues of a given process is invariably a difficult task, if at all analytically possible. Numerically, it is often of high computational cost. Such expense may be avoided for a certain class of signals if a transform can be said to approximate a KL transform. Elements of the Fourier basis [340] have been shown to be eigenfunctions for a large class of stationary covariance functions. Finite observation time causes signals to be nonstationary, frequently making the Fourier basis inadequate as a KL approximation. In the search for other bases for the efficient representation of nonstationary signals, wavelets receive special attention for being time-localized as well as for not being a fixed set of functions. The ability to choose a wavelet from a library is usually a welcome advantage. In the next section, we characterize the classes of signals for which wavelets can be eigenfunctions.

6.2.3 Covariance functions with eigenwavelets

For $f(t) \in L^2(\mathbf{R})$, represented in terms of the wavelet series in (6.15), the statistical independence of the representation coefficients means

$$E[d_{jk}d_{lm}] = \lambda_{jk}\delta_{jl}\delta_{km} \tag{6.29}$$

and we have

$$E[d_{jk}^2] = \lambda_{jk} \geq 0.$$

In order for (6.29) to be true, $\psi_{jk}(t)$ must be eigenfunctions of the covariance function of $f(t)$, i.e.

$$\lambda_{jk}\psi_{jk}(t) = \int r(t,u)\psi_{jk}(u)\,du.$$

A covariance function generated by an orthonormal set of wavelets can be written using Mercer's formula as[4]

$$r(t,u) = \sum_{jk} \lambda_{jk}\psi_{jk}(t)\psi_{jk}(u). \tag{6.30}$$

For ease of reference, wavelets that satisfy (6.30) will be called eigenwavelets of $r(t,u)$. Given that a wavelet basis is the KL basis function of $r(t,u)$ on $L^2(\mathbf{R})$, the following remarks are true:

Remark 1 *The autocovariance function in (6.30) can be written as the sum of its projections onto* $\mathbf{W}_j$.

We have

$$r(t,u) = \sum_j r_j(t,u) \tag{6.31}$$

where

$$r_j(t,u) = \sum_k \lambda_{jk}\psi_{jk}(t)\psi_{jk}(u). \tag{6.32}$$

[4]Unless especially noted, all summations in the wavelet series are over $\mathbf{Z} \times \mathbf{Z}$, where $\mathbf{Z}$ is the space of all integers.

Figure 6.1 shows an example of $r_j(t,u)$ for $j = 0, 1$ and 2 for the Haar wavelet.

Remark 2 *All non-white random processes with finite energy satisfy*

$$\xi = \int_{-\infty}^{\infty} r(t,t)\,dt = \sum_{jk} \lambda_{jk} < \infty. \tag{6.33}$$

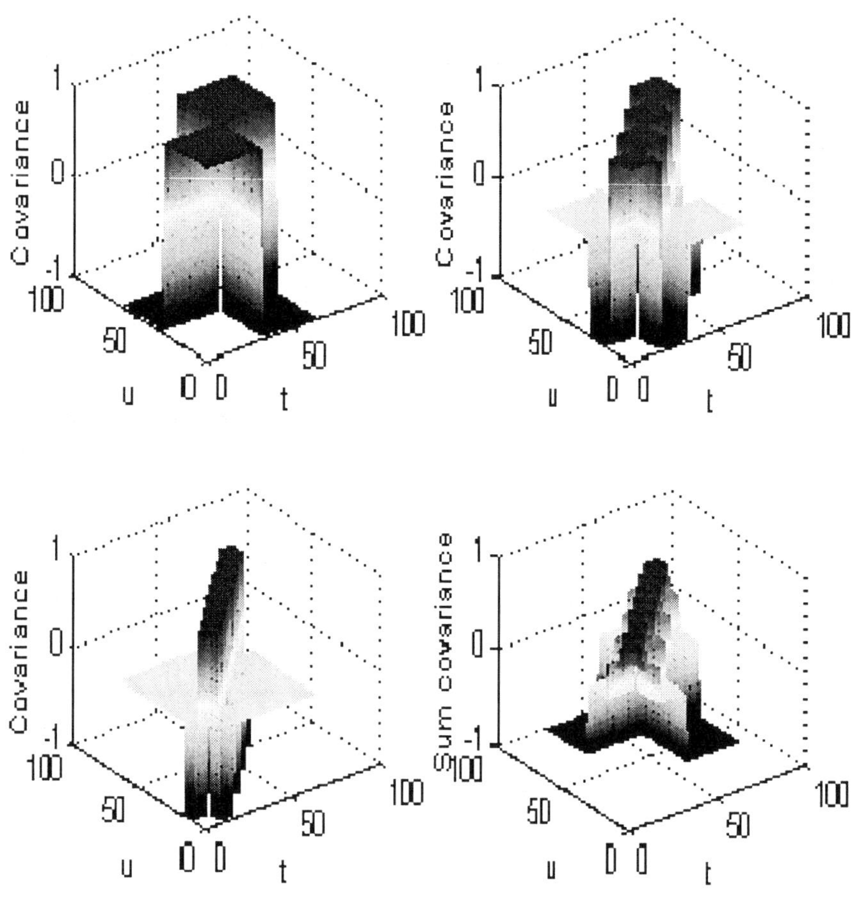

Figure 6.1 Covariance function $r_j(t,u)$ per (6.32) for the Haar wavelet with $\lambda_{jk} = 2^{-|j|}$ for three consecutive scales and their sum.

A trivial implication of (6.33) is

$$\xi_j = \int_{-\infty}^{\infty} r_j(t,t)\,dt = \sum_k \lambda_{jk} < \infty. \tag{6.34}$$

Remark 3 *If there are no more than a finite number of zero eigenvalues, then the doubly infinite sum of the eigenvalues in (6.33) can be finite only by imposing the decay rate that $\lambda_{jk} < c\alpha^{-|j|}\beta^{-|k|}$ for some positive constant c and $|\alpha|, |\beta| < 1$. It implies that $\sum_k \lambda_{jk} < A\alpha^{-|j|}$ and $\sum_j \lambda_{jk} < B\beta^{-|k|}$ for positive constants A and B.. This is a stronger condition on the single sums $\sum_k \lambda_{jk}$ and $\sum_j \lambda_{jk}$ than is required for the convergence of each one alone.*

Remark 4 *The two dimensional Fourier transform of $r(t,u)$ is*

$$\hat{r}(\omega_t, \omega_u) = \sum_j 2^{-j}\hat{\psi}\left(2^{-j}\omega_t\right)\hat{\psi}\left(2^{-j}\omega_u\right)\Lambda_j\left(2^{-j}(\omega_t+\omega_u)\right) \tag{6.35}$$

where Λ_j is the Fourier transform of the sequence λ_{jk}. It is $2^j 2\pi$- periodic in both ω_t and ω_u, and is given by

$$\Lambda_j(\omega) = \sum_k \lambda_{jk}\, e^{-ik\omega}.$$

When $\omega_t + \omega_u = 0$, (6.35) becomes

$$\hat{r}\left(\omega_t, -\omega_t\right) = \sum_j \xi_j 2^{-j}\left|\hat{\psi}\left(2^{-j}\omega_t\right)\right|^2$$

where ξ_j in (6.34) is the energy in scale j and $\hat{\psi}(\omega)$ is the Fourier transform of $\psi(t)$. We note the quantity $A_j = \xi_j 2^{-j}\hat{\psi}\left(2^{-j}\omega_t\right)$. Its energy is ξ_j. Its effective bandwidth, β_j, and center frequency, ω_j, are proportional to 2^j. ξ_j may be thought of as the energy of A_j concentrated at ω_j. Since by Remark 3, ξ_j are bounded above by a multiple of 2^{-j}, then for large values of ω_t, we have an exponentially decreasing energy distribution. One may also define the average spectral energy distribution of A_j as $2^{-j}\xi_j$ over a bandwidth of β_j. Again for large values of ω_t, we would have an exponentially decaying energy density envelope. These processes are said to have $1/f$ spectra and are effectively represented by wavelets [366].

Remark 5 *Projections, $f_j(t)$, defined in (6.16), of the process $f(t)$ to $\mathbf{W}_j$ and $\mathbf{W}_l$ are not correlated, i.e. $E\left[f_j(t)f_l(u)\right] = 0$ if $j \neq l$.*

Remark 6 *Treated as a function of two variables, $r_j(t,u)$ and $r_l(t,u)$ are orthogonal, i.e. the two dimensional inner product[5] $\langle r_j(t,u), r_l(t,u)\rangle_2 = \delta_{jl}\sum_k \lambda_j k^2$.*

[5] $\langle r_j, r_l\rangle_2 = \iint r_j(t,u) r_l(t,u)\,dt\,du$

Remark 7 $\langle r(t,u), r(t,u)\rangle_2 = \sum_{jk} \lambda_{jk}^2$.

Covariance functions that can be expressed as in (6.31) can be designed using a choice of wavelets and some appropriate eigenvalues. These pathological expressions, may be used to develop new models for physical random signals. Existing models may be tested for representability as well.

In Section 6.2.1, it was shown that satisfying (6.32) for every j does not imply the satisfaction of (6.31). The structure of (6.32) is, however, less restrictive than that of (6.31). We will show that if Haar wavelets are used, then (6.32) is satisfied by the nonstationary Wiener process [247]. Developed as a model for Brownian motion, its covariance function is

$$r_w(t,u) = \sigma^2 \min(t,u). \tag{6.36}$$

The covariance of its wavelet coefficients, $E[d_{jk}d_{jm}]$, is zero whenever $r_w(t,u) = t$ or u over the entire the support of $\psi_{jk}(t)\psi_{jm}(u)$. This is the case when the rectangular support of $\psi_{jk}(t)\psi_{jm}(u)$ does not intersect the line $t = u$ in the (t,u) plane. For Haar wavelets, that means $E[d_{jk}d_{jm}] = 0$ if $k \neq m$ and (6.32) is satisfied. For the compactly supported wavelets of Daubechies, the result depends on the order of the wavelet. For the Daubechies-3 wavelet, which overlaps its nearest four orthogonal translates, $E[d_{jk}d_{jm}] \neq 0$ for $m = k, k \pm 1$ and $k \pm 2$. (6.32) is satisfied for all but a small number of coefficients around the diagonal. In most applications this amount of decorrelation is satisfactory . Its sparse nature makes it easy to transform any finite section of the correlation matrix to a diagonal matrix. Since a large class of processes can be transformed into the Wiener process, this result is significant in establishing the usefulness of wavelets as a tool for nonstationary signal analysis.

The analysis of random processes over MRA is incomplete without the knowledge of the properties of wavelet coefficients for stationary signals. In the absence of a derived mathematical model, one relies on ergodicity for statistical information. Ergodicity [246] is the property that allows ensemble averages to be replaced by time averages and is meaningful only for stationary signals. In the next section, we analyze the behavior of the wavelet coefficients of stationary processes and find the properties of covariance functions that can be synthesized using stationary coefficients.

6.2.4 Stationarity of wavelet coefficients

The random process $f(t)$ is stationary if its covariance function can be written as $r(|t-u|)$. The wavelet series coefficients have the joint expected value

$$E[d_{jk}d_{lm}] = \int r(t)\mathcal{A}(j,k,l,m,t)\, dt$$

where

$$\mathcal{A}(j,k,l,m,t) = \int \psi_{jk}(\tau)\psi_{lm}(\tau - t)\, d\tau. \tag{6.37}$$

(6.37) can be expressed in the frequency domain as

$$\mathcal{A}(j,k,l,m,t) = \frac{1}{2\pi}\int \hat{\psi}(-\omega)\hat{\psi}(2^{j-l}\omega)e^{i\omega b}\,d\omega$$

where $b = k - 2^j\tau - 2^{j-l}m$. Clearly, the coefficients are not stationary. If, however, we consider the case of $j = l$, then we get

$$E\left[d_{jk}d_{jm}\right] = \frac{1}{2\pi}\int \left|\hat{\psi}(\omega)\right|^2 \hat{r}(2^j\omega)e^{i\omega(k-m)}\,d\omega \tag{6.38}$$

where $\hat{r}(\omega)$ is the Fourier transform of $r(t)$. The right hand side of (6.38) is the inverse Fourier transform of a non-negative definite autocovariance function and therefore is symmetric in $k-m$. For a given j, the covariance of the coefficients at the same scale is a function of $|k-m|$ and therefore is stationary.

The above result can be extended to covariance functions that have the form

$$r(t,u) = a(t) + b(u) + c(|t-u|) \tag{6.39}$$

where $c(t)$ is stationary. The covariance of the coefficients,

$$E\left[d_{jk}d_{lm}\right] = \iint c(|t-u|)\psi_{jk}(t)\psi_{lm}(u)\,du\,dt,$$

is not affected by $a(t)$ or $b(u)$ since the average (DC) value of a wavelet is zero. An example of this type of autocovariance is that of fractional Brownian motion with parameter H given by

$$r_{FBM}(t,u) = \frac{\sigma^2}{2}\left(|t|^{2H} + |u|^{2H} - |t-u|^{2H}\right). \tag{6.40}$$

Detailed results on the statistics of their wavelet coefficients can be found in [92]. Other examples are the Ornstein-Uhlenbeck process with the covariance function

$$r_{o-u}(t,u) = \frac{1}{2\alpha}\exp\left(-\alpha\,|t-u|\right), \tag{6.41}$$

and the sinusoidal covariance function such as

$$r_s(t,u) = \cos(t-u).$$

If wavelets $\psi_{jk}(t)$ are eigenfunctions of a stationary process, the eigenvalues of the process are given by

$$\lambda_{jk} = \int r(t)\int \psi_{jk}(u-t)\psi_{jk}(u)\,du\,dt. \tag{6.42}$$

Rewriting (6.42) in the frequency domain, we have

$$\begin{aligned}\lambda_{jk} &= \frac{1}{2\pi}\int \left|\hat{\psi}(\omega)\right|^2 \hat{r}(2^j\omega)\,d\omega \qquad (6.43)\\ &= \frac{2^{-j}}{2\pi}\int \left|\hat{\psi}(2^{-j}\omega)\right|^2 \hat{r}(\omega)\,d\omega.\end{aligned}$$

We note that λ_{jk} are independent of k and vanish only if $\hat{r}(2^j\omega)$ vanishes entirely over the bandwidth (support) of $\hat{\psi}(\omega)$. We also note that λ_{jk} are also independent of j if $\hat{r}(\omega)$ is a constant, representing the spectrum of white noise. As given in (6.43), the coefficients, $\{d_{jk}\}_k$, of the random process $f_j(t)$ are whitened and its covariance is given by

$$r_j(t, t+\tau) = \lambda_{j0} \sum_k \psi_{jk}(t)\psi_{jk}(t+\tau).$$

For a given τ, $r_j(t, t+\tau)$ is periodic with period 2^{-j}, hence it is cyclostationary. As a function of τ, when t is fixed, it is non-zero only for $|\tau| < 2^{-j}$.

This section's results can be summarized as follows. Random processes exist for which wavelets form KL bases. To find the KL basis of a random process, one must consider the whole process on L^2, which remains a difficult task. A KL basis can be found for a scale limited restriction of a random process only by considering all the scales in question as one unit. The characteristic sparsity of wavelet coefficients for a large class of signals may lead to approximate solutions that are not computationally expensive. Decorrelation of same scale wavelet coefficients is not sufficient to remove the cross-covariance between coefficients of different scales. Requirement of the decorrelation of same scale wavelet coefficients is, however, much less restrictive and can be performed on the isolated subspace alone. Stationarity of same scale wavelet coefficients is satisfied by a class of covariance functions, as represented in (6.39), that include stationary processes and FBMs. We conclude that for such purposes as detection, it is much more convenient to work on each subspace (scale) alone. In the next section, we show that detectability need not be compromised using a detection strategy that involves the specification of eigenfunctions and values in each subspace alone.

6.3 DETECTION WITH WAVELETS

In Section 6.1, the fundamental elements of detection of a known signal in Gaussian noise were given as whitening (6.10), matched filtering (6.11) and the computation of a sufficient statistic (6.12). All of these operations were given in terms of the eigenvalues and eigenfunctions or the KL basis of the noise covariance function $r_\eta(t, u)$. In Section 6.2, various methods were discussed to transform a process over MRA subspaces so that it can be represented in terms of uncorrelated coefficients. It was shown that for a class of random processes, as expressed in (6.39), that include stationary processes, fractional Brownian motion, as in (6.40), and the Ornstein-Uhlenbeck process given in (6.41), the wavelet coefficients of the same scale formed a stationary sequence; the nonstationary Wiener process in (6.36), gave rise to the same-scale Haar wavelet coefficients that were white. It was also shown that finding KL bases for the projection of a process on subspaces did not guarantee the same for the whole process. The reduced size of projections onto MRA subspaces as well as the above mentioned properties make it reasonable to consider detection on each subspace alone. In this section, the advantages of transposing the detection problem to the wavelet subspaces are discussed and it is shown that in the process detectability need not be sacrificed.

The binary detection problem for a known signal, $s(t)$, in additive Gaussian noise, $\eta(t)$, was defined in (6.1). $r_\eta(t,u)$, given in (6.2), is assumed to be the square-integrable covariance function of the zero-mean second-order noise process. In a method similar to that introduced in Section 6.2.1, we consider the KL basis for $r_\eta(t,u)$ on $L^2(\mathbf{R})$ and denote it $\{\theta_k(t)\}_k$. The corresponding eigenvalues are λ_k, and x_k and s_k are the respective coefficients of $x(t)$ and $s(t)$ with respect to $\phi_k(t)$. We have

$$s(t) = \sum_k s_k \theta_k(t)$$

and

$$r_\eta(t,u) = \sum_k \lambda_k \theta_k(t)\theta_k(u).$$

The binary detection problem in this context was given in (6.1). Its detectability, d^2, and sufficient statistic, g, are given by

$$d^2 = \sum_k \frac{s_k^2}{\lambda_k} \quad \text{and} \quad g = \sum_k \frac{x_k s_k}{\lambda_k}.$$

We also consider the projection, $\eta_j(t)$, of $\eta(t)$ onto the wavelet subspace $\mathbf{W}_j$ and its covariance function $r_j(t,u)$ described by its KL basis $\{\theta_{jk}(t)\}_k$ as

$$r_j(t,u) = \sum_k \lambda_{jk} \theta_{jk}(t)\theta_{jk}(u).$$

The projection of the signal $s(t)$ onto $\mathbf{W}_j$ expressed in terms of $\theta_\bullet$ is

$$s_j(t) = \sum_k s_{jk}\theta_{jk}(t).$$

We note that the double versus single subscripts are the only indicators that the λ_{jk} and λ_k are different and represent related but different covariance functions. A similar comment holds for the coefficients. For a given j, we may state the detection problem of (6.1) as

$$\begin{aligned} H_1 &: x_j(t) = s_j(t) + \eta_j(t), \\ H_0 &: x_j(t) = \eta_j(t) \qquad t \in T_o. \end{aligned} \tag{6.44}$$

Detectability of (6.6) and the sufficient statistic of (6.9) become, respectively,

$$d_j^2 = \sum_k \frac{s_{jk}^2}{\lambda_{jk}} \quad \text{and} \quad g_j = \sum_k \frac{x_{jk} s_{jk}}{\lambda_{jk}}.$$

The above decomposition of the detection problem requires a decision strategy that, for each j, integrates the comparison of the sufficient statistic, g_j, to a threshold which depends on d_j^2. If, however, we define the noise process $\bar{\eta}(t)$, which is characterized by its covariance $\bar{r}(t,u) = E\left[\bar{\eta}(t)\bar{\eta}(u)\right]$ given by

$$\bar{r}(t,u) = \sum_j r_j(t,u) = \sum_{jk} \lambda_{jk}\theta_{jk}(t)\theta_{jk}(u),$$

then detection in each subspace as given in (6.44) can be restated as

$$\begin{aligned} H_1 &: \bar{x}(t) = s(t) + \bar{\eta}(t), \\ H_0 &: \bar{x}(t) = \bar{\eta}(t) \qquad t \in T_o. \end{aligned} \tag{6.45}$$

The detectability and sufficient statistic are given, respectively by,

$$\bar{d}^2 = \sum_j d_j^2 = \sum_{jk} \frac{s_{jk}^2}{\lambda_{jk}} \quad \text{and} \quad \bar{g} = \sum_j g_j = \sum_{jk} \frac{x_{jk}s_{jk}}{\lambda_{jk}}. \tag{6.46}$$

If (6.46) defines the optimal detector for the detection of $s(t)$ in additive Gaussian noise $\bar{\eta}(t)$, what happens when the real problem is the detection of the same signal in additive noise $\eta(t)$? In Section 6.2.1, it was shown that $\{\theta_{jk}(t)\}_{jk}$ form a complete ON basis on $L^2(\mathbf{R})$, therefore $s(t)$ is completely described by s_{jk}. Further, by Parseval's rule, we have

$$\sum_k s_k^2 = \sum_{jk} s_{jk}^2, \tag{6.47}$$

and

$$\sum_k \lambda_k = \sum_{jk} \lambda_{jk}.$$

$\{\theta_{jk}(t)\}_k$ forms a KL basis for $\bar{r}(t,u)$ but not for $r_\eta(t,u)$, and we have

$$r_\eta(t,u) = \bar{r}(t,u) + \sum_{j \neq l} E\left[\eta_j(t)\eta_l(u)\right].$$

The cross-covariance terms can be written as

$$E\left[\eta_j(t)\eta_l(u)\right] = \sum_{km} E\left[\eta_{jk}\eta_{lm}\right]\theta_{jk}(t)\theta_{lm}(u) \quad \text{for } j \neq l.$$

Their contribution to the eigenvalues, and therefore to the expected value of the energy of the noise process is zero. Changing the detection strategy from solving the detection problem of (6.1) to that in (6.45), changes the eigenvalues of the noise process while preserving the expected value of the noise energy. Although detectability is a meaningful measure of goodness, an *a priori* comparison of d^2 and $\bar{d}^2$ is very difficult. It is possible to achieve high values of detectability in either case by a judicious choice of the signal if possible. The pathological singular detection condition (case II in Section 6.1), where the signal has a component along the eigenfunction corresponding to a zero eigenvalue, is possible for d_j^2 or d^2 independently of one another. If the noise process is white so that all the eigenvalues in all the bases are identical and equal to the noise variance, σ^2, then according to (6.47), $d^2 = \bar{d}^2$. Due to its simplicity, we choose an optimal receiver based on the sufficient statistic $\bar{g}$ of (6.46).

The sufficient statistic, $\bar{g}$, and its components, g_j, are given in (6.46) in terms of the coefficients of $x(t)$ and $s(t)$ with respect to the eigenfunctions $\theta_{jk}(t)$. They can be calculated from their wavelet coefficients by noting that $\{\theta_{jk}(t)\}_k$ and $\{\psi_{jk}(t)\}_k$ are ON bases of the same subspace $\mathbf{W}_j$. One can be written in terms of the other using the coefficients defined by their inner product

$$\alpha^j_{km} = \langle \theta_{jk}(t), \psi_{jm}(t) \rangle .$$

We have

$$\theta_{jk}(t) = \sum_m \alpha^j_{km} \psi_{jm}(t), \quad \text{and} \quad \psi_{jm}(t) = \sum_k \alpha^j_{km} \theta_{jk}(t).$$

Orthonormality of each set of functions implies the orthonormality of the coefficient sequences and we have

$$\sum_n \alpha^j_{kn} \alpha^j_{ln} = \delta_{kl} \quad \text{and} \quad \sum_k \alpha^j_{km} \alpha^j_{kn} = \delta_{mn}.$$

The coefficients, x_{jk}, of $\theta_{jk}(t)$, can be written in terms of the wavelet coefficients, d^x_{jk}, of $x(t)$. Signal, $s(t)$, and noise, $\eta(t)$, can be expressed similarly. Thus we have

$$x_{jk} = \sum_m \alpha^j_{km} d^x_{jm}, \; s_{jk} = \sum_m \alpha^j_{km} d^s_{jm}, \; \text{and} \; \eta_{jk} = \sum_m \alpha^j_{km} d^\eta_{jm}. \tag{6.48}$$

Now the sufficient statistic, g_j, as in (6.46), can be expressed in terms of the wavelet coefficients as

$$g_j = \sum_{mn} d^x_{jm} d^s_{jn} \beta^j_{mn} \tag{6.49}$$

where

$$\beta^j_{mn} = \sum_k \frac{\alpha^j_{km} \alpha^j_{kn}}{\lambda_{jk}}.$$

(6.49) can be represented using a symmetric matrix $\mathbf{B}^j$ with elements $[\mathbf{B}^j]_{mn} = \beta^j_{mn}$. With vectors $\mathbf{d}^\bullet_j = [\ldots, d^\bullet_{jm}, d^\bullet_{jm+1}, d^\bullet_{jm+2}, \ldots]^T$, we can write

$$g_j = \mathbf{d}^{xT}_j \mathbf{B}^j \mathbf{d}^s_j. \tag{6.50}$$

We also have

$$\mathbf{B}^j = \mathcal{A}^{jT} \Lambda_j^{-1} \mathcal{A}^j$$

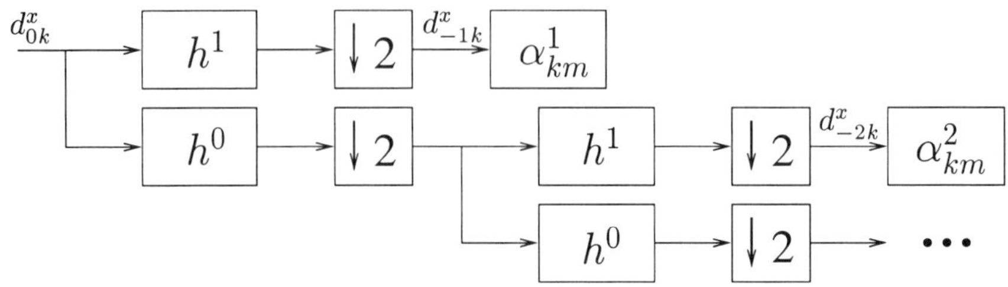

Figure 6.2 Whitening of the noise as accomplished by passing the wavelet coefficients of the input through a time varying discrete time filter.

where $\left[\mathcal{A}^j\right]_{mn} = \alpha^j_{mn}$ and $\Lambda_j^{-1} = \text{diag}(\ldots 1/\lambda_{jm}, 1/\lambda_{jm+1}, 1/\lambda_{jm+2}, \ldots)$. Using the matrix form of the noise coefficients in (6.48), and the vector $\mathbf{n}_j = [\ldots, \eta_{jm}, \eta_{jm+1}, \eta_{jm+2}, \ldots]^T$ we can write the noise covariance matrix relationships:

$$E\left[\mathbf{n}_j \mathbf{n}_j^T\right] = \Lambda_j = \mathcal{A}^j E\left[\mathbf{d}_j^\eta \mathbf{d}_j^{\eta T}\right] \mathcal{A}^{jT}.$$

In light of the above equation, we see that $\mathbf{B}^j = \left[E\left[\mathbf{d}_j^\eta \mathbf{d}_j^{\eta T}\right]\right]^{-1}$. The "whitening" process mentioned in Section 6.1, i.e. (6.10), can be carried out by the linear transformation described by its action on the noise wavelet coefficients by

$$\gamma_j^\eta = \Lambda_j^{-1/2} \mathcal{A}^{jT} \mathbf{d}_j^\eta$$

where γ_j^η is a white noise sequence. Defining the transformed vectors of the known and received signals by

$$\gamma_j^s = \Lambda_j^{-1/2} \mathcal{A}^{jT} \mathbf{d}_j^s \quad \text{and} \quad \gamma_j^x = \Lambda_j^{-1/2} \mathcal{A}^{jT} \mathbf{d}_j^x,$$

we obtain an alternate expression for the sufficient statistic as

$$g_j = \gamma_j^{xT} \gamma_j^s.$$

A realization based on the whitening of the noise process is depicted in Figure 6.2.

The sufficient statistic, g_j, in (6.50) requires the calculation of the wavelet coefficients of the known signal. It is well known that this quantity is affected by the time of origin of the signal relative to the analyzing wavelet. The effect of shift variance has been studied in [13] and [11]. To correct the problem, design of maximally shift invariant wavelets have been proposed [21, 72, 294]. Alternatively, in [12], it is suggested that the unsubsampled frame be searched for the critically sampled ON path which corresponds to the optimal initial phase. This search reduces the amount of leakage of the signal energy to other scales.

6.4 CONCLUSION

The problem of detection of a known signal in Gaussian noise has been projected to wavelet defined multiresolution subspaces. Properties of wavelets to be KL bases for signals over $L^2(\mathbf{R})$ and over each subspace as well as the process of finding the KL bases that decorrelate wavelet coefficients have been discussed. Detection strategies that involve whitening and matched filtering of same scale wavelet coefficients have been proposed. Sufficient statistics and detectability measures have been derived.

7 MPEG AUDIO CODING

James D. Johnston, Schuyler R. Quackenbush, Grant A. Davidson, Karlheinz Brandenburg and Jurgen Herre

AT&T Laboratories – Research
Florham Park, NJ
http://www.research.att.com/info/srq

7.1 INTRODUCTION

7.1.1 The MPEG Committee and Its Standards

The International Standards Organization - Moving Picture Experts Group (ISO-MPEG) standards committee has for many years supported an audio coding group for the purpose of serving its charter, "Video with Associated Audio." For a variety of reasons, much of the international standards effort for audio coding has concentrated there, leading to a wide-ranging set of standards. There are currently three sets of MPEG-Audio coding algorithms, first the MPEG-1 coders, referred to as MPEG-1 Layer 1/2/3, then the 5-channel extensions of the MPEG-1 coders, called the MPEG-2 BC coders, and finally the latest MPEG-Audio effort, MPEG-2 AAC, "Advanced Audio Coding." Approximately a year from press time, another, more flexible and diverse coding standard, MPEG-4 Audio, will appear, incorporating MPEG-2 AAC as well as other voice and low-rate coding algorithms.

In this chapter, we describe the general nature of the MPEG-Audio encoders, briefly discuss some features of the human auditory system, and detail the filter bank and associated processing for each of the MPEG-Audio standards, with particular focus on the newest standard, MPEG-2 AAC. We discuss the needs of a perceptual coder in terms of filter bank design and flexibility, and finally provide some suggestions for further enhancement of perceptual coding via filter bank improvements.

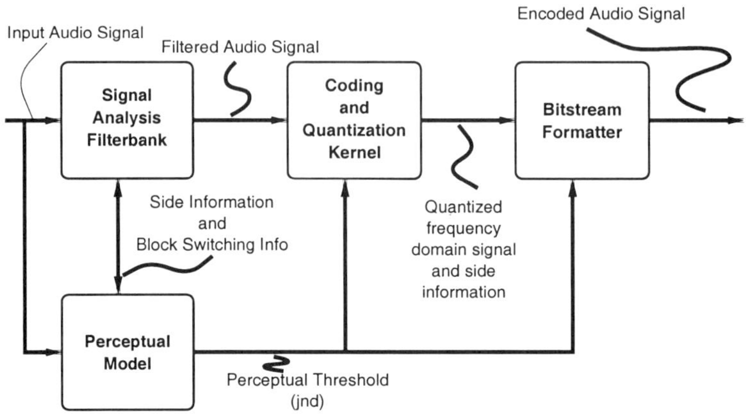

Figure 7.1 Block diagram of a perceptual audio coder.

7.1.2 Perceptual Audio Coding

All of the MPEG audio coders are perceptual coders, i.e. they work by the process of separating and eliminating information from the signal that the human ear cannot detect (coined "irrelevant") as opposed to the usual approach for speech coding, where a source model is used to reduce redundancy in the signal. The more efficient MPEG coders use both a model of perception to extract irrelevancy and filter bank gain to extract redundancy. A block diagram of the basic perceptual coder is shown in Figure 7.1. In the figure, there are four blocks: a filter bank that converts the time domain signal to a frequency-domain form, a perceptual model that calculates a set of noise-injection levels that are inaudible (i.e. quantization step-sizes), a coding kernel that implements the results of the perceptual model on the filter bank signals, and a bitstream formatter, sometimes containing a noiseless coding section, that converts the quantized filter bank values into a form suitable for storage and/or transmission.

A discussion on source coding versus perceptual coding for audio purposes can be found in several places. In particular [161], [159], and [107] provide a great deal of detail on the subject.

7.1.3 The Human Auditory System

The human auditory system (HAS) is described in several books, either as physiology of the auditory system, [377], psychology of the auditory system [225], or as phenomena [94]. A full treatment of interesting parts of the HAS would require several times the size of this book, so we will limit ourselves to a short discussion here.

The Cochlear Filter Mechanism. A number of researchers, as cited above, also [283] and [5], among others, have observed that the cochlea implements a mechanical filter bank with some unusual properties. One such property is that the center frequency of this filter bank varies continuously along the length of the basilar membrane, and that the bandwidth of the filter about that center frequency varies strongly from low to high frequencies. The bandwidth of such a filter at a given point on the cochlea has been dubbed a "critical bandwidth," and a standardized scale of critical bandwidths has been dubbed the "Bark" scale. While [283] and others report fixed points on this scale, the approximation $b = 13\tan^{-1}(0.76f/1000) + 3.5\tan^{-1}((f/7500)^2)$ provides a fairly accurate mapping of frequency, f, to the Bark value, b.

The filters that realize these critical bandwidths vary by about 40:1 in bandwidth, and, as one would expect, their time response also varies by about 1:40 in time, with the longest (time-wise) cochlear filters at low frequencies, where the critical bandwidth is approximately 100 Hz, and the shortest filters in the 15 kHz to 20 kHz range, where the critical bandwidth rises to approximately 4kHz. This variation in filter time response creates one of the most significant difficulties in perceptual coding, in that one must simultaneously control time/frequency artifacts over a 40:1 variation in time/frequency resolution, while simultaneously maintaining good control over coding gain and other source coding issues.

Auditory Masking. The other interesting part of the HAS, from the point of view of the audio coder, is the phenomenon of auditory masking. This refers to the masking, or making inaudible, of one signal by another signal that is proximate in frequency (on the Bark scale). There are several classic results known for auditory masking, from the basic work of [94] to the well known report on tone masking noise, [283] and summaries of noise masking tone [121]. Hellman's work is particularly interesting in that the author contrasts masking ability of different kinds of maskers.

The principle of masking is used extensively in perceptual audio coders, as it allows one to calculate the parts of a signal that are self-masked. It is these parts of the signal that are "irrelevant" and hence can be removed. A summary of masking results appears in [161] and are not be repeated here.

7.2 MPEG 1 CODERS

7.2.1 Basic Block Diagram

The block diagram of MPEG-1 audio [143] follows the basic paradigm of perceptual coding systems, as shown in Figure 7.1. The main blocks are the filter bank, perceptual model, quantization and coding and bitstream formatting. MPEG-1 was devised as a generic audio coding scheme addressing a large number of potential applications. Therefore MPEG-1 audio covers a wide range of bit rates (from 32 kbit/s up to 448 kbit/s) and modes. To enable a trade-off between complexity and quality at a given bit rate, MPEG-1 audio consists of three different coding schemes called Layers. These Layers all follow the same basic paradigm but use different filter bank and quantization tools. The MPEG-1 audio Layers share the definition of the basic bitstream format

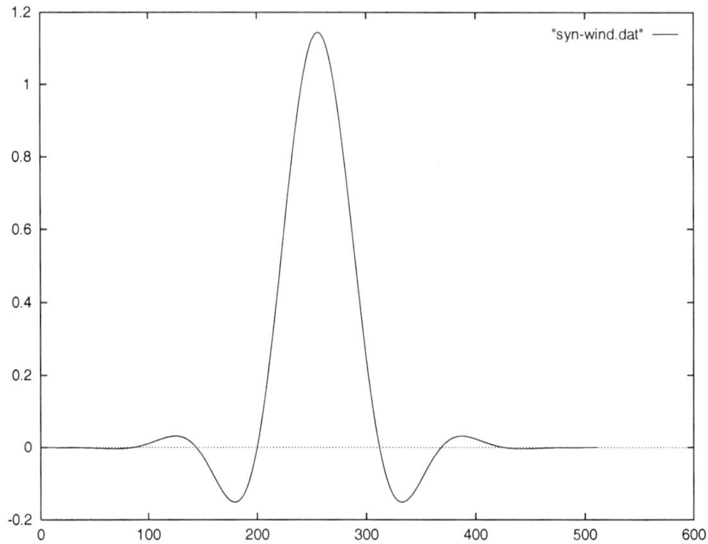

Figure 7.2 Window function of the MPEG-1 polyphase filter bank.

including a 4 byte header with synchronization information and other parameters necessary for successful decoding (like sampling frequency, bit rate, stereo or mono modes etc.). Beyond the basic quantization and coding tools, all three Layers allow for joint stereo coding techniques (intensity stereo for all Layers, combined intensity/broadband mid/side coding for Layer 3). These techniques are explained below in the context of MPEG-2 Advanced Audio Coding.

7.2.2 Layer 1

This coding scheme uses a polyphase filter bank which maps the digital audio input into 32 subbands, a fixed segmentation to format the data into blocks, a psychoacoustic model to determine the adaptive bit allocation, and quantization using block companding and frame coding. The following description follows the lines of the basic block diagram of a perceptual coding system as in Figure 7.1.

Filter Bank. The filter bank is a 32 band polyphase filter bank. This type of filter bank was introduced by [273]. The filter bank is equally spaced. Polyphase filter banks are useful for audio coding because they combine the design flexibility of generalized quadrature mirror filter (QMF) banks with low computational complexity. In MPEG-1 audio, a 511 tap prototype filter is used. Figure 7.2 shows the prototype filter (window function). It has been optimized for a very steep filter response and a stop band attenuation of better than 96 dB. While the impulse response has a length of 10.6 ms at 48 kHz sampling frequency, the form of the prototype filter keeps pre-echo artifacts (see the description in

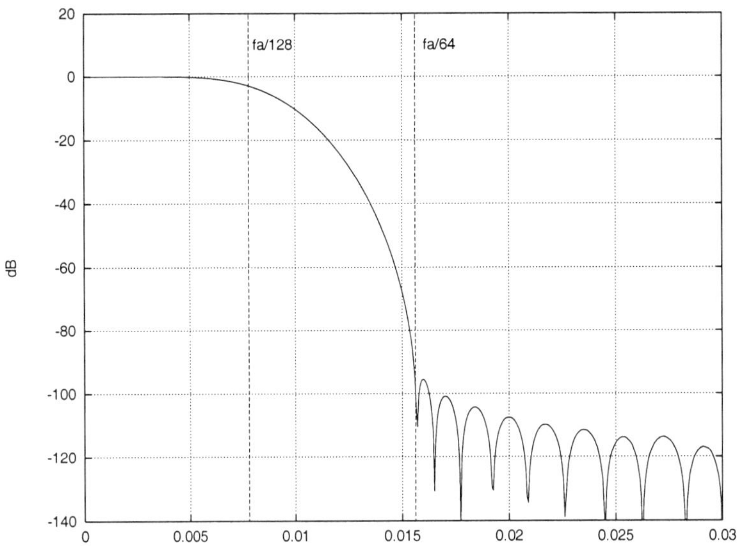

Figure 7.3 Frequency response of the MPEG-1 polyphase filter bank. The sampling frequency is normalized to 2.0, and the horizontal scale to the Nyquist limit of 1.0.

[162]) due to the filter bank itself near the threshold of audibility. Its time resolution is 0.66 ms at 48 kHz sampling frequency.

Figure 7.3 shows the frequency response of the filter bank. In addition to the attenuation requirements, it was designed as a reasonable trade-off between time behavior and frequency localization [71]. The stop band attenuation is more than 96 dB, equivalent to a 16 bit resolution of an input signal. The filter response is down to the stop band attenuation at twice the bandwidth of a single band, i.e. there is negligible aliasing outside the adjacent bands. The filter bank is not of the perfect reconstruction type, but the reconstruction errors in the absence of quantization are on the order of one LSB for 16 bit resolution.

Other Features. The quantization and coding step in Layer 1 uses block companding (block floating point) for groups of 12 subband samples. The basic block length is $32 \times 12 = 384$ samples. The bit allocation is signalled by a four bit field for every subband, which specifies an allocation of between zero and sixteen bits for each subband. For each band with a non-zero bit allocation a six bit scale factor is transmitted which is the exponent of the block floating point quantization. The subband data are transmitted according to the bit allocation field. The main idea behind the separate bit allocation field in the side information is to retain flexibility and the possibility of upgrading the bit allocation algorithm without the need to upgrade decoders.

Different perceptual models can be used for Layer 1 encoding. The standard [143] contains an example in which the perceptual model is based on the output of a 512 line fast Fourier transform (FFT). The perceptual model results in the estimation of signal-to-mask ratios (SMR) for each coder subband which then serve as an input to the bit allocation procedure. The bit allocation works using the subband signals and the SMRs delivered by the perceptual model. A more detailed description on this and other features of MPEG-1 Audio can be found in [27].

7.2.3 Layer 2

Filter Bank. In Layer 2, exactly the same filter bank as in Layer 1 is used.

Other Features. Layer 2 gains additional compression efficiency over Layer 1 by employing more advanced techniques to code scale factors and main information. The basic bitstream frame is three time the size of Layer 1, i.e. 24 ms at 48 kHz sampling frequency (1152 samples in the time domain).

- Side information
  Just as in Layer 1, the scale factors (block floating point exponents) are coded with 6 bit accuracy. With the so-called "scale factor select information," a scale factor can apply to one, two or all three sub-blocks of 12 subband samples within a frame (corresponding to 36 subband samples or 1152 time domain samples). This permits reduction of the number of bits spent on the scale factors. Another difference is in the coding of the bit allocation information. In Layer 2, the bit allocation field can have a length between zero bits (values are always zero) and four bits (16 possible bit allocations). The Layer 2 bit allocation tables contain the possible quantizers per subband (possible bit allocation) for each subband. They are different for different bit rates and sampling frequencies.

- Coding of subbands
  Layer 2 adds radix coding of the quantized values. Rather than an integer number of bits, three quantized values with 3, 5, 7 or 9 quantization steps can be grouped and coded using 5, 7, 9 or 10 bits thus allowing allocation of fractional bits for small quantized values.

Again, examples for perceptual models are given in the standard. A more detailed description can be found in [27].

7.2.4 Layer 3

Layer 3 adds a number of concepts to MPEG audio. It exhibits much more flexibility and yields better compression efficiency. This comes at the cost of higher implementation complexity.

Filter Bank. The Layer 3 filter bank was designed to allow the greatest possible commonality with Layer 1 and 2 and, at the same time, increase coding efficiency. Since the coding gain of a subband coding scheme (or equivalent filter bank) increases with the number of channels, Layer 3 was designed as a high

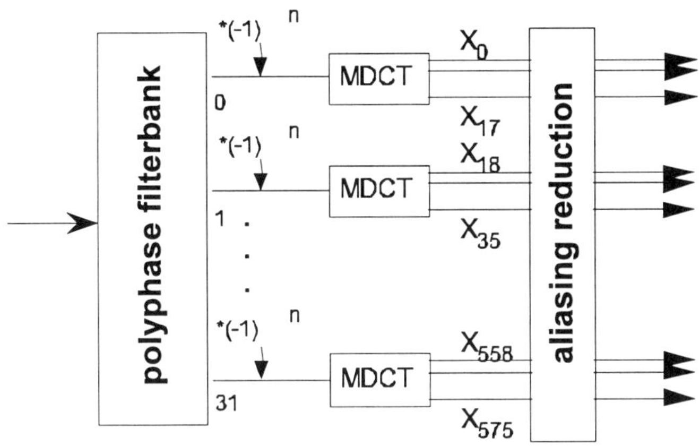

Figure 7.4 Block diagram of the MPEG Layer 3 hybrid filter bank.

frequency resolution subband coding scheme. To achieve both the objectives, commonality and coding gain, Layer 3 comprises a cascaded filter bank scheme (hybrid filter bank). Figure 7.4 shows the block diagram of the analysis filter. The first filter bank is the 32 channel polyphase filter bank from Layers 1 and 2. Each output channel is again subdivided into 18 bands via a modified discrete cosine transform (MDCT) which is a lapped orthogonal transform using time domain aliasing cancellation (TDAC) as proposed in [255]. The total number of frequency lines is 576.

One problem of cascaded filter banks including this hybrid filter bank must be mentioned. Since the frequency selectivity of the complete filter bank can be derived as the product of a single filter with the alias components folded in for all other filters, there are spurious responses (aliasing components) created about the edges of the polyphase subband frequencies. Crosstalk between subbands over a distance of several times the bandwidth of the final channel separation occurs and the overall frequency response shows peaks within the stop bands.

In [81] a solution to this problem was proposed. It is based on the fact that every frequency component of the input signal influences two subbands of the cascaded filter bank, one as a signal component and the other as an aliasing component. Since this influence is symmetric, a compensation can be achieved using a butterfly structure with the appropriate weighting factors. Complete cancellation of the additional alias terms can not be achieved, but an optimization for overall frequency response flatness is possible. The resulting frequency response of the hybrid filter banks shows an improvement in the aliasing side lobes by about 5 to 10 dB.

Adaptive Block Switching. The Layer 3 filter bank allows for dynamic switching of the time-frequency decomposition (filter bank resolution). This is necessary to ensure that the time spread of the filter bank does not exceed

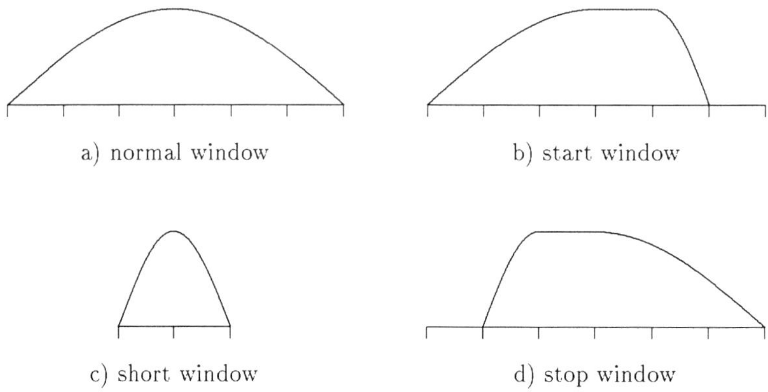

Figure 7.5 Window forms used in Layer 3.

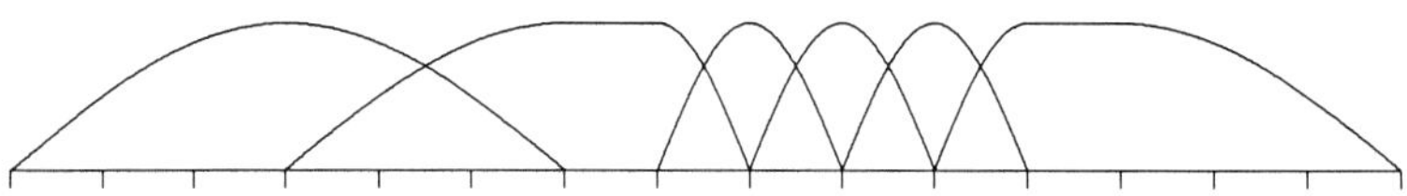

Figure 7.6 Example sequence of window forms.

the premasking period, thus avoiding the pre-echos. The technique is based on the fact that alias terms which are caused by subsampling in the frequency domain of the MDCT are constrained to either half of the window. Adaptive window switching as used in Layer 3 is based on [80]. Due to the nature of the hybrid filter bank, which does not allow for "optimum" time-frequency resolution, the amount of pre-echo protection achieved can not match that of a single filter bank. Window switching, however, can achieve reasonable performance under most operating conditions. Figure 7.5 shows the different windows used in Layer 3. Figure 7.6 shows a typical sequence of window types if adaptive window switching is used. The function of the different window types is explained as follows:

- Long window
  The normal window type used for stationary signals.

- Short window
  The short window has the same form as the long window, but with 1/3 of the window length. It is followed by an MDCT of 1/3 length. The time resolution is enhanced to 4 ms at 48 kHz sampling frequency. The combined frequency resolution of the hybrid filter bank in the case of short windows is 192 lines compared to 576 lines for the normal windows used in Layer 3.

- Start window
  In order to switch between the long and the short window type, this hybrid window is used. The left half has the same form as the left half

of the long window type. The right half has the value one for 1/3 of the length and the shape of the right half of a short window for 1/3 of the length. The remaining 1/3 of the window is zero. Thus, alias cancellation can be obtained for the part which overlaps the short window.

- Stop window
  This window type enables the switching from short windows back to normal windows. It is the time reverse of the start window.

Other Features. As in the case of Layer 1 and 2, the perceptual model in MPEG-1 Layer 3 is a feature of the encoder. An example is given in the standard text [143] and described in [27].

The quantization and coding structure of Layer 3 is quite different from the one used in Layers 1 and 2. Layer 3 employs non-uniform quantization, variable rate (Huffman) noiseless coding and an analysis-by-synthesis iteration loop structure in the encoder. Unlike traditional coders, no bit direct allocation is employed, rather an indirect scheme, often called noise allocation, is used. Two nested iteration loops do the encoding:

- Rate Loop
  This inner iteration loop does the quantization and coding of the spectral values. The data are quantized and coded using one (for each segment of data) of a number of fixed Huffman code tables. After this is done, the bits are counted. If the number of bits is larger than the number of available bits, the quantization step size is increased. This leads to smaller quantized values and fewer bits. This process is repeated iteratively until the required bit rate is met.

- Distortion Control Loop
  The outer iteration loop is designed to keep the quantization noise below the masking threshold as determined by the perceptual model. The nonuniform quantizer is designed to do some noise coloration (larger values exhibit larger quantization noise), and the means for controlling the quantization noise are the scale factors, i.e. quantizer step size. If the injected noise is too large in any scale factor band, the step size is reduced. The process is repeated until the quantization noise is below the masking threshold (allowed noise) for each scale factor band. A more detailed description of this technique can be found in the example encoder in the international standard [143].

7.2.5 Filter Bank Implications

In the MPEG-1 systems, there are two filter banks, a 32 band filter bank and a 576 band filter bank. In the case of the 32 band filter bank, the system is primarily perceptual, i.e. the filter bank does not provide much extraction of redundancy, or at least enough where coding gain becomes an issue along with perceptual concerns. With the 576 band filter bank, the coder offers substantial source coding gain and enables greater use of perception in the encoding process. This is primarily due to greater source coding gain and better frequency resolution at low frequencies. Although more complicated rate loop and block length switching are required to meet the demanding conditions, the coding gain is enhanced by the extraction of additional redundancy. As a result, the Layer 3 coder is more efficient.

Unfortunately, due to the hybrid nature of the filter bank, the Layer 3 filter bank has several properties that reduce the total coding gain. As mentioned above, the aliasing from the initial filter bank is reduced, but not eliminated. Furthermore, the length of the combined filter bank is longer than a single-stage filter bank, and this leads to difficulties in controlling pre-echo situations. This leads to the Layer 3 coder showing an increased coding gain more typical of a 350-450 tap uniform filter bank, which is still a substantial advantage compared to the Layer 1/2 32 band filter bank.

7.3 MPEG 2 BACKWARDS COMPATIBLE CODING

At the start of MPEG-2 deliberations, it was decided that the two channel MPEG-1 standard coders should be extended to multichannel coding. In particular, "backwards compatibility" (BC) was required, where BC was defined as the ability of the MPEG-1 coder to parse and reproduce an acceptable two channel sound stream from the MPEG-2 bitstream. While this was an admirable goal, it required the use of the MPEG-1 syntax and filter banks, as well as the use of a "mixdown" channel for the BC channels. This "mixdown channel" was required to carry the full 2-channel mix created from a 5-channel source. With those requirements, MPEG-2 BC Audio was created to carry other formats, specifically the 5 or 5.1 channel formats known in home theater. As is shown later, the cost of backwards compatibility is substantial.

A would-be successor to the traditional two-channel sound format is the 3/2+1 multi-channel system, sometimes called the 5.1 channel system. The MPEG-2 BC audio coding standard (International Standard 13818-3 [145]) contains the description of a backward compatible low bit rate multi-channel encoding system. The MPEG-1 L and R channels are replaced by the matrixed signals L_C and R_C and encoded with an MPEG-1 encoder, and therefore an MPEG-1 decoder can reproduce a comprehensive mixdown of the full 5 channel information, if it is able to parse the "extension part" of the MPEG-1 bitstream. Denoting the left, right, center, left surround and right surround channels of the multichannel signal as L, R, C, L_S and R_S, respectively, the following equations describe the mixdown of five to two channels:

$$L_C = \frac{1}{1+1/\sqrt{2}+a}[L + C/\sqrt{2} + aL_S]$$
$$R_C = \frac{1}{1+1/\sqrt{2}+a}[R + C/\sqrt{2} + aR_S].$$

L_C and R_C are the compatible left and right channels generated from the five channel signal. The matrix-mixdown coefficient a is usually selected to be one of $1/\sqrt{2}$, $1/2$, $1/(2\sqrt{2})$, or 0. The basic frame format is identical to the MPEG-

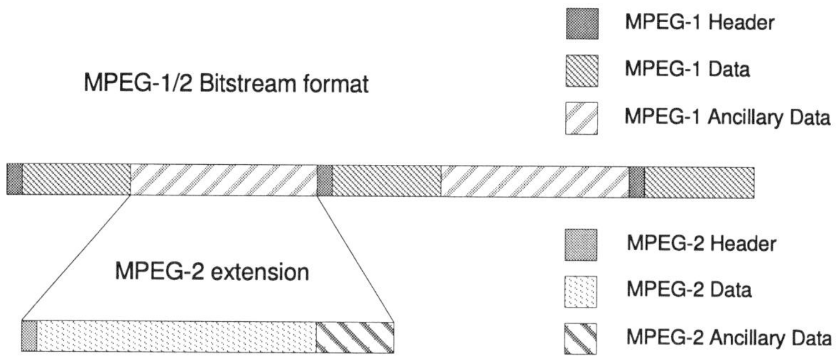

Figure 7.7 Transmission of MPEG-2 multichannel information within an MPEG-1 bitstream.

1 bitstream. Channels C, L_S, and R_S are transmitted in the MPEG-1 ancillary data field as shown in Figure 7.7.

As in the case of MPEG-1, there are three versions of the multichannel (MC) extension called Layer 1, Layer 2 and Layer 3. Layer 1 and Layer 2 BC MC extensions use a bitstream syntax similar to Layer 2. As in the case of MPEG-1, Layer 3 is the most flexible system. In MPEG-2, Layer 3 allows a flexible number of extension channels rather than enforcing exactly three. While the original idea behind this was to alleviate the dematrixing artifacts for some worst case material, this feature can also be used to do simulcast of two-channel stereo and 5-channel extension without the artistic restrictions of a fixed compatibility matrix.

7.3.1 Layer 1 and Layer 2

The multichannel extensions of Layer 1 and Layer 2 both use Layer 2 coding in the extension part. Generalized joint stereo coding (*dynamic crosstalk*) and, optionally, prediction between different channels can be used to increase coding efficiency.

7.3.2 Layer 3

The extension part for Layer 3 multichannel coding uses a structure similar to MPEG-1 Layer 3 to code the spectral values. The number of additional channels within the multichannel extension is flexible. This allows for simulcast (the transmission of a full five channel signal in the extension) as a special case of Layer 3 multichannel coding.

7.3.3 The Problem with "Backwards Compatibility"

While backwards compatibility as implemented in the first version of MPEG-2 audio appears to be a very convenient feature, it introduces disadvantages. The following have been noted:

- Cross-feed of quantization noise
  The dematrixing procedure in the decoder can produce the effect that most of the sound in a channel is canceled, but the corresponding quantization noise is not canceled. This leads to dematrixing artifacts wherein the quantization noise generated by coding of some channels can become audible in the dematrixed channels.

- Inadequate sound quality for the stereo or multichannel sound
  The optimum mix of a multichannel signal and the corresponding mix of a two channel stereo signal are quite different in most cases. If automatic mixdown is employed, either the two channel or the five channel signal will exhibit a sound stage which may be far inferior to an optimum mix.

It is these limitations of the BC model that resulted in the establishment of the MPEG-2 NBC (Non-Backwards Compatible), now known as MPEG-2 AAC, coding effort. In particular, for signals with time-domain audio imaging information, it is possible to show that the resolution in the "matrixed mode" must approach the original signal's dynamic range around sudden transitions if noise imaging problems are to be mitigated, let alone eliminated.

While there are ways around these problems, they must translate into an increased bit rate demand for very high quality multi-channel sound. Before we describe this next generation MPEG audio standard, here are some ways to alleviate the problems of matrixing:

- Matrix artifacts
  Predistortion techniques have been proposed to get rid of the cross-feed of quantization noise. It can be shown that it is possible to precalculate the quantization noise and subtract it in the encoder so that it is reduced in the dematrixing step. The requirements for this are tight control of the quantization step size and substantial additional requirements from a carefully calculated, time aware perceptual model.

- Inadequate sound stage
  The only way to render the desired sound stage and dialog level for both 2 and 5-channel presentation is to simulcast the 2 and 5-channel program mixes. With newer, more efficient coding systems like MPEG-2 AAC becoming available, this becomes reasonably efficient. In the case of simulcast, MPEG-2 BC multichannel cannot provide efficiency approaching that of the MPEG-2 AAC standard.

However, all these techniques require additional bits to be transmitted, and do not eliminate the inefficiencies associated with "backwards compatibility."

7.3.4 Filter Bank Implications

Since the filter banks are the same in MPEG-1 and MPEG-2 BC, the filter bank implications are essentially unchanged. The additional difficulties of MPEG-2 BC are related mostly to the requirement of backwards compatibility, where the

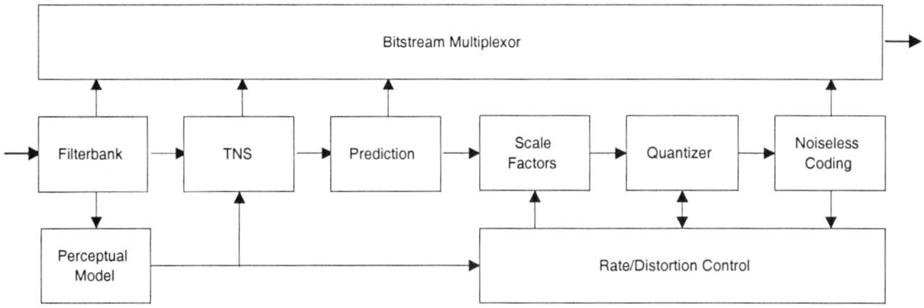

Figure 7.8 AAC encoder block diagram.

cross-feed and mixdown problems are effectively impossible to conquer in the case of most signals. Despite these handicaps and its limited ability to render good audio quality at a low bit rate, MPEG-2 BC has been chosen for several hardware applications, including some high-definition television (HDTV) and digital versatile disk (DVD) applications. The performance of BC versus AAC coders are discussed below.

7.4 MPEG-2 AAC - ADVANCED AUDIO CODING

The ISO/IEC MPEG-2 AAC (Advanced Audio Coding) technology delivers unsurpassed audio quality at rates at or below 64 kbps/channel. It has a very flexible bitstream syntax that supports multiple audio channels, subwoofer channels , embedded data channels and multiple programs consisting of multiple audio, subwoofer, and embedded data channels. AAC combines the coding efficiencies of a high resolution filter bank, backward-adaptive prediction, joint channel coding, and Huffman coding with a flexible coding architecture to permit application-specific functionality while still delivering excellent signal compression.

AAC supports a wide range of sampling frequencies (from 8 kHz to 96 kHz) and a wide range of bit rates. This permits it to support applications ranging from professional or home theater sound systems through internet music broadcast systems to low (speech) rate speech and music preview systems.

A block diagram of the AAC encoder is shown in Figure 7.8. The blocks are:

- Filter Bank
  AAC uses a resolution-switching filter bank which can switch between a high frequency resolution mode of 1024 (for maximum statistical gain during intervals of signal stationarity), and a high time resolution mode of 128 bands (for maximum time-domain coding error control during intervals of signal non-stationarity).

- Temporal Noise Shaping
  The temporal noise shaping (TNS) tool modifies the filter bank charac-

teristics so that the combination of the two tools is better able to adapt to the time/frequency characteristics of the input signal.

- Perceptual Model
  A model of the human auditory system that sets the quantization noise levels based on the loudness characteristics of the input signal.

- MS, Intensity and Coupling
  These two blocks actually comprise three tools, all of which seek to protect the stereo or multichannel signal from noise imaging, while achieving coding gain based on correlation between two or more channels of the input signal.

- Prediction
  A backward adaptive recursive prediction that removes additional redundancy from individual filter-bank outputs.

- Scale Factors
  Scale factors set the effective step sizes for the non-uniform quantizers.

- Quantization, Noiseless Coding
  These two tools work together. The first quantizes the spectral components and the second applies Huffman coding to vectors of quantized coefficients in order to extract additional redundancy from the non-uniform probability of the quantizer output levels. In any perceptual encoder, it is very difficult to control the noise level accurately, while at the same time achieving an "optimum quantizer." It is, however, quite efficient to allow the quantizer to operate unconstrained, and to then remove the redundancy in the PDF of the quantizer outputs through the use of entropy coding.

- Rate/Distortion Control
  This tool adjusts the scale factors such that more (or less) noise is permitted in the quantized representation of the signal which, in turn, requires fewer (or more) bits. Using this mechanism the rate/distortion control tool can adjust the number of bits used to code each audio frame and hence adjust the overall bit rate of the coder.

- Bitstream Multiplexor
  The multiplexor assembles the various tokens to form a bitstream.

7.4.1 Profiles

There are three profiles in the AAC standard: Main, Low Complexity (LC) and Scalable Sampling Rate (SSR). Each profile specifies a different set of coding tools and a different set of parameters for those tools.

The Main Profile provides the greatest compression at a given quality, and is appropriate when processing power and storage are not of significant concern.

With the exception of the SSR Profile's gain control tool, all tools may be used in order to provide the best data compression possible.

The LC Profile provides nearly the same compression as the Main Profile for all but the most demanding signals, but requires significantly less processing power and storage than the Main Profile. The features in this profile are a subset of those in the Main Profile such that any LC Profile bitstream can be decoded by a Main Profile decoder. The most significant reduction in complexity comes from restricting the LC Profile to not use the prediction tool.

The SSR Profile is significantly different from the other two profiles in that it does not inter-operate with them, but rather is in itself a set of four interoperable specifications that permit bandwidth scalability and hence sampling rate scalability. This is appropriate for applications that require a family of AAC decoders, each of which achieves a different price/performance trade-off. Specifically, SSR Profile is similar to LC Profile but with a modified time to frequency mapping tool that splits the signal into four equal bands and applies gain control to each band prior to further processing. The bitstream has a matching embedded structure such that a full-bandwidth SSR bitstream can be decoded by a 1-band, 2-band, 3-band or 4-band SSR decoder to yield 6 kHz, 12 kHz, 18 kHz or 24 kHz bandwidth output signals, respectively. As the number of bands supported by the decoder decreases, its complexity (and cost) decreases.

The three profiles have limited interoperability. A Main Profile decoder can decode both Main Profile and LC Profile bitstreams. A LC decoder can decode only LC bitstreams. As mentioned in the previous paragraph, the SSR Profile is not interoperable with either the Main or LC Profiles, but rather has a scalable interoperability within its own profile. The SSR Profile can do a limited decoding of the LC Profile for the first (lowest) polyphase quadrature-mirror filter (PQMF) bank.

7.4.2 Analysis/Synthesis Filter Bank

From the beginnings of the MPEG-2 AAC coder in 1995, the objective was to significantly exceed the multichannel coding performance of the MPEG-2 backwards-compatible Layer II audio coder. The frequency and time resolution of the filter bank in an audio coder are critical to determining the ultimate achievable coding performance. Of significant importance as well are the properties of critical sampling and overlap-add reconstruction. This section discusses these and other design issues, performance characteristics, and implementation efficiencies of the filter bank employed in the AAC multichannel audio coding system.

The objective of significantly improving upon the performance of MPEG-2 Layers 2 and 3 suggested the use of a filter bank with more capability than both the 32-band uniform polyphase filter bank of Layers I and II and the 576-band hybrid filter bank of Layer III. The design goals outlined in this section ultimately resulted in a coder satisfying the desired subjective quality performance [146].

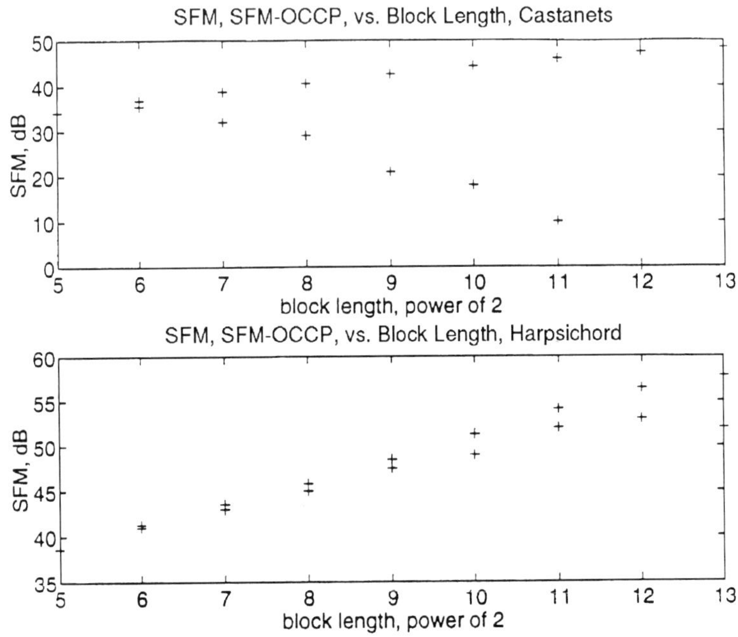

Figure 7.9 Two examples: theoretical and actual gain versus filter bank length.

Since AAC was targeted for use at very low bit rates, a primary design goal of the AAC filter bank was to provide high coding gain. Furthermore, since overall assessments of audio coders are frequently determined by performance on the worst-case (not average) audio test sequence, the filter bank should provide high coding gain across a wide variety of audio signals. Our approach to this problem was to employ a filter bank with adaptive time-frequency resolution (block switching) and adaptive window shape (window switching).

Of the many considerations involved in filter bank design, two of the most important for achieving high coding gain are the impulse response length and the window shape. These key filter bank parameters determine the achievable time and frequency resolution. For transform coders, the impulse response length is directly related to the window length.

A long window length is most suitable for input signals whose spectrum remains stationary, or varies slowly in time relative to the window length. Since frequency resolution increases in proportion to window length, and coding gain most often follows frequency resolution, long window lengths will provide greater coding gain for most signals. On the other hand, a shorter window length, which has better time resolution, is more effective for coding non-stationary or pitchy signals, due to the ear's ability to resolve the time-structure of the signal, regardless of coding gain.

A good illustration of these two effects is Figure 7.9, which shows two curves for each of two signals. The upper curve in each case is the abstract "coding gain" that can be achieved with a filter bank of uniform frequency resolution of 2^n where n is shown on the x-ordinate. The coding gain is specified by the spectral flatness measure (SFM) [157]. The lower curve is the amount of coding gain that can actually be realized, given the perceptual constraints. The upper panel in the figure is for the "castanets" signal, a very impulsive signal and hence prone to pre-echo distortion. The lower curve falls off rapidly in this case. The lower panel is for the "harpsichord" signal, which, although it has several abrupt attacks, also has a very rich harmonic structure, and is amenable to the predictive/rate gain of a high-resolution filter bank. For this signal, the lower curve continues to rise with the resolution until between 4096 and 8192 taps in the filter bank frequency response. The two graphs make it very clear that there is no one ideal length for a filter bank, in particular the optimum for the two signals in the figure are 64 taps and 4096 taps, hence the choice of two window lengths, and a switching mechanism.

In principle, a variable-resolution filter bank can be constructed by tiling the time-frequency plane in any arbitrary fashion. The height (bandwidth) and width (time duration) of the rectangles can be controlled depending upon signal or perception characteristics. One criterion for time-frequency tiling is to layout each tile to encompass a region where the psychoacoustic masking threshold must be constant, i.e. by using a filter bank with roughly the same time-frequency tiling as the auditory system. This variable time-frequency characteristic (essentially implying a critical band filter bank) is similar to that known to occur in human hearing; however, it is not necessarily optimal for coding, unless only the irrelevancy of the signal is to be extracted [254]. In the case where a joint gain in irrelevancy and redundancy is desired, a more complex, adaptive filter bank is necessary.

For many audio signals, we can achieve a higher coding gain than the nonuniform time-frequency tiling scheme, and with lower computational complexity, by employing a dual-resolution uniform filter bank (block switching). In this approach, the encoder filter bank assumes one of two resolution states, each uniform in frequency. As we will see in a later section, the two states in AAC differ in resolution by a ratio of 8:1.

By an argument similar to the case for block switching, above, we can illustrate that there is no single window shape which is optimal for all input signals. The closely-space frequency lines present in audio signals with dense harmonic structure are better resolved using a window with a narrow main-lobe. Conversely, windows with higher ultimate rejection (at the expense of a relatively wide main lobe) typically provide higher coding gains for signals with a sparse harmonic structure. We conclude that, other parameters being equal, a filter bank with variable window shape provides higher coding gain than one employing a fixed window.

Another consideration is minimizing inter-band leakage. In general, the separation between frequency bands, measured as the leakage from a tonal component occurring in one band into the adjacent bands, is not perfect in any filter bank. The degree of incomplete separation has a direct consequence on the number of bits required to transparently encode the audio signal. As separation increases, fewer bits are required. The high leakage which typically occurs in hybrid filter banks compared to single filter banks of identical order (and somewhat lower complexity) makes them less attractive from a coding perspective. However, hybrid schemes are useful in the design of scalable coding architectures, such as the AAC SSR Profile.

Finally, we summarize three filter bank characteristics which are commonly employed in audio coding, and are relevant to AAC as well: critical sampling, overlap/add synthesis, and perfect reconstruction. In a critically-sampled filter bank, the number of time samples input to the analysis filter per second equals the number of frequency coefficients generated per second. Critically-sampled filter banks minimize the number of frequency coefficients which must be quantized and transmitted in the bitstream. discrete Fourier transform (DFT) and discrete cosine transform (DCT) based filter banks are not critically-sampled unless the overlap between adjacent input blocks is zero.

This brings us to the next characteristic, overlap/add synthesis. Continuously time-overlapped filter banks are commonly employed in speech, audio and, more recently as spatial overlap, in image coding. The overlap feature is very useful for reducing artifacts caused by block-to-block variations in signal quantization. In a "transform" coder, in which the blocks do not overlap, the frequency meaning of any given bin in the transform is uncertain unless analyzed strictly on the block boundaries. This means that inter-frequency leakage is quite high because of the discontinuities at block boundaries. Using a transform, it is possible to overlap/add, but this adds a penalty of some extra samples in the transform domain due to the overlap. The use of an orthonormal filter bank, such as the MDCT, allows the analysis/synthesis to remain 1:1 and onto while at the same time it remains critically sampled, i.e. there are exactly the same number of samples in both the time and filtered domains. The filter bank overlap problem is particularly acute for stationary input signals, since to a large degree human perception mechanisms detect changes in signal features. Block artifacts are readily perceived since they add time-varying features to time-invariant input signals, or in the case of video, provide discernible and recognizable shapes that the eye can extract.

A final desired feature of the AAC filter bank is perfect reconstruction (PR). PR implies that in the absence of frequency coefficient quantization, the synthesis filter output will be identical, within numerical error, to the analysis filter input. Near perfect-reconstruction filter banks can also be useful in coding applications, but only if aliasing or frequency-shaping errors added by the filters are sufficiently small.

7.4.3 Transform (Filter Bank)

The filter bank adopted for use in AAC is a modulated, overlapped filter bank — the MDCT. Despite its name, the MDCT is in fact a filter bank, and not a transform, because it does not, and can not, locally meet the requirements of Parseval's theorem. Principles of operation of the non-switched filter bank are described in [255], with the block switching extensions from [80] and the window switching extensions from [26]. A good general discussion of such types of filter banks can be found in [207]. In the next section, we define the MDCT and inverse MDCT used in the AAC Main and LC Profiles. The AAC SSR Profile employs a hybrid filter bank consisting of the MDCT in cascade with a polyphase quadrature filter (PQF).

Definitions. The AAC Main and LC Profiles employ only the MDCT filter bank. In the encoder, a block of N audio samples are modulated by one of two window functions prior to time-to-frequency conversion. Each block of input samples is overlapped by 50% with the preceding block. The window length N can be dynamically selected as 2048 or 256 samples depending on signal characteristics within the block, and the shift length follows as half of the window length. The window lengths and shapes for adjoining blocks are selected in a manner which preserves both critical sampling and perfect reconstruction during switching intervals.

The analytical definition of the forward MDCT (analysis filter bank) is

$$X_{ik} = 2\sum_{n=0}^{N-1} z_{in} \cos\left(\frac{2\pi}{N}(n+n_0)(k+\frac{1}{2})\right), \quad 0 \leq k < N/2 \tag{7.1}$$

where z_{in} is the windowed input sequence, n, k and i are the the sample index, spectral coefficient index and block index, respectively, N is the length of one transform window (256 or 2048 samples) and $n_0 = (N/2+1)/2$. The transform sequence X_{ik} is periodic with period N. However, since the sequence is also odd-symmetric, the coefficients from 0 to $N/2-1$ uniquely specify X_{ik}. Hence $N/2$ unique nonzero transform coefficients are generated for each new block of N samples.

The analytical expression of the inverse MDCT (synthesis filter bank) is

$$x_{in} = \frac{2}{N}\sum_{k=0}^{\frac{N}{2}-1} X_{ik} \cos\left(\frac{2\pi}{N}(n+n_0)(k+\frac{1}{2})\right), \quad 0 \leq n < N. \tag{7.2}$$

Following the inverse transform in (7.2), the N-length output sample block x_{in} is windowed with a synthesis window, and then overlap-added with the preceding windowed sample block $x_{(i-1)n}$. The overlap factor in the synthesis filter is identical to that in the analysis filter (50%).

The factor of two undersampling inherent in the forward MDCT transform leads to time-aliasing in the inverse transform output sequence, x_{in}, in much the way that the aliasing quantization in the QMF filter bank in [85] works as the frequency-domain dual. However, in the absence of transform coefficient quantization, alias terms from the overlapping segments of adjoining blocks will exactly cancel during window-overlap-add synthesis.

To achieve perfect reconstruction with a unity-gain MDCT transform, the shape of the analysis and synthesis windows must satisfy two design constraints. First, the analysis/synthesis windows for two overlapping transform blocks must be related by

$$a_i(n+N/2)s_i(n+N/2) + a_{i+1}(n)s_{i+1}(n) = 1, \quad 0 \leq n < N/2 \tag{7.3}$$

where n, N, and i are defined as above, $a(n)$ is the analysis window, and $s(n)$ is the synthesis window. This is the well-known condition that the analysis/synthesis window products must add so that the result is flat [255]. The second design constraint is

$$a_i(N/2-n-1)s_i(n) - a_i(n)s_i(N/2-n-1) = 0, \quad 0 \le n < N/2. \qquad (7.4)$$

This constraint must be satisfied for the time-domain aliasing introduced by the analysis filters to be canceled during synthesis.

Note that (7.3) and (7.4) do not preclude the use of asymmetric windows. (7.4) is satisfied in AAC by choosing the analysis and synthesis windows to be equal within each block (although they vary from block-to-block). A sufficient condition for (7.3) is then to ensure that window samples in the second-half of one block are a mirror-image reflection of those in the first-half of the succeeding block. These observations are used to advantage in the window switching implementation.

In summary, we see that the AAC filter bank simultaneously satisfies the properties of critical sampling, perfect reconstruction and overlap. A fundamental advantage of this approach is that 50% frame overlap is achieved without increasing the required bit rate. Any non zero overlap used with conventional transforms (such as the DFT or standard DCT) precludes critical sampling, resulting in a higher bit rate for the same level of subjective quality.

The AAC SSR Profile employs a hybrid filter bank similar to the one described in [28]. The encoder input signal is first subdivided by a PQF into four uniform-width frequency bands. The impulse response length of each band is 96, providing a high degree of separation between adjacent bands. The coefficients of each PQF band are

$$h_i(n) = 4\cos\left(\frac{(2i+1)(2n+5)\pi}{16}\right) Q(n),$$

with $0 \le n < 96$ and $0 \le i < 4$, where $Q(n), 0 \le n < 48$, are real-valued prototype coefficients from a table [148] and

$$Q(n) = Q(95-n), \quad 48 \le n < 96.$$

Each of the four bands is followed by a 256/32 switchable MDCT of the type defined in (7.1) and (7.2). The decoder contains a mirror-symmetric inverse PQF (IPQF). In the absence of coefficient quantization, aliasing components introduced into adjacent PQF bands by the PQF in the encoder are canceled by IPQF, resulting in a near perfect reconstruction structure.

The hybrid structure provides a means for decoders of both high and low computational capability to decode the same audio bitstream. Lower-cost decoders simply ignore coded audio information in one or more of the higher PQF bands, with a commensurate reduction in audio bandwidth.

7.4.4 Transform Length Selection

The selection of the transform lengths in AAC was based in part on statistics of audio signals, and in part on experiments that evaluate the performance of one

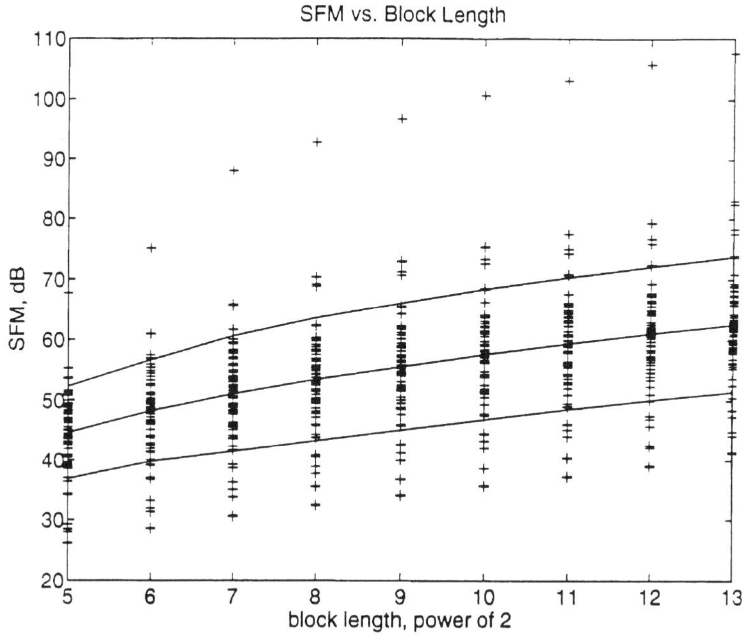

Figure 7.10 Transform gain versus transform length.

block length versus another. The statistics are presented in Figure 7.10, where the coding gain of a large number of signals has been calculated. The figure plots many data at each block length, with the vertical axis showing coding gain in dB, as measured by the SFM, and the horizontal axis showing $\log_2$ of the shift length. In this data, as in other figures in this section, the influence of zero and near-zero blocks is removed. The lines are the mean, and $\pm$ sigma of the data at each block length. The figure, from [161], shows that the average coding gain is still going up at a shift of 1024, and is growing slowly even beyond that point. However, as shown in Figure 7.11, which plots the average non-stationarity of the signal spectrum (i.e. the absolute dB difference in level between successive blocks for a given frequency band), the non-stationarity of the signal is starting to increase substantially at a shift of 1024. The amount of coding gain that can be used is approximately equal to the difference between the coding gain and the signal non-stationarity. This leads to a broad peak in coding gain between 700 and 1200 or so in shift length. It is this set of statistics that was used in order to determine the trial values for the AAC coder. Subsequently, various shift lengths and number of frequency bands were tested in experiments, and the shift length of 1024 for "long blocks" was selected.

As we have shown previously in Figure 7.9, a transform length of 64 or 128 shift points is necessary for transient signals. In order to avoid the problems that come about due to highly unsymmetric start/stop blocks, we chose the 128 shift length for transient signals.

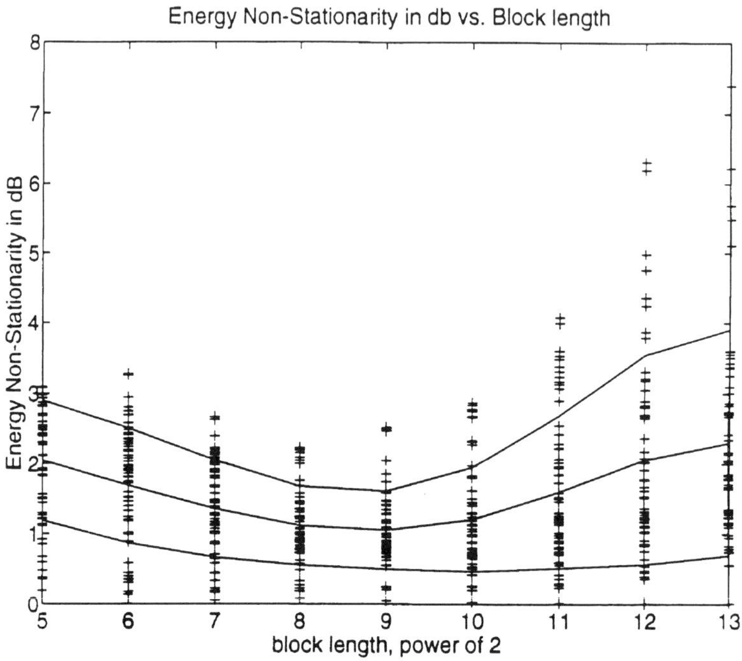

Figure 7.11 Nonstationarity in frequency versus filter bank length.

7.4.5 Block Switching

The use of 2048 time-domain samples transform improves coding efficiency for signals with complex spectra, but creates problems for transient signals. Quantization errors preceding a transient by more than a millisecond or so are not masked by the transient itself. This leads to a phenomenon called pre-echo in which quantization error from one transform block is spread in time and becomes audible. Long transforms are inefficient for coding signals which are transient in nature. Transient signals are best encoded with relatively short transform lengths. Unfortunately, short transforms produce inefficient coding of steady-state signals due to poorer frequency resolution [63, 161].

AAC circumvents this problem by allowing the block length of the transform to vary as a function of the signal conditions as in [80]. Signals that are short-term stationary are best accommodated by the long transform, while transient signals are generally reproduced more accurately by short transforms.

Transitions between long and short transforms are seamless in the sense that the overlap/add properties (OLA) of the synthesis filter bank are retained during block switch events. Furthermore, aliasing is completely canceled in the absence of transform coefficient quantization. A complete switching event consists of one or more short-block frames bracketed on the front end by an asymmetric start window of length $N = 2048$ and on the back end by a stop window (time-reversed version of the start window). The short-block frames

Table 7.1 Allowed window sequence transitions.

From	To
ONLY_LONG	ONLY_LONG or LONG_START
LONG_START	LONG_STOP or EIGHT_SHORT
LONG_STOP	ONLY_LONG or LONG_START
EIGHT_SHORT	EIGHT_SHORT or LONG_STOP

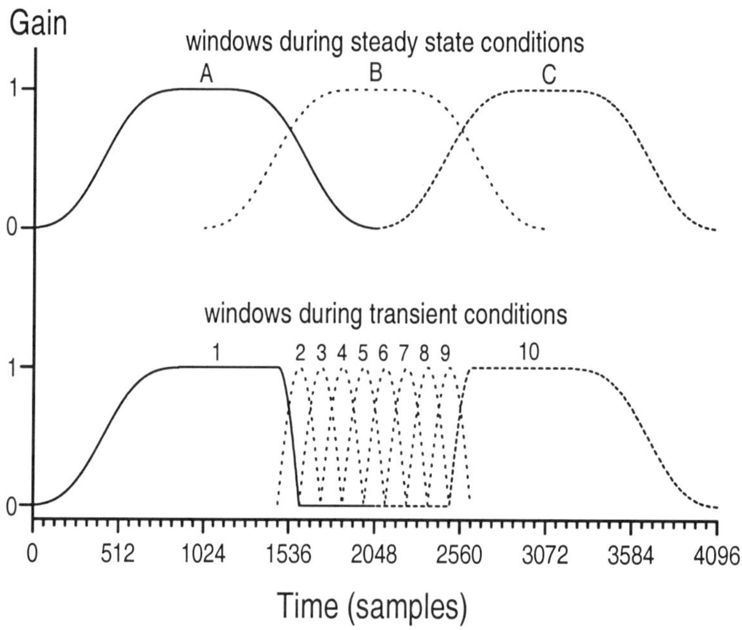

Figure 7.12 Example of block switching during stationary and transient signal conditions.

consists of eight windows of size 256; therefore, every frame processed in the encoder contains N/2 = 1024 transform coefficients prior to quantization and formatting. A two-bit per frame bitstream code is used to signal the window sequence type to the decoder. The four AAC window sequence types, and the allowed state transitions between them, are presented in Table 7.1.

Figure 7.12 displays the window-overlap-add process for both stationary and transient input signals. The top window sequence represents the case when block switching is not employed and all windows have a length of 2048 samples. The bottom window sequence depicts how block switching transitions to and from the shorter $N = 256$ time sample transforms. For transients which are closely separated in time, a single EIGHT_SHORT window sequence can be extended with another sequence of the same type.

7.4.6 *Window Switching*

As we have discussed earlier, window shape influences achievable coding gain, and there is no single window length which is optimal for all signals. For these reasons, the AAC filter bank has provision for seamless switching between two different window shapes. Perfect reconstruction, critical sampling, and overlap are preserved during window shape adaptation. The shape of the window is selected by the encoder and conveyed to the decoder via a single unity-valued bit per block code. This mechanism allows the filter bank to efficiently discriminate and code spectral components over a wider variety of input signals.

The sine window is a natural and popular choice for use in MDCT filter banks because it has a rather narrow main-lobe and satisfies the overlap-add constraint in (7.3). The sine window has the additional benefit that DC offsets present in the input signal block will be localized to the first transform coefficient only. Other windows will spread DC offsets across more than one coefficient, sometimes leading to increased bit rate or coding noise.

Alternatively, frequency selectivity of the sine window is suboptimal for certain classes of signals. A second window was designed for AAC using a numerical procedure which allows optimization of the transition bandwidth and the ultimate rejection of the filter bank, and simultaneously guarantees perfect reconstruction. This window is called a Kaiser-Bessel Derived (KBD) window, as introduced in [91].

Figure 7.13 shows the relative amplitude response or spectral leakage of one MDCT band as a function of frequency for both sine and KBD windows. A comparison with a minimum masking threshold from [91] is also shown in the figure. The figure shows that the sine window leakage is approximately 20 dB greater than the minimum masking threshold for frequency offsets greater than 110 Hz, while the KBD window response lies below the masking threshold. This indicates that fewer bits are necessary to transparently encode a harmonic signal using the KBD window when the spectral components are spaced more than 220 Hz apart. Note that the increased rejection of the KBD window comes at the price of poorer rejection of components for frequency offsets less than $\pm$ 70 Hz. Therefore, when perceptually significant frequency components are separated by less than 140 Hz, the sine window will generally yield a more bit efficient filter bank.

Window shape decisions made by the encoder are applicable to the second half of the window only since the first half is constrained to use the time-reversed window shape from the preceding frame. Figure 7.14 shows an overlap-add sequence of blocks for the two situations. The sequence of windows labeled A-B-C all employ the KBD window, while the sequence D-E-F shows the transition to and from a single block employing the sine function window. In actual use, the window shape selector typically produces window shape run-lengths greater than that shown in the figure.

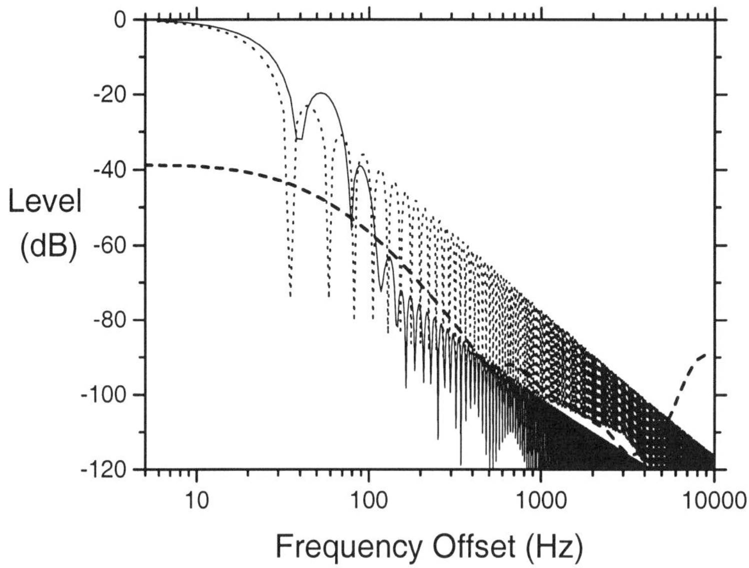

Figure 7.13 Comparison of the frequency selectivity of 2048-point sine and KBD windows to the minimum masking threshold. The dotted line represents the sine window, the solid line represents the KBD window, and the dashed line represents the minimum masking threshold.

Window coefficients for the sine window are derived from the following equations:

$$W_{SIN\_LEFT,N}(n) = \sin\left(\frac{\pi}{N}(n+\frac{1}{2})\right), \quad 0 \le n < \frac{N}{2}$$

$$W_{SIN\_RIGHT,N}(n) = \sin\left(\frac{\pi}{N}(n+\frac{1}{2})\right), \quad \frac{N}{2} \le n < N.$$

Coefficients for the KBD window are derived from

$$W_{KBD\_LEFT,N}(n) = \sqrt{\frac{\sum_{p=0}^{n}[W'(n,\alpha)]}{\sum_{p=0}^{N/2}[W'(p,\alpha)]}}, \quad 0 \le n < \frac{N}{2}$$

$$W_{KBD\_RIGHT,N}(n) = \sqrt{\frac{\sum_{p=0}^{N-n}[W'(n,\alpha)]}{\sum_{p=0}^{N/2}[W'(p,\alpha)]}}, \quad \frac{N}{2} \le n < N$$

where W', a Kaiser-Bessel kernel window [115], is defined as

$$W'(n,\alpha) = \frac{I_0\left[\pi\alpha\sqrt{1.0-(\frac{n-N/4}{N/4})^2}\right]}{I_0[\pi\alpha]}, \quad 0 \le n \le \frac{N}{2}$$

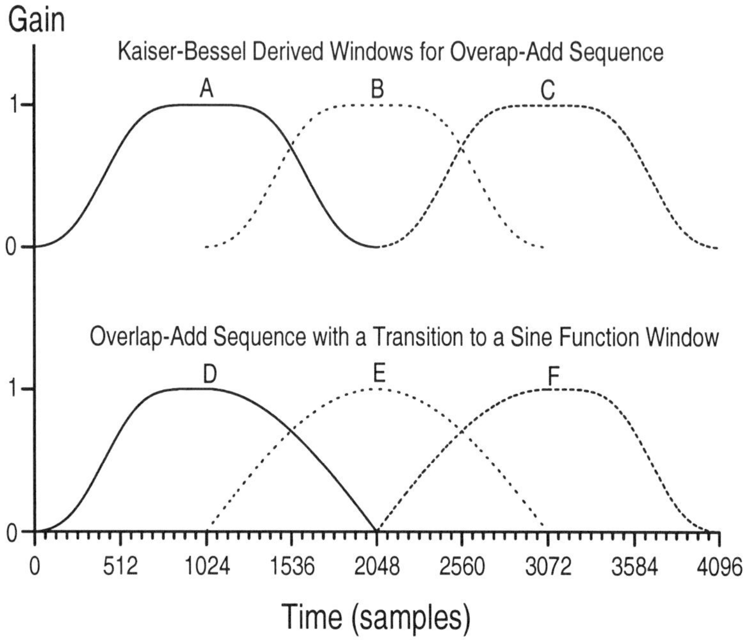

Figure 7.14 (Top) Window OLA sequence for KBD window. (Bottom) Window shape switching example from KBD window to sine window and back again.

with

$$I_0[x] = \sum_{k=0}^{\infty} \left[\frac{(\frac{x}{2})^k}{k!} \right]^2 .$$

The Kaiser window parameter α has been selected to provide a compromise in the trade-off between main lobe width and ultimate rejection. For the long window, $\alpha = 4$, and for the short window, $\alpha = 6$, i.e.

$$\alpha = \begin{cases} 4 & \text{for} \quad N = 2048 \\ 6 & \text{for} \quad N = 256. \end{cases}$$

7.4.7 Filter Bank Implementation

In many of today's emerging digital audio transmission applications, decoder cost is a central issue in the process of coding algorithm selection. Therefore, a computation and memory-efficient synthesis filter bank implementation is highly desirable. Toward this end, several memory and computation-efficient techniques are available for implementing the AAC forward and inverse transforms (for example, see [287]).

In the decoder, one of the more efficient techniques is derived by rewriting (7.2) in the form of an N-point inverse DFT (IDFT) combined with two complex vector multiplies. The IDFT can be efficiently computed using an inverse

FFT (IFFT). Two properties of the output sequence x_{in} further reduce the IFFT length. Firstly, the N-length sequence x_{in} contains only $N/2$ unique samples, and secondly, the output signal is purely real. By observing symmetries implied by these properties, one can show that an N-point inverse MDCT (IMDCT) can be computed using an $N/4$-point complex IFFT in combination with the two complex vector multiplies.

The computation rate for the fast inverse MDCT filter bank, including the window-overlap-add, is approximately $3N(1 + 0.25 \log_2(N/4))$ real multiply-adds per sample. This results in an overall computation rate of about 20 multiply-add operations per sample when $N = 2048$, and 15 multiply-add operations per sample when $N = 256$. In some AAC profiles, the synthesis computation can be further reduced when the decoder produces fewer output channels than are contained in the bitstream. In this case, the decoder generates a mixdown program by combining frequency coefficients from appropriate channels immediately prior to synthesis.

Filter Bank Implementations on Joint Coding. The time duration of the filter bank also has an impact on the ability of a coder to use various joint coding strategies. In general, approaches such as intensity stereo are advantaged by short filter banks and approaches such as mid/side (MS) coding benefit from long filter banks. For either kind of joint coding, however, the filter bank length should not exceed the point at which signal non-stationarity begins to increase.

7.4.8 Temporal Noise Shaping

A novel concept in perceptual audio coding is represented by the temporal noise shaping (TNS) tool in AAC [128]. This tool is motivated by the fact that, despite the advanced state of today's perceptual audio coders, the handling of transient and pitched input signals still presents a major challenge. This is mainly due to the problem of maintaining time domain details of the masking effect in the reproduced audio signal. In particular, coding is difficult because of the temporal mismatch between masking threshold and quantization noise (also known as the "pre-echo problem" [162], or as the "German male speech" problem, when time structure of the audio signal permits unmasking between the pitch pulses of the audio signal).

The TNS technique permits the coder to exercise a control over the temporal structure of the quantization noise within a filter bank window. The concept of this technique can be outlined as follows:

- Time/Frequency Duality Considerations
  The concept of TNS uses the duality between time and frequency domain to extend the known coding techniques by a new variant. It is well-known that signals with an "un-flat" spectrum can be coded efficiently either by directly coding spectral values, "transform coding," or by applying predictive coding methods to the time signal [157]. Consequently, the corresponding dual statement relates to the coding of signals with an "un-

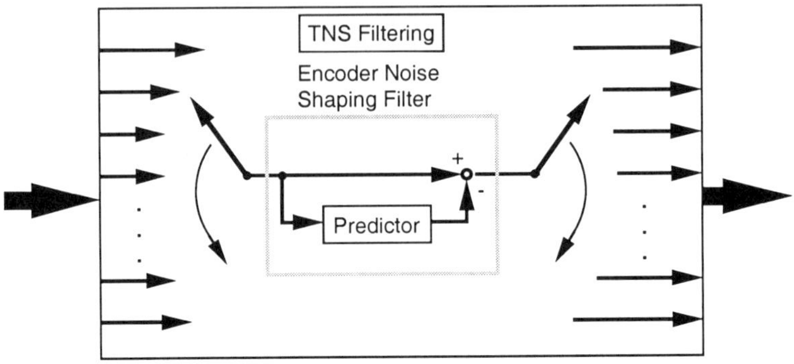

Figure 7.15 Diagram of encoder TNS filtering stage.

flat" time structure, i.e. transient signals. Efficient coding of transient signals can thus be achieved by either directly coding time domain values or by employing predictive coding methods to the spectral data. Such a predictive coding of spectral coefficients over frequency in fact constitutes the dual concept to the intra-channel prediction tool in AAC. While intra-channel prediction over time increases the coder's spectral resolution, prediction over frequency enhances its temporal resolution.

- Noise Shaping by Predictive Coding
  If an open-loop predictive coding technique is applied to a time signal, the quantization error in the final decoded signal is known to be adapted in its power spectral density (PSD) to the PSD of the input signal [157]. Dual to this, if predictive coding is applied to spectral data over frequency, the temporal shape of the quantization error signal will appear adapted to the temporal shape of the input signal at the output of the decoder. This effectively puts the quantization noise under the actual signal and in this way avoids problems of temporal masking, either in transient or pitched signals. This type of predictive coding of spectral data is therefore referred to as the TNS method.

Since the TNS processing can be applied either for the entire spectrum, or for only part of the spectrum, the time-domain noise control can be applied in any necessary frequency-dependent fashion. In particular, it is possible to use several predictive filters operating on distinct frequency regions.

Implementation in the AAC Encoder/Decoder. The predictive encoding/decoding process over frequency can be realized easily by adding one building block to the standard structure of a generic perceptual encoder and decoder. This is shown for the encoder in Figure 7.8 as the block "TNS." This performs an in-place filtering operation on the spectral values, i.e. replaces the target spectral coefficients (set of spectral coefficients to which TNS should be applied) with the prediction residual. This is symbolized by a "rotating switch"

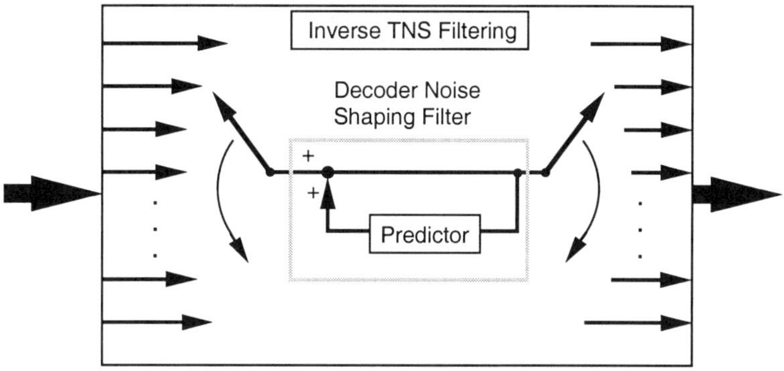

Figure 7.16 Diagram of decoder inverse TNS filtering stage.

circuitry in Figure 7.15. TNS prediction in both increasing and decreasing frequency direction is possible.

Similarly, the TNS decoding process is done immediately before the synthesis filter bank in the AAC decoder. An inverse in-place filtering operation is performed on the residual spectral values so that the target spectral coefficients are replaced with the decoded spectral coefficients by means of the inverse prediction (all-pole) filter as seen in Figure 7.16.

The TNS operation is signaled to the decoder via a dedicated part of the side information that includes a TNS on/off flag, the number of TNS filters applied in each transform window (maximum of 3), the order of the prediction filter (maximum of 12 or 20, depending on the profile) and the filter data itself.

Properties of TNS Processing. The properties of the TNS technique can be described as follows:

- The combination of filter bank and adaptive prediction filter can be interpreted as a continuously signal adaptive filter bank as opposed to the classic "switched filter bank" approach. In fact, this type of adaptive filter bank dynamically provides a continuum in its behavior between a high-resolution filter bank (for stationary signals) and a low-resolution filter bank (for transient signals) thus approaching the optimum filter bank structure for a given signal.

- The TNS approach permits a more efficient use of masking effects by adapting the temporal structure of the quantization noise to that of the masker (signal). In particular, it enables a better encoding of "pitch-based" signals such as speech, which consist of a pseudo-stationary series of impulse-like events where traditional block switching schemes do not offer an efficient solution.

- The method reduces the peak bit demand of the coder for transient signal segments by exploiting irrelevancy. As a side effect, the coder can more

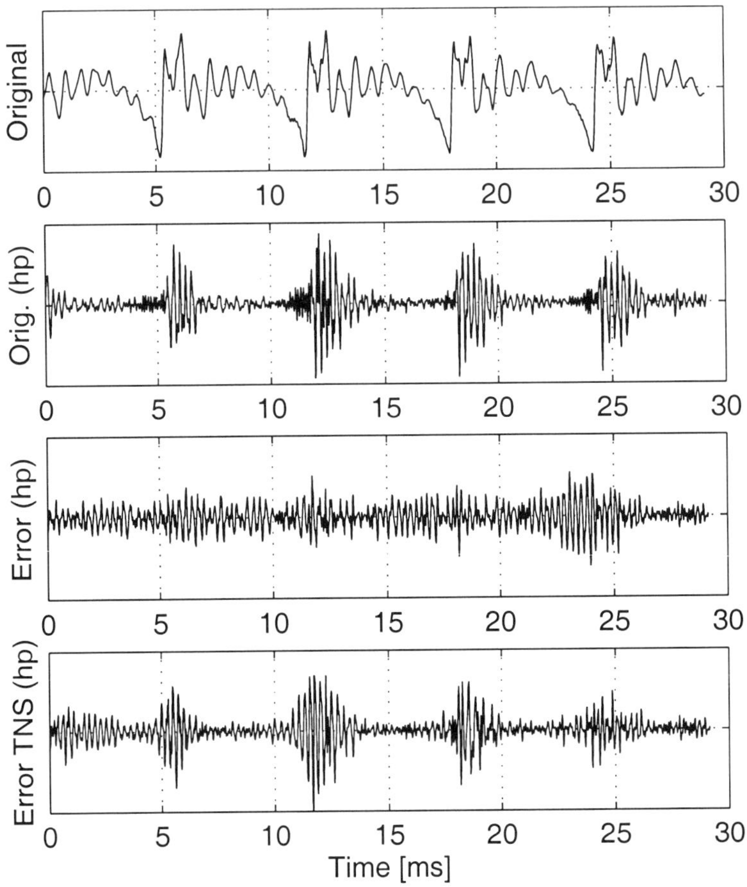

Figure 7.17 TNS operation: (Top) time waveform of the signal, (Second) high-passed time waveform, (Third) the waveform of the quantization noise at high frequencies without TNS operating and (Bottom) the waveform of the quantization noise with TNS operating.

often stay in the more efficient "long block" mode so that use of the less efficient "short block" mode is minimized.

- The technique can be applied in combination with other methods addressing the temporal noise shaping problem, such as block switching and pre-echo control.

During the course of the standardization process of MPEG-2 AAC, the TNS tool demonstrated a significant increase in coder performance for speech stimuli (approximately a 0.9 increase in CCIR impairment grade for the most critical speech item, "German Male Speech") as well as advantages for transient segments of stimuli (e.g. "Glockenspiel").

As Figure 7.17 shows, TNS is quite effective in changing the time envelope of the quantization noise. The final plot in the figure shows quite clearly that the quantization noise, which was previously flat in the time domain, has been shaped to match the high frequency energy in the signal very closely. This strongly mitigates the so-called "German male speech" effect, by moving quantization noise from between the pitch periods to being synchronous with the pitch periods, eliminating the co-articulation unmasking problem.

The only difficulty with TNS is that the time aliasing in the MDCT filter bank applies to the PSD noise shaping as well, and thus the process can not arbitrarily shape the noise PSD. For some signals, this can lead to incomplete noise shaping and some audible temporal unmasking.

7.4.9 Perceptual Model

The perceptual model estimates the threshold of masking, which is the level of noise that is subjectively just noticeable given the current input signal. Because models of auditory masking are primarily based on frequency domain measurements [284, 383], these calculations typically are based on the short term power spectrum of the input signal, and threshold values are adapted to the time/frequency resolution of the filter bank outputs. The threshold of masking is calculated relative to each frequency coefficient for each audio channel for each frame of input signal, so that it is signal-dependent in both time and frequency. When the high-time-resolution filter bank is used, it is calculated for the spectra associated with each of the sequence of eight windows used in the time/frequency analysis. In intervals in which pre-echo degradation is likely, more than one frame of signal is considered such that the threshold in frames just prior to a non-stationary event is depressed to insure that leakage of coding noise is minimized. Within a single frame, calculations are done in a granularity of approximately 1/3 critical band, following the critical band model in psychoacoustics. The model calculations are similar to those in psychoacoustic model II in the MPEG-1 audio standard [27]. The following steps are used to calculate the monophonic masking threshold of an input signal:

- Calculate the power spectrum of the signal in 1/3 critical band partitions.
- Calculate the tonal or noiselike nature of the signal in the same partitions, called the tonality measure.
- Calculate the spread of masking energy, based on the tonality measure and the power spectrum.
- Calculate the time domain effects on the masking energy in each partition.
- Relate the masking energy to the filter bank outputs.

Once the masking threshold is known, it is used to set the scale factor values in each scale factor band such that the resulting quantizer noise power in each band is below the masking threshold in that band. When coding audio channel pairs that have a stereo presentation, binaural masking level depression must be considered, [225].

7.4.10 Prediction

Prediction is used for improved redundancy reduction and is especially effective in case of stationary parts of a signal which belong to the most demanding signals in terms of bit rate. Because the use of a short window in the filter bank indicates signal changes, i.e. non-stationary signal characteristic, prediction only applies to long windows.

For each channel, prediction is applied to the spectral components resulting from the spectral decomposition of the filter bank. For each spectral component up to 16 kHz, there is one corresponding predictor, resulting in a bank of predictors, where each predictor exploits the auto-correlation between the spectral component values of consecutive frames.

To achieve high coding efficiency, the overall coding structure using a filter bank with high spectral resolution implies the use of backward adaptive predictors with coefficients calculated from quantized spectral components in the encoder as well as in the decoder. In contrast to forward adaptive predictors, in this case no additional side information for predictor coefficients need be transmitted. A second order backward-adaptive lattice structure predictor is used for each spectral component, so that each predictor is working on the spectral component values of the two preceding frames. The predictor parameters are adapted to the current signal statistics on a frame by frame basis, using a least-mean-square based adaptation algorithm. If prediction is activated, the quantizer is fed with a prediction error instead of the original spectral component, resulting in a coding gain. A more detailed description of the principles can be found in [99].

In order to guarantee that prediction is only used if this results in a coding gain, an appropriate predictor control is required and a small amount of predictor control information has to be transmitted to the decoder. For the predictor control, the predictors are grouped into scale factor bands. The predictor control information for each frame is determined in two steps. First, it is determined for each scale factor band whether or not prediction gives a coding gain and all predictors belonging to a scale factor band are switched on/off accordingly. Then it is determined whether the overall coding gain by prediction in the current frame compensates at least the additional bit need for the predictor side information. Only in this case, prediction is activated and the side information is transmitted. Otherwise, prediction is not used in the current frame and only one signaling bit is transmitted.

In order to increase the stability by re-synchronizing the predictors of the encoder and the decoder and to allow defined entry points in the bitstream, a cyclic reset mechanism is to be applied in the encoder, where all predictors are initialized again during a certain time interval in an interleaved way. The whole set of predictors is subdivided into 30 so-called reset groups (Group 1: P1, P31, P61, ...; Group 2: P2, P32, P62, ...; ...; Group 30: P30, P60,...) which are then periodically reset, one after the other with a certain spacing. For example, if one group is reset every eighth frame, then all predictors are reset within an interval of $8 \times 30 = 240$ frames. The reset mechanism is controlled by a reset

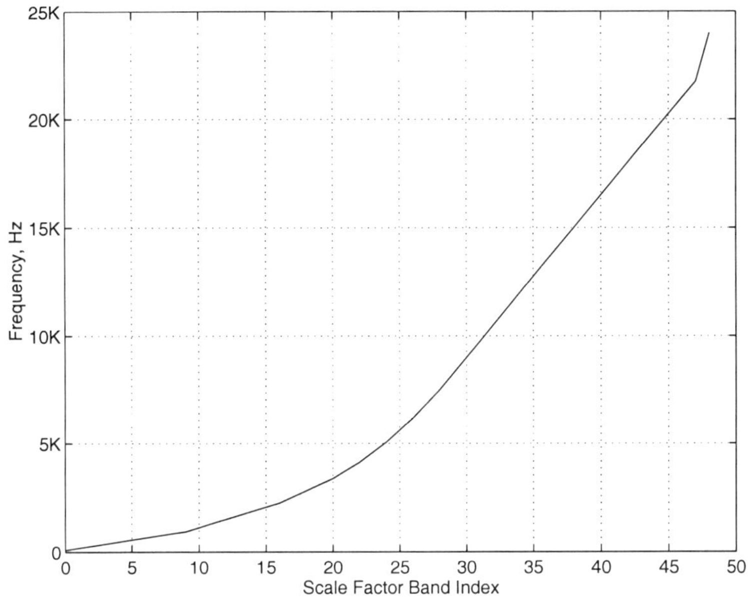

Figure 7.18 AAC scale factor band boundaries at 48kHz sampling rate.

on/off bit, which always has to be transmitted as soon as prediction is enabled and a conditional 5 bit index specifying the group of predictors to be reset. In case of short windows prediction is always disabled and a full reset, i.e. all predictors at once, is carried out.

Listening tests conducted on the AAC standard (see Section 7.4.16) have shown that statistically significant improvement in sound quality is achieved by the prediction tool when coding locally stationary signals, such as "Pitch Pipe" and "Harpsichord."

The lattice structure of the predictors is quite robust to quantization of predictor coefficients, predictor state variables and adaptation variables. Therefore variables associated with a predictor that must be saved from frame to frame are truncated to a 16-bit IEEE floating point precision (i.e. the 16 least significant bits of the floating point word is set to zero) prior to storage. An investigation of prediction gain as a function of variable quantization indicated that the loss in prediction gain for the 16-bit quantization was negligible (less than 10 percent) while it provides 50 percent reduction in the storage associated with the prediction tool.

7.4.11 Quantization and Coding

The spectral coefficients are coded using one quantizer per scale factor band, which is a fixed division of the spectrum. For high-resolution blocks there are 49 scale factor bands whose frequency boundaries are shown in Figure 7.18.

The psycho-acoustic model specifies the quantizer step size (inverse of scale factor) per scale factor band. An AAC encoder is an instantaneously variable rate coder, but if the coded audio is to be transmitted over a constant rate channel then the rate/distortion module adjusts the step sizes and number of quantization levels so that a constant rate is achieved.

7.4.12 Quantizer

AAC uses a nonlinear quantizer for spectral component x_i to produce $\hat{x}_i$,

$$\hat{x}_i = \text{sign}(x_i)\text{nint}\left[\left(\frac{|x_i|}{(\sqrt[4]{2})^{stepsize}}\right)^{3/4}\right]$$

where nint[·] represents the nearest integer function. The main advantage of the nonlinear quantizer is that it shapes the noise as a function of the amplitude of the coefficients, such that the increase of the signal to noise ratio with raising signal energy is much lower compared to that of a linear quantizer. The range of the quantized values, $\hat{x}_i$, is limited to ±8191. The exponent *stepsize* represents the quantizer step size in a given scale factor band. Thus the quantizer can be changed in steps of 1.5 dB.

Scale Factors and Noise Shaping. The use of a nonlinear quantizer is not sufficient to achieve maximum compression while rendering the coding noise perceptually inaudible. In addition AAC can selectively amplify the set of spectral coefficients within a given scale factor band, i.e, change the quantizer step size, so as to alter the quantizer signal to noise ratio in that band. It is desirable to be able to shape the quantization noise in units similar to the critical bands of the human auditory system. Since the AAC system offers a relatively high frequency resolution for long blocks (23.43 Hz/line at 48 kHz sampling frequency), it is possible to build groups of spectral values which reflect the bandwidth of the critical bands very closely.

The AAC system offers the possibility of individual amplification using the scale factor bands in steps of 1.5 dB. Noise shaping is achieved because coefficients that have larger amplitudes will generally get a higher signal to noise ratio after quantization. On the other hand larger amplitudes normally require more bits in coding, so that the distribution of bits among the scale factor bands is implicitly controlled. The amplification information, stored in the scale factors (in units of 1.5 dB steps) is transmitted to the decoder.

Rate/Distortion Control. The quantized coefficients created by the quantizer are coded using Huffman Codes. A highly flexible coding method allows to use several Huffman tables for one spectrum. Two- and four-dimensional tables with and without sign are available. The noiseless coding process is described in detail in Section 7.4.13. To calculate the number of bits needed to code a spectrum of quantized data, the coding process has to be performed and the number of bits needed for the spectral data and the side information has to be accumulated.

7.4.13 Noiseless Coding

The noiseless coding module in AAC is nearly identical to that of AT&T's Perceptual Audio Coder [164]. The input to the noiseless coding is the set of 1024 quantized spectral coefficients and their associated scale factors. If the high time resolution filter bank is selected then the 1024 coefficients are actually a matrix of 8×128 coefficients representing the time/frequency evolution of the signal over the duration of the eight short-time spectra.

Grouping and Interleaving. The filter bank typically switches to high time resolution when there is a non-stationary event in the time domain such that some of the eight spectra are very different from the others. In such circumstances, it is usually true that the required scale factors for several sets of short blocks will be very similar. Consequently, similar short-block spectra are grouped in order to reduce the scale factor transmission costs. For each of the groups that shares scale factors, all of the spectral data are interleaved in frequency, such that the effective scale factor bandwidth is multiplied by the number of short blocks in the scale factor group. Thus, interleaving also has the advantage of combining the high-frequency zero-valued coefficients (due to band-limiting or quantization) within each group such that they form a single, longer run of zeros which is more efficient to code.

Spectral Clipping. The first step in noiseless coding is a method of dynamic range compression that may be applied to the spectrum. Up to four coefficients can be coded separately as magnitudes in excess of one, with a value of ± 1 left in the quantized coefficient array to carry the sign. The "clipped" coefficients are coded as integer magnitudes and an offset from the base of the coefficient array to mark their location. Since the side information for carrying the clipped coefficients costs some bits, this noiseless compression is applied only if it results in a net savings of bits.

Huffman Coding. In AAC the spectral coefficients are raised to the 3/4 power and quantized using one of several uniform quantizers. Although the power provides some dynamic range compression, the probability distribution for the levels in any given quantizer is far from uniform. To compensate for this, variable-length (Huffman) coding is used to represent the quantizer levels. However, the variable length code must still be an integral number of bits long, such that if $H(p)$ is the entropy of the codeword representing a given quantizer level (the level being a member of the code alphabet) and l is the length of this codeword, the bound on l is

$$H(p) \leq l < H(p) + 1.$$

To circumvent this inefficiency of as much as one bit, the Huffman code can be extended such that N letters are coded at once, which is to say that codes are designed that represent a sequence of spectral coefficients. In this case, the

Table 7.2 Huffman Codebooks.

Codebook index	Tuple size	Max absolute value	Signed values
0		0	
1	4	1	yes
2	4	1	yes
3	4	2	no
4	4	2	no
5	2	4	yes
6	2	4	yes
7	2	7	no
8	2	7	no
9	2	12	no
10	2	12	no
11	2	16 (ESC)	no

bound on l is

$$\frac{H(p)}{N} \leq l < \frac{H(p)+1}{N}$$

and the codeword inefficiency is reduced to at most $1/N$ bits.

In AAC an extended Huffman code is used to represent n-tuples of quantized coefficients, with the Huffman codewords drawn from one of 11 codebooks. The spectral coefficients within n-tuples are ordered (low to high) and the n-tuple size is two or four coefficients. The maximum absolute value of the quantized coefficients that can be represented by each Huffman codebook and the number of coefficients in each n-tuple for each codebook is shown in Table 7.2. There are two codebooks for each maximum absolute value, with each representing a distinct probability distribution function, however the best overall fit is always chosen from any applicable codebook. In order to save on codebook storage (an important consideration in a mass-produced decoder), most codebooks represent unsigned values. For these codebooks the magnitude of the coefficients is Huffman coded and the sign bit of each non-zero coefficient is appended to the codeword.

Two codebooks require special note: Codebook 0 and Codebook 11. As mentioned in Table 7.2, Codebook 0 indicates that all coefficients within a section are zero. In addition to requiring no transmission of the spectral values, all scale factor information is omitted. Codebook 11 can represent quantized coefficients that have an absolute value greater than or equal to 16. If the magnitude of one or both coefficients is greater than or equal to 16, a special escape coding mechanism is used to represent those values. The magnitude of the coefficients is limited to no greater than 16 and the corresponding 2-tuple is Huffman coded. The sign bits, as needed, are appended to the codeword.

For each coefficient magnitude greater or equal to 16, an escape code is also appended as follows:

```
escape code = <escape_prefix><escape_separator><escape_word>
```

where <escape_prefix> is a sequence of N binary "1's," <escape_separator> is a binary "0," <escape_word> is an $N+4$ bit unsigned integer, msb first, and N is a count that is just large enough so that the magnitude of the quantized coefficient is equal to

$$2^{N+4} + \texttt{<escape\_word>}$$

Sectioning. Another form of noiseless coding segments the set of 1024 quantized spectral coefficients into sections, such that a single Huffman codebook is used to code each section. For reasons of coding efficiency, section boundaries can only be at scale factor band boundaries so that for each section of the spectrum one must transmit the length of the section, in scale factor bands, and the Huffman codebook number used for the section.

Sectioning is dynamic and typically varies from block to block, such that the number of bits needed to represent the full set of quantized spectral coefficients is minimized. This is done using a greedy merge algorithm starting with the maximum possible number of sections, each of which uses the Huffman codebook with the best fit. Sections are merged if the new sectioning results in a lower total bit count. Merges that yield the greatest bit count reduction are done first. If the sections to be merged do not use the same Huffman codebook then the codebook with an index equal to at least the higher index of the two must be used, although codebooks with an even higher index can be used if they result in a lower total bit count. Sections often contain only coefficients whose value is zero. For example, if the audio input is bandlimited to 20 kHz then the highest coefficients will be zero and coded using Huffman Codebook 0. The sectioning is the only side information that is required for the noiseless coding. No explicit bit allocation is done and therefore no bit allocation information need be sent. Instead, the sectioning indirectly indicates the number of spectral coefficients that are sent (section lengths for sections coded with other than the zero codebook), and the number of coefficients represented by each codeword is known from the codebook index. Since the Huffman codewords are uniquely decodable (i.e. self-terminating and prefix-free), the sequence of codewords uniquely determines the number of bits used to represent the spectral coefficients.

Scale Factors. The coded spectrum uses one quantizer per scale factor band. The step sizes of each of these quantizers is specified as a set of scale factors and a global gain which normalizes these scale factors. In order to increase compression, scale factors associated with sections that have only zero-valued coefficient are not transmitted. If the entire section is not zero, then scale factors associated with scale factor bands that have only zero-valued coefficients are a free parameter in the coding process and are coded with the value that

Table 7.3 Huffman Compression.

Codebook index	Number of codewords	Relative frequency	Compression for codebook
1	81	0.16	1.92
2	81	0.08	1.24
3	81	0.10	2.20
4	81	0.08	1.42
5	81	0.08	2.20
6	81	0.06	1.34
7	64	0.08	2.06
8	64	0.04	1.18
9	169	0.17	2.65
10	169	0.05	1.31
11	289	0.10	1.29
Average			1.85

results in the lowest total bit count. Both the global gain and the scale factors are quantized in 1.5 dB steps. The global gain is coded as an 8-bit unsigned integer and the scale factors are differentially encoded relative to the previous scale factor (or global gain for the first scale factor) and then Huffman coded. The dynamic range of the global gain is sufficient to represent values from a 24-bit PCM audio source.

Compression. For each spectral Huffman codebook, Table 7.3 lists the codebook index, the number of codewords, and the relative frequency with which the codebooks is selected. The last column indicates the compression C provided by the individual codebook, calculated as

$$C = \frac{\lceil \log_2(m) \rceil}{\bar{l}}$$

where $\bar{l}$ is the expected value of the codeword length in the book and m is the number of codewords in the book. The table shows that the individual Huffman codebooks provide a maximum compression of up to 2.65, which is for the most likely codebook. The average compression realized by using the set of spectral codebooks is 1.85.

7.4.14 Joint Channel Coding

AAC includes two well-known techniques for joint stereo coding of signals: MS stereo coding (also known as "Sum/Difference coding") and intensity stereo coding. Both joint coding strategies can be combined by selectively applying them to different frequency regions. By using MS stereo coding, intensity stereo coding, and left/right (independent) coding as appropriate, it is possible to avoid the expensive overcoding due to "binaural masking level depression," [23, 225], to correctly account for noise imaging, and, very frequently, to achieve a significant savings in bit rate. The concept of joint stereo coding in AAC is discussed in greater detail in [163]

MS Coding. MS stereo is used to control the imaging of coding noise, as compared to the imaging of the original signal. In particular, this technique is capable of addressing the issue of binaural masking level depression [23, 225], where a signal at lower frequencies (below 2 kHz) can show up to 20 dB difference in masking thresholds depending on the phase of the signal and noise present (or lack of correlation in the case of noise). A second important issue is that of high-frequency time domain imaging on transient or pitched signals. In either case, the properly coded stereo signal can require more bits than two transparently coded monophonic signals.

In AAC, MS stereo coding is applied within each channel pair of the multi-channel signal, i.e. between a pair of channels that are arranged symmetrically on the left/right listener axis. In this way, imaging problems due to spatial unmasking is avoided to a large degree.

MS stereo coding can be used in a flexible way by selectively switching in time (on a block-by-block basis), as well as in frequency (on a coder band by coder band basis) [158]. The switching state (MS stereo coding "on" or "off") is transmitted to the decoder as an array of signaling bits. This can accommodate short time delays between the L and R channels and still accomplish both image control and some signal-processing gain. While the amount of time delay that it allows is limited, the time delay is greater than the inter-aural time delay, and allows for control of the most critical imaging issues [158].

Intensity Stereo Coding. The second important joint stereo coding strategy for exploiting inter-channel irrelevance is the well-known generic concept of "intensity stereo coding" [350, 27]. This idea has been widely utilized in the past for stereophonic and multi-channel coding under various names ("dynamic crosstalk" [306], "channel coupling" [67]). Intensity stereo exploits the fact that the perception of high frequency sound components mainly relies on the analysis of their energy-time envelopes [23]. Thus, it is possible for signals with stationary and similar envelopes in all source channels to transmit a single set of spectral values that is shared among several audio channels with virtually no loss in sound quality. The original energy-time envelopes of the coded channels are preserved approximately by means of a scaling operation.

AAC provides two mechanisms for applying generic intensity stereo coding:

- The first is based on the "channel pair" concept as used for MS stereo coding and implements an straight-forward coding concept that covers most of the normal needs without introducing noticeable signaling overhead into the bitstream. For simplicity, this mechanism is referred to as the "intensity stereo coding" tool. While the intensity stereo coding tool only implements joint coding within each channel pair, it may be used for coding of both 2-channel as well as multi-channel signals.

- In addition, a second, more sophisticated mechanism is available that is not restricted by the channel pair concept and allows better control over the coding parameters. This mechanism is called the "coupling channel" element and is discussed in the next section.

Thus, AAC provides appropriate coding tools for many types of stereophonic material from traditional two channel recordings to five or seven channel surround sound material.

Coupling Channel. The coupling channel element (CCE) is a monophonic audio channel that provides both compression and post-processing flexibility [163]. As a compression tool, coupling channels may be used to implement generalized intensity stereo coding, where channel spectra can be shared across channel boundaries. Alternatively, coupling channels may be used to dynamically perform a downmix of a known and isolated sound object into the multichannel sound image. The latter case is particularly important for material (such as movie soundtracks) that may be distributed with multi-language dialogue tracks in which the user can select the language to be used.

There are two kinds of CCE: "independently switched" and "dependently switched." An independently switched CCE is one in which the window state (i.e. window_sequence and window_shape) of the CCE does not have to match that of any of the audio channels to which the CCE is coupled (target channels). This requires that the independently switched CCE must be decoded all the way to the time domain before it is scaled and added to the target audio channels. A dependently switched CCE, on the other hand, must have a window state that matches all of the target audio channels. Therefore the CCE only needs to be decoded as far as the frequency domain prior to being added to the target audio channels.

Independently switched CCEs are most appropriate for encoding multi-language programming in that the dialogue typically is not correlated with the "background" audio and hence should not be forced to switch in synchrony with the target audio channels. Carrying dialogue in dependently switched CCEs would typically result in lower compression since the coupling channel would require a very high signal-to-noise ratio during intervals in which it is non-stationary and subject to "pre-echo" degradation but the target signals are stationary and processed using the high-frequency resolution filter bank. Dependently switched CCEs are most appropriate for generalized intensity stereo compression, or for applications in which decoder complexity is an issue (since coupling is done in the frequency domain so that the CCE need not be processed by the synthesis filter bank).

Table 7.4 Bitstream Elements.

Token	Meaning
prog_config_ele	Program configuration element
audio_ele	Audio element
coupling_ele	Multi-channel coupling
data_ele	Data element – segment of data stream
fill_ele	Fill element – adjusts bit rate for constant rate channels
term_ele	Terminator – signals end of block

7.4.15 Bitstream Syntax

General Structure. AAC has a very flexible bitstream syntax. Two layers are defined: the lower specifies the "raw" audio information while the higher specifies a recommended transport mechanism. The raw data stream contains all data which belong to the audio (including ancillary data). Since any one transport cannot be appropriate for all applications, the raw data layer is designed to be parsable on its own, and in fact is entirely sufficient for applications such as compression to computer storage devices.

The composition of a bitstream is as follows:

```
<stream> -> {<transport>}<block>{<transport>}<block>...
<block> -> [<prog_config_ele>]<audio_ele>[<audio_ele>]
           [<coupling_ele>][<data_ele>][<fill_ele>]<term_ele>
```

where -> indicates a production rule and tokens in the bitstream are indicated by angle brackets (< >). Braces ({ }) indicate an optional token and brackets ([ ]) indicate that the token may appears zero or more times. The bitstream is indicated by the token <stream> and is a series of <block> tokens each containing all information necessary to decode 1024 audio frequency samples. Furthermore, each <block> token begins on a byte boundary relative to the start of the first <block> in the bitstream. Between <block> tokens there may be transport information, indicated by <transport>, such as would be needed for synchronization on break-in or for error control.

Raw Data Block. Since AAC has a bit buffer that permits its instantaneous bit rate to vary as required by the audio signal, the length of each <block> is not constant. In this respect the AAC bitstream uses variable rate headers (header being the <transport> token). The byte-aligned constant rate headers used in the AAC coder permit editing of bitstreams at any block boundary. Tokens within a <block> are shown in Table 7.4.

Bitstream Elements. The prog_config_ele is a configuration element that specifies the audio channel to output speaker assignment so that multi-channel coding can be a flexible as possible. It can specify the correct voice tracks for multi-lingual programming and specifies the analog sampling rate.

There are three possible audio elements (`audio_ele`): `single_channel_ele` is a single monophonic audio channel, `channel_pair_ele` is a stereo pair and `low_freq_effects_ele` is a sub-woofer (low frequency effects) channel. Each of the audio elements is named with a 4-bit tag such that up to 16 of any one element can be represented in the bitstream and assigned to a specific output channel. At least one audio element must be present.

`coupling_ele` is a mechanism to code signal components common to two or more audio channels. `data_ele` is a tagged data stream that can continue over an arbitrary number of blocks. Unlike other elements, the data element contains a length count such that an audio decoder can strip it from the bitstream without knowledge of its meaning. As with the audio elements, up to 16 distinct data streams are supported.

The `fill_ele` is a bit stuffing mechanism that enables an encoder to increase the instantaneous rate of the compressed audio stream such that it fills a constant rate channel. Such mechanisms are required as, first, the encoder has a region of convergence for its target bit allocation so that the bits used may be less than the bit budget, and second, the encoder's representation of a digital zero sequence is so much less than the average coding bit budget that it must resort to bit stuffing. `term_ele` signals the end of a block. It is mandatory as this makes the bitstream parsable. Padding bits may follow the `term_ele` such that the next `<block>` begins on a byte boundary.

To illustrate the bitstream syntax, a 5.1 channel bitstream, where the .1 indicates the low frequency effects channel, `<block>` is

```
<block> -> <single_channel_ele><channel_pair_ele>
           <channel_pair_ele><low_freq_effects_ele><term_ele>
```

Although discussion of the syntax of each element is beyond the scope of this chapter, all elements make frequent use of conditional components. This increases flexibility while keeping bitstream overhead to a minimum. For example, a one-bit field indicates whether prediction is used in an audio channel in a given block. If set to one, then the set of bits indicating which scale factor bands use prediction follows. Otherwise the bits are not sent [150].

7.4.16 AAC Audio Quality

Test Conditions. The MPEG-Audio committee has tested AAC for subjective audio quality in both 2-channel (stereo) and 5-channel (surround) presentation formats. In addition, the Communications Research Centre (CRC) has tested AAC, two commercial software codecs, Dolby AC-3 and Lucent Technologies' Perceptual Audio Coding (PAC), and one commercial hardware codec. MPEG-Audio tested AAC in 5-channel format in the fall of 1996 at the British Broadcasting Company's (BBC) Research and Development Department at Kingswood Warren, UK and Nippon Hoso Kyokai (NHK) Science and Technical Research Labs, Tokyo, Japan [146]. In this test there were a total of 56 listeners, 32 at the BBC and 24 at NHK. TMPEG-Audio tested AAC in

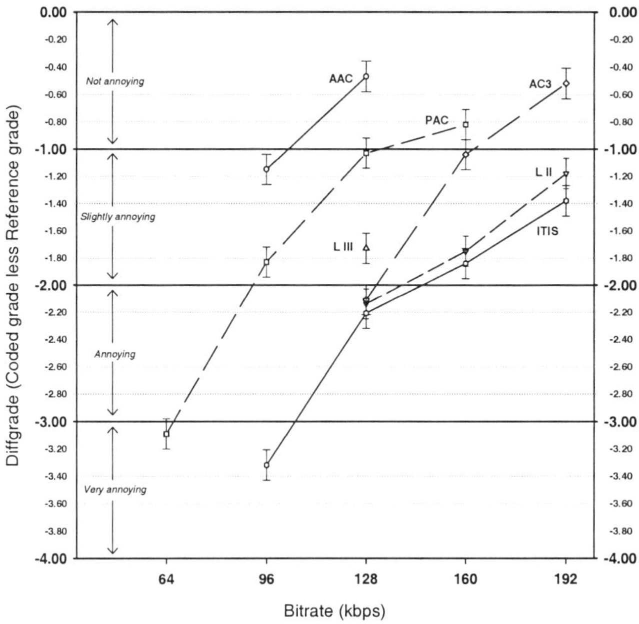

Following the completion of the tests, Lucent Technologies indicated that a programming bug might exist in the PAC codec which could affect the performance for one of the test sequences. However, at the time of publication, the existence of a bug had not been shown. It is important to note that, regardless of how PAC may have performed for this sequence in the absence of the presumed bug, the relative overall ranking of the codecs would remain unchanged.

Figure 7.19 CRC subjective audio quality test results for 2-channel coding including the required disclaimer.

2-channel format in the fall of 1997 at NHK Science and Technical Research Labs [214]. In this test there were a total of 31 listeners.

The 5-channel MPEG presentations tested four coding systems: the Main Profile AAC at both 256 and 320 kb/s, the LC Profile AAC at 320 kb/s and MPEG-2 BC Layer II at 640 kb/s. The 2-channel MPEG presentations tested seven coding systems: Main Profile AAC at both 96 kb/s and 128 kb/s, LC Profile AAC at both 96 kb/s and 128 kb/s, SSR Profile AAC at 128 kb/s, MPEG-2 Layer II at 192 kb/s and MPEG-2 Layer III at 128 kb/s. The 2-channel CRC presentations tested codecs at various rates, as shown in Figure 7.19.

As much as possible, test methodologies adhered to recommendation ITU-R BS 1116. The tests were conducted according to the triple-stimulus/hidden-reference/double-blind method, and the grading scale used was the ITU-R 5-point impairment scale. Descriptors associated with grades on this scale are represented in Table 7.5

In order for a coder to be considered ITU-R broadcast quality, decoded signals must be "indistinguishable" from the reference (uncoded) signal, as indicated by the fact that the 95% confidence intervals of the test and reference subjective assessments overlap. Ideally, all test items should be indistinguishable from the reference. However, ITU-R broadcast quality requires only that at least 70% of the total number of test items be judged indistinguishable (in the case of both tests, 7 items), and that the remaining test items have the ratio of the upper confidence interval limits of test and reference greater than 0.85.

Test Results. The results from the MPEG tests are presented on an inverted scale of *difference grades* (diff grades), in which 5.0 is subtracted from all scores so that larger negative scores correspond to lower grades on the impairment scale. Test result mean scores and 95% confidence intervals on the diff grade scale are shown in Figures 7.20 and 7.21.

The most significant outcome of the MPEG tests is that AAC is the first coding system that satisfies the requirements of ITU-R broadcast quality for both 2-channel presentation at 128 kb/s and 5-channel presentation at 320 kb/s. Specifically, in the 2-channel presentation, the AAC Main Profile at 128 kb/s and AAC LC Profile at 128 kb/s qualify as ITU-R broadcast quality, with AAC SSR Profile at 128 kb/s misses qualifying by a very slight margin. In the 5-channel presentation, the AAC Main Profile at 320 kb/s qualifies and the AAC LC Profile at 320 kb/s misses qualifying by a small margin.

Another significant outcome is that in the 2-channel presentation both the AAC Main Profile and AAC LC Profile at the rates of both 128 kb/s and 96 kb/s score better than MPEG-2 Layer II at 192 kb/s for all program item scores combined. Additionally, the Main Profile, LC Profile and the SSR Profile at 128kb/s, and the Main Profile at 96kb/s all showed statistically significant better performance than the MPEG-1 Layer 3 codec at 128 kb/s. Finally, the lower confidence interval for the LC Profile at 96kb/s just barely overlapped the upper confidence bound of the MPEG-1 Layer 3 codec at 128 kb/s.

Similarly, in the 5-channel presentation both the AAC Main Profile and AAC LC Profile at the rate of 320 kb/s score better than MPEG-2 Layer II at 640 kb/s for all program item scores combined. This clearly indicates that AAC has a factor of 2 advantage in compression relative to the MPEG-2 Layer II coding system.

The CRC results are quite interesting, as well. They show that the AAC Main Profile at 128 kb/s and Dolby AC-3 at 192 kb/s have an encoded quality

Table 7.5 ITU-R 5-point impairment scale.

Grade	Degradation
5.0	Imperceptible
4.0	Perceptible but not annoying
3.0	Slightly annoying
2.0	Annoying
1.0	Very annoying

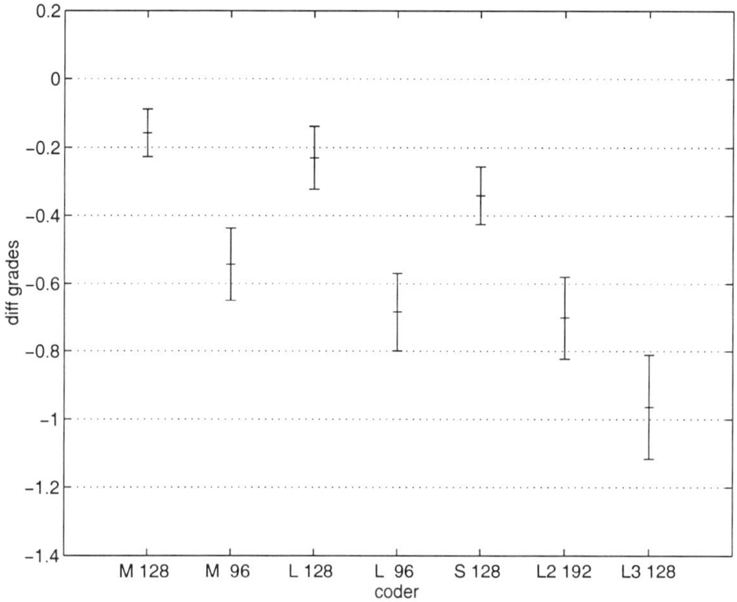

Figure 7.20 MPEG subjective audio quality test results for 2-channel coding. The figure indicates mean scores as the middle tick of each vertical stroke, with the 95% confidence interval indicated by the length of the stroke. Each stroke indicates the mean score for each of seven coders for the set of ten stimuli.

advantage over all of the included codecs, regardless of bit rate. These results clearly show the superiority of the AAC standard over both the previous MPEG standards and the de facto standards established prematurely by industry and government groups.

7.5 SUMMARY

From the filter bank perspective, the various MPEG coders are "middle of the road" filter bank users in that they do not use any esoteric or extremely new techniques. Their filter banks are well known, and the characteristics of the filter banks are important in that they provide a decent joint optimization of the redundancy and irrelevancy considerations The authors note that there have been a number of other filter banks proposed for audio coding, most notably fixed "wavelet" structures that use critical bandwidth arrangements. While these coders provide very good quality for transient signals, or for other signals in which the time structure of the signal is the primary consideration, they do not in general provide enough source coding gain to reduce the signal redundancy to a useful level, and require many more bits, on average, for high quality encoding of the average musical signal.

A perceptual coder, in its efficient form, is not only good at removing irrelevant material, but also at reducing signal redundancy. As the results for AAC

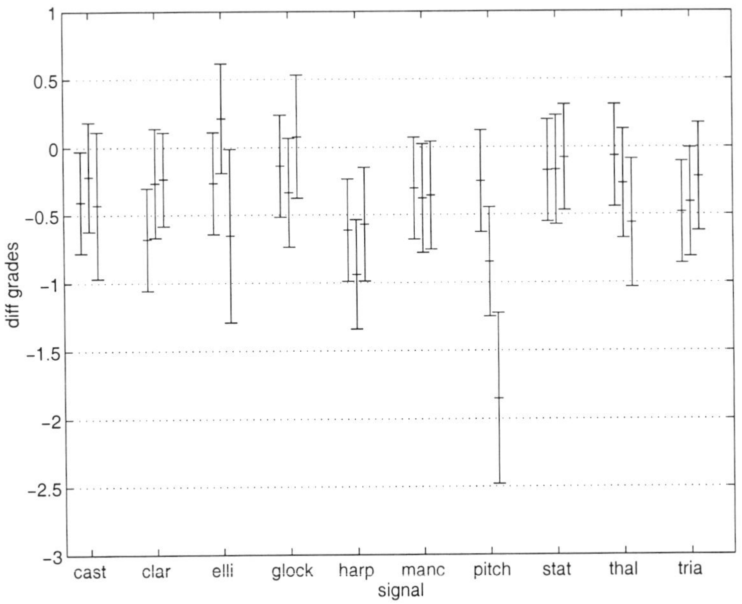

Figure 7.21 MPEG subjective audio quality test results for 5-channel coding. The figure indicates mean scores as the middle tick of each vertical stroke, with the 95% confidence interval indicated by the length of the stroke. For each of ten stimuli, a trio of strokes is show, the leftmost being AAC Main Profile at 320 kb/s, the middle AAC LC Profile at 320 kb/s and the rightmost MPEG-2 Layer II at 640 kb/s.

show, even with a 1024 line filter bank there is substantial signal redundancy remaining on some signals, enough to make backward adaptive prediction of the signals at the analysis filter bank output a useful technique.

In the future, advances in filter banks that allow the filter bank to actively change its time/frequency tiling as a function of the signal statistics and perceptual threshold, in a much more flexible way than the current long/short or long/wavelet paradigm, may permit more efficient audio coding. It is interesting to note that at the present, filter banks of simple construction (MDCT/OBT (MDCT/overlapped block transform)) [207] have shown the best performance in audio coders, and more sophisticated filter banks, such as wavelet/MDCT, have not yet shown verified performance at the level of the simple length-switched MDCT coders.

As final food for thought, an audio signal analyzed with a set of cochlear filters followed by an envelope detector is shown in Figure 7.22. At the figure's onset, the signal has significant transient high-frequency content, and it is clear that a filter bank with critical-band tiling will provide good coding for this part. At the same time, at low frequencies, the signal is so slowly varying as to be nearly stationary, and would be best coded with narrow frequency resolution. In the future, a filter bank that can adapt to optimally analyze

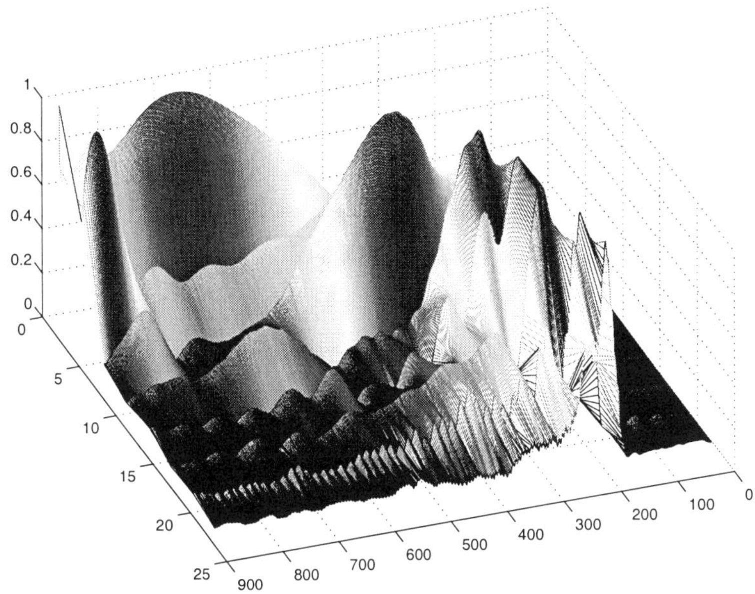

Figure 7.22 Time (x-axis) versus frequency (y-axis), in barks, versus level (z-axis) for a real audio signal.

both signal types simultaneously may provide a needed advance in audio coding. While the TNS tool of AAC provides a start to this ability, it cannot avoid the difficulties with time-aliasing intrinsic in the MDCT, and it is clear that even more ability to dynamically adapt between time and frequency resolution, in a signal responsive fashion, will lead to better control of co-articulation and stereo imaging artifacts.

Acknowledgments

The authors would like to acknowledge the aid of Jont Allen, Alan Docef, Stanley Kuo and Lynda Feng for their help in creating, assembling, editing, and organizing this document.

We would also like to expressly thank Gilbert Soloudre of the CRC and the Audio Engineering Society for allowing us to use Figure 7.19, which appears in the March, 1998 issue of the Journal of the Audio Engineering Society.

8 SUBBAND IMAGE COMPRESSION

Aria Nosratinia[1], Geoffrey Davis[2],
Zixiang Xiong[3] and Rajesh Rajagopalan[4]

[1] Rice University
Houston, TX
aria@ece.rice.edu

[2] Dartmouth College
Hanover, NH
geoffd@microsoft.com

[3] University of Hawaii
Honolulu, HI
zx@lena.eng.hawaii.edu

[4] Lucent Technologies
Murray Hill, NJ
rajesh@ee.princeton.edu

8.1 INTRODUCTION

Digital imaging has had an enormous impact on industrial applications and scientific projects. It is no surprise that image coding has been a subject of great commercial interest. The JPEG image coding standard has enjoyed widespread acceptance and the industry continues to explore issues in its implementation.

In addition to being a topic of practical importance, the problems studied in image coding are also of considerable theoretical interest. The problems draw upon and have inspired work in information theory, applied harmonic analysis, and signal processing. This chapter presents an overview of subband image coding, arguably one of the most fruitful and successful directions in image coding.

8.1.1 Image Compression

An image is a positive function on a plane. The value of this function at each point specifies the luminance or brightness of the picture at that point.[1] Digital images are sampled versions of such functions, where the value of the function is specified only at discrete locations on the image plane, known as *pixels*. The value of the luminance at each pixel is represented to a pre-defined precision, M. Eight bits of precision for luminance is common in imaging applications. The eight-bit precision is motivated by both the existing computer memory structures (1 byte = 8 bits) as well as the dynamic range of the human eye.

The prevalent custom is that the samples (pixels) reside on a rectangular lattice which we will assume for convenience to be $N \times N$. The brightness value at each pixel is a number between 0 and $2^M - 1$. The simplest binary representation of such an image is a list of the brightness values at each pixel, a list containing N^2M bits. Our standard image example in this paper is a square image with 512 pixels on a side. Each pixel value ranges from 0 to 255, so this canonical representation requires $512^2 \times 8 = 2,097,152$ bits.

Image coding consists of mapping images to strings of binary digits. A good image coder is one that produces binary strings whose lengths are on average much smaller than the original canonical representation of the image. In many imaging applications, exact reproduction of the image bits is not necessary. In this case, one can perturb the image slightly to obtain a shorter representation. If this perturbation is much smaller than the blurring and noise introduced in the formation of the image in the first place, there is no point in using the more accurate representation. Such a coding procedure, where perturbations reduce storage requirements, is known as *lossy coding*. The goal of lossy coding is to reproduce a given image with minimum distortion, given some constraint on the total number of bits in the coded representation.

But why can images be compressed on average? Suppose for example that we seek to efficiently store any image that has ever been seen by a human being. In principle, we can enumerate all images that have ever been seen and represent each image by its associated index. We generously assume that some 50 billion humans have walked the earth, that each person can distinguish on the order of 100 images per second, and that people live an average of 100 years. Combining these figures, we estimate that humans have seen some 1.6×10^{22} images, an enormous number. However, $1.6 \times 10^{22} \approx 2^{73}$, which means that the entire collective human visual experience can be represented with a mere 10 bytes (73 bits, to be precise)! This collection includes any image that a modern human eye has ever seen, including artwork, medical images, and so on, yet the collection can be conceptually represented with a small number of bits. The remaining vast majority of the $2^{512 \times 512 \times 8} \approx 10^{600,000}$ possible images in the canonical representation are not of general interest, because they contain little or no structure, and are noise-like.

[1] Color images are a generalization of this concept, and are represented by a three-dimensional vector function on a plane. In this paper, we do not explicitly treat color images. However, most of the results can be directly extended to color images.

While the above conceptual exercise is intriguing, it is also entirely impractical. Indexing and retrieval from a set of size 10^{22} is completely out of the question. However, we can see from the example the two main properties that image coders exploit. First, only a small fraction of the possible images in the canonical representation are likely to be of interest. *Entropy coding* can yield a much shorter image representation on average by using short code words for likely images and longer code words for less likely images.[2] Second, in our initial image gathering procedure we sample a continuum of possible images to form a discrete set. The reason we can do so is that most of the images that are left out are visually indistinguishable from images in our set. We can gain additional reductions in stored image size by discretizing our database of images more coarsely, a process called *quantization*. By mapping visually indistinguishable images to the same code, we reduce the number of code words needed to encode images, at the price of a small amount of distortion.

8.1.2 Chapter Overview

It is possible to quantize each pixel separately, a process known as *scalar quantization*. Quantizing a group of pixels together is known as *vector quantization*, or VQ. Vector quantization can, in principle, capture the maximum compression that is theoretically possible. In Section 8.2 we review the basics of quantization, vector quantization, and the mechanisms of gain in VQ.

VQ is a very powerful theoretical paradigm, and can asymptotically achieve optimality. But the computational cost and delay also grow exponentially with dimensionality, limiting the practicality of VQ. Due to these and other difficulties, most practical coding algorithms have turned to *transform coding* instead of high-dimensional VQ. Transform coding usually consists of scalar quantization in conjunction with a linear transform. This method captures much of the VQ gain, with only a fraction of the effort. In Section 8.3, we present the fundamentals of transform coding. We use a second-order model to motivate the use of transform coding.

The success of transform coding depends on how well the basis functions of the transform represent the features of the signal. At present, one of the most successful representations is the *subband/wavelet transform*. A complete derivation of fundamental results in subband signal analysis is beyond the scope of this chapter, and the reader is referred to excellent existing references such as [348, 337]. The present discussion focuses on compression aspects of subband transforms.

Section 8.4 outlines the key issues in subband coder design, from a general transform coding point of view. However, the general transform coding theory is based only on second-order properties of a random model of the signal. While subband coders fit into the general transform coding framework, they also go beyond it. Because of their nice temporal properties, subband decompositions

[2]For example, mapping the ubiquitous test image of Lena Sjööblom (cf. Figure 8.12) to a one-bit codeword would greatly compress the image coding literature.

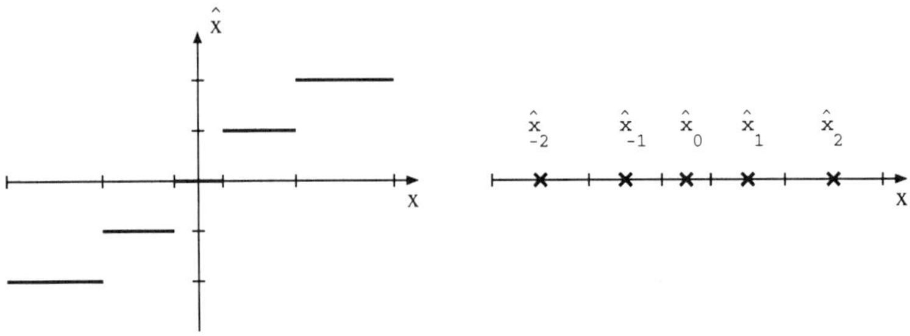

Figure 8.1 (Left) Quantizer as a function whose output values are discrete. (Right) because the output values are discrete, a quantizer can be more simply represented only on one axis.

can capture redundancies beyond general transform coders. We describe these extensions in Section 8.5, and show how they have motivated some of the most recent coders, which we describe in Sections 8.6, 8.7 and 8.8. We conclude by a summary and discussion of future directions.

8.2 QUANTIZATION

At the heart of image compression is the idea of quantization and approximation. While the images of interest for compression are almost always in a digital format, it is instructive and more mathematically elegant to treat the pixel luminances as being continuously valued. This assumption is not far from the truth if the original pixel values are represented with a large number of levels.

The role of quantization is to represent this continuum of values with a finite — preferably small — amount of information. Obviously this is not possible without some loss. The quantizer is a function whose set of output values are discrete and usually finite (cf. Figure 8.1). Good quantizers are those that represent the signal with minimum distortion.

Figure 8.1 also indicates a useful view of quantizers as a concatenation of two mappings. The first map, the *encoder*, takes partitions of the x-axis to the set of integers $\{-2, -1, 0, 1, 2\}$. The second, the *decoder*, takes integers to a set of output values $\{\hat{x}_k\}$. We need to define a measure of distortion in order to characterize "good" quantizers. We need to be able to approximate any possible value of x with an output value $\hat{x}_k$. Our goal is to minimize the distortion on average, over all values of x. For this, we need a probabilistic model for the signal values. The strategy is to have few or no reproduction points in locations at which the probability of the signal is negligible, whereas at highly probable signal values, more reproduction points need to be specified. While improbable values of x can still happen — and will be costly — this strategy pays off *on average*. This is the underlying principle behind all signal compression, and will be used over and over again in different guises.

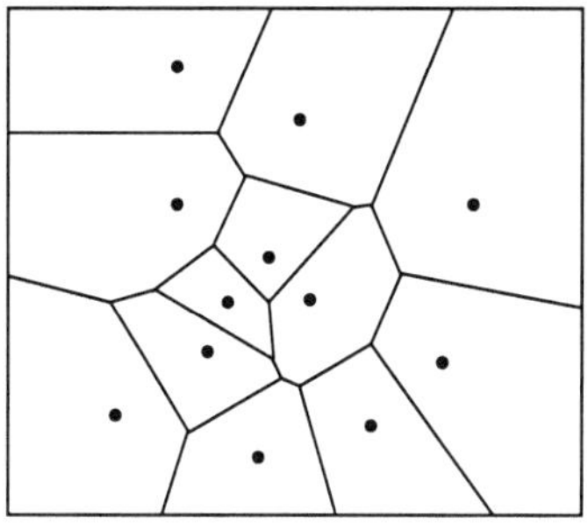

Figure 8.2 A Voronoi diagram.

The same concepts apply to the case where the input signal is not a scalar, but a vector. In that case, the quantizer is known as a *vector* quantizer.

8.2.1 Vector Quantization

Vector quantization (VQ) is the generalization of scalar quantization to the case of a vector. The basic structure of a VQ is essentially the same as scalar quantization, and consists of an encoder and a decoder. The encoder determines a partitioning of the input vector space and to each partition assigns an index, known as a *codeword.* The set of all codewords is known as a *codebook.* The decoder maps each index to a reproduction vector. Combined, the encoder and decoder map partitions of the space to a discrete set of vectors.

Vector quantization is a very important concept in compression: In 1959 Shannon [288] delineated the fundamental limitations of compression systems through his "Source coding theorem with a fidelity criterion." While this is not a constructive result, it does indicate, loosely speaking, that fully effective compression can only be achieved when input data samples are encoded in blocks of increasing length, i.e. in large vectors.

Optimal vector quantizers are not known in closed form except in a few trivial cases. However, two optimality conditions are known for VQ (and for scalar quantization as a special case) which lead to a practical algorithm for the design of quantizers. These conditions were discovered independently by Lloyd [198, 197] and Max [213] for scalar quantization, and were extended to VQ by Linde, Buzo, and Gray [194]. An example of cell shapes for a two-dimensional optimal quantizer is shown in Figure 8.2. We state the result here and refer the reader to [103] for proof.

Let $p_{\mathbf{X}}(\mathbf{x})$ be the probability density function for the random variable $\mathbf{X}$ we wish to quantize. Let $D(\mathbf{x}, \mathbf{y})$ be an appropriate distortion measure. Like scalar quantizers, vector quantizers are characterized by two operations, an encoder and a decoder. The encoder is defined by a partition of the range of $\mathbf{X}$ into sets $\mathcal{P}_k$. All realizations of $\mathbf{X}$ that lie in $\mathcal{P}_k$ are encoded to k and decoded to $\hat{\mathbf{x}}_k$. The decoder is defined by specifying the reproduction value $\hat{\mathbf{x}}_k$ for each partition $\mathcal{P}_k$.

A quantizer that minimizes the average distortion D must satisfy the following conditions:

1. *Nearest neighbor condition:* Given a set of reconstruction values $\{\hat{\mathbf{x}}_k\}$, the optimal partition of the values of $\mathbf{X}$ into sets $\mathcal{P}_k$ is the one for which each value $\mathbf{x}$ is mapped by the encoding and decoding process to the nearest reconstruction value. Thus,

$$\mathcal{P}_k = \{\mathbf{x} : D(\mathbf{x}, \hat{\mathbf{x}}_k) \leq D(\mathbf{x}, \hat{\mathbf{x}}_j) \text{ for } j \neq k\}.$$

2. *Centroid condition:* Given a partition of the range of $\mathbf{X}$ into sets $\mathcal{P}_k$, the optimal reconstruction values values $\hat{\mathbf{x}}_k$ are the generalized centroids of the sets $\mathcal{P}_k$. They satisfy

$$\hat{\mathbf{x}}_k = \arg\min \int_{\mathcal{P}_k} p_{\mathbf{X}}(\mathbf{z}) D(\mathbf{z}, \hat{\mathbf{x}}_k) d\mathbf{z}.$$

With the squared error distortion, the generalized centroid corresponds to the $p_{\mathbf{X}}(\mathbf{x})$-weighted centroid.

8.2.2 *Limitations of VQ*

Although vector quantization is a very powerful tool, the computational and storage requirements become prohibitive as the dimensionality of the vectors increase. The complexity of VQ has motivated a wide variety of constrained VQ methods. Among the most prominent are tree structured VQ, shape-gain VQ, classified VQ, multistage VQ, lattice VQ, and hierarchical VQ [103].

There is another important consideration that limits the practical use of VQ in its most general form: the design of the optimal quantizer requires knowledge of the underlying probability density function for the space of images. While we may claim empirical knowledge of lower order joint probability distributions, the same is not true of higher orders. A training set is drawn from the distribution we are trying to quantize, and is used to drive the algorithm that generates the quantizer. As the dimensionality of the model is increased, the amount of data available to estimate the density in each bin of the model decreases, and so does the reliability of the probability density function (pdf) estimate.[3] The issue is commonly known as "the curse of dimensionality."

Instead of accommodating the complexity of VQ, many compression systems opt to move away from it and employ techniques that allow them to use sample-wise or scalar quantization more effectively. To design more effective scalar quantization systems one needs to understand the source of the compression efficiency of the VQ. Then one can try to capture as much of that efficiency as possible, in the context of a scalar quantization system.

[3] Most existing techniques do not estimate the pdf. to use it for quantization, but rather use the data directly to generate the quantizer. However, the reliability problem is best pictured by the pdf estimation exercise. The effect remains the same with the so-called direct or data-driven methods.

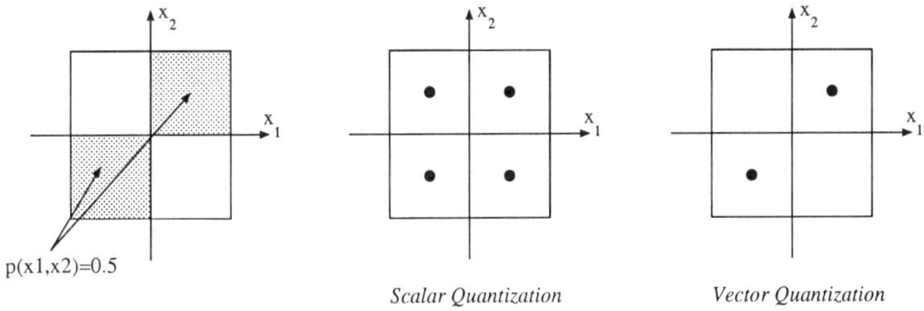

Figure 8.3 The leftmost figure shows a probability density for a two-dimensional vector $\mathbf{X}$. The realizations of $\mathbf{X}$ are uniformly distributed in the shaded areas. Center figure shows the four reconstruction values for an optimal scalar quantizer for $\mathbf{X}$ with expected squared error $\frac{1}{12}$. The figure on the right shows the two reconstruction values for an optimal vector quantizer for $\mathbf{X}$ with the same expected error. The vector quantizer requires 0.5 bits per sample, while the scalar quantizer requires 1 bit per sample.

8.2.3 Why VQ Works

The source of the compression efficiency of VQ is threefold: (a) exploiting correlation redundancy, (b) sphere covering and density shaping, and (c) exploiting fractional bitrates.

Correlation Redundancy. The greatest benefit of jointly quantizing random variables is that we can exploit the dependencies between them. Figure 8.3 shows a two-dimensional vector $\mathbf{X} = (X_1, X_2)$ distributed uniformly over the squares $[0,1] \times [0,1]$ and $[-1,0] \times [-1,0]$. The marginal densities for X_1 and X_2 are both uniform on $[-1,1]$. We now hold the expected distortion fixed and compare the cost of encoding X_1 and X_2 as a vector, to the cost of encoding these variables separately. For an expected squared error of $\frac{1}{12}$, the optimal scalar quantizer for both X_1 and X_2 is the one that partitions the interval $[-1,1]$ into the subintervals $[-1,0)$ and $[0,1]$. The cost per symbol is 1 bit, for a total of 2 bits for $\mathbf{X}$. The optimal vector quantizer with the same average distortion has cells that divides the square $[-1,1] \times [-1,1]$ in half along the line $y = -x$. The reconstruction values for these two cells are $\hat{\mathbf{x}}_a = (-\frac{1}{2}, -\frac{1}{2})$ and $\hat{\mathbf{x}}_b = (\frac{1}{2}, \frac{1}{2})$. The total cost per vector $\mathbf{X}$ is just 1 bit, only half that of the scalar case.

Because scalar quantizers are limited to using separable partitions, they cannot take advantage of dependencies between random variables. This is a serious limitation, but we can overcome it in part through a preprocessing step consisting of a linear transform.

Sphere Covering and Density Shaping. Even if random components of a vector are *independent*, there is some gain in quantizing them jointly, rather than independently. This may at first seem surprising, but is universally true and is due to the geometries of multidimensional spaces. We demonstrate by an example:

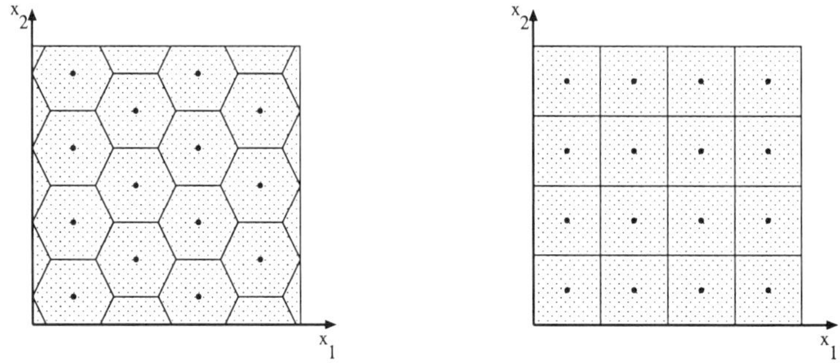

Figure 8.4 Tiling of the two-dimensional plane. The hexagonal tiling is more efficient, leading to a better rate-distortion.

Assume we intend to quantize two uniformly distributed, independent random variables $\mathbf{X}_1$ and $\mathbf{X}_2$. One may quantize them independently through two scalar quantizers, leading to a rectangular tiling of the $x_1 - x_2$ plane. Figure 8.4 shows this, as well as a second quantization strategy with hexagonal tiling. Assuming that these rectangles and hexagons have the same area, and hence the same rate (we disregard boundary effects), the squared error from the hexagonal partition is 3.8% lower than that of the square partition due to the extra error contributed by the corners of the rectangles.

In other words, one needs to cover the plane with shapes that have maximal ratio of area to moment-of-inertia. It is known that the best two-dimensional shape in that respect is the circle. It has also been shown that in 2-D the best polygon in that respect is the hexagon (so our example is in fact optimal).

Generally, in n-dimensional spaces, the performance of vector quantizers is determined in part by how closely we can approximate spheres with n-dimensional convex polytopes [102]. When we quantize vector components separately using scalar quantizers, the resulting Voronoi cells are all rectangular prisms, which only poorly approximate spheres. VQ makes it possible to use geometrically more efficient cell shapes. The benefits of improved spherical approximations increase in higher dimensions. For example, in 100 dimensions, the optimal vector quantizer for uniform densities has an error of roughly 0.69 times that of the optimal scalar quantizer for uniform densities, corresponding to a PSNR gain of 1.6 dB [102].

This problem is closely related to the well-studied problem of sphere covering in lattices. The problem remains largely unsolved, except for the uniform density at dimensions 2, 3, 8, and 24. Another noteworthy result is due to Zador [378], which gives asymptotic cell densities for high-resolution quantization.

Fractional Bitrates. In scalar quantization, each input sample is represented by a separate codeword. Therefore, the minimum bitrate achievable is one bit per sample, because symbols cannot be any shorter than one bit.

Since each symbol can only have an integer number of bits, one can generate fractional bitrates per sample by coding multiple samples together, as done in vector quantization. A vector quantizer coding N-dimensional vectors using a K-member codebook can achieve a rate of $(\log_2 K)/N$ bits per sample. For example, in Figure 8.3 scalar quantization cannot have a rate lower than one bit per sample, while vector quantization achieves the same distortion with 0.5 bits per sample.

The problem with fractional bitrates is especially acute when one symbol has very high probability and hence requires a very short code length. For example, the zero symbol is very common when coding the high-frequency portions of subband-transformed images. The only way of obtaining the benefit of fractional bitrates with scalar quantization is to jointly re-process the codewords after quantization. Useful techniques to perform this task include arithmetic coding, run-length coding (as in JPEG), and zerotree coding.

Finally, the three mechanisms of gain noted above are not always separable and independent of each other; processing aimed at capturing one form of gain may capture others as well. For example, run-length coding and zerotree coding are techniques that enable the attainment of fractional bitrates as well as the partial capture of correlation redundancy.

8.3 TRANSFORM CODING

The advantage of VQ over scalar quantization is primarily due to VQ's ability to exploit dependencies between samples. Direct scalar quantization of the samples does not capture this redundancy, and therefore suffers. However, we have seen that VQ presents severe practical difficulties, so the usage of scalar quantization is highly desirable. Transform coding is one mechanism by which we can capture the correlation redundancy, while using scalar quantization (Figure 8.5).

Transform coding does not capture the geometrical "packing redundancy," but this is usually a much smaller factor than the correlation redundancy. Scalar quantization also does not address fractional bitrates by itself, but other post-quantization operations can capture the advantage of fractional bitrates with manageable complexity (e.g. zerotrees, run-length coding, arithmetic coding).

To illustrate the exploitation of correlation redundancies by transform coding, we consider a toy image model. Images in our model consist of two pixels, one on the left and one on the right. We assume that these images are realizations of a two-dimensional random vector $\mathbf{X} = (X_1, X_2)$ for which X_1 and X_2 are identically distributed and jointly Gaussian. The identically distributed assumption is a reasonable one, since there is no *a priori* reason that pixels on the left and on the right should be any different. We know empirically that adjacent image pixels are highly correlated, so let us assume that the autocorrelation matrix for these pixels is

$$E[\mathbf{X}\mathbf{X}^T] = \begin{bmatrix} 1 & 0.9 \\ 0.9 & 1 \end{bmatrix}.$$

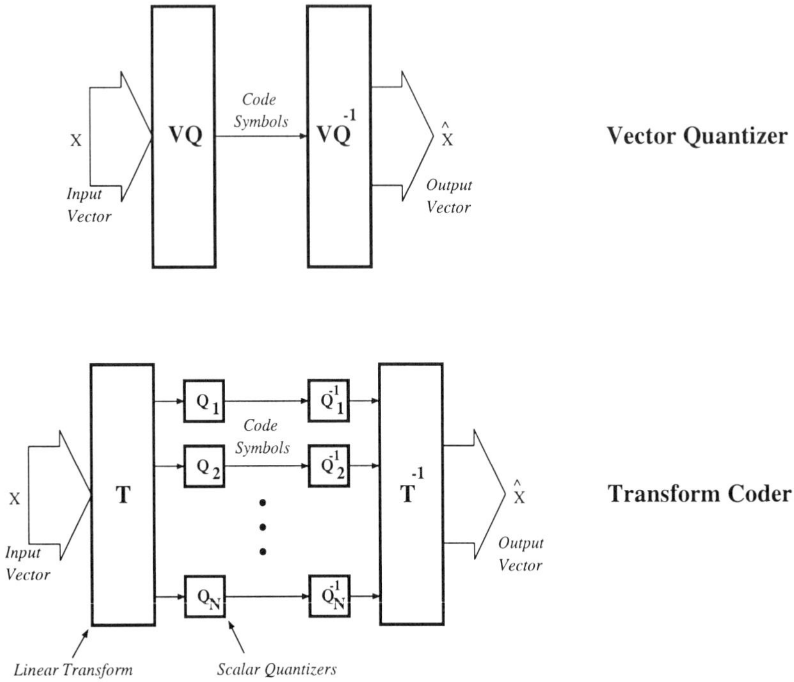

Figure 8.5 Transform coding simplifies the quantization process by applying a linear transform.

By symmetry, X_1 and X_2 will have identical quantizers. The Voronoi cells for this scalar quantization are shown on the left in Figure 8.6. The figure clearly shows the inefficiency of scalar quantization: most of the probability mass is concentrated in just five cells. Thus a significant fraction of the bits used to code the bins are spent distinguishing between cells of very low probability. This scalar quantization scheme does not take advantage of the coupling between X_1 and X_2.

We can remove the correlation between X_1 and X_2 by applying a rotation matrix. The result is a transformed vector $\mathbf{Y}$ given by

$$\mathbf{Y} = \frac{1}{\sqrt{2}} \begin{bmatrix} 1 & 1 \\ 1 & -1 \end{bmatrix} \begin{bmatrix} X_1 \\ X_2 \end{bmatrix}.$$

This rotation does not remove any of the variability in the data. Instead it packs that variability into the variable Y_1. The new variables Y_1 and Y_2 are independent, zero-mean Gaussian random variables with variances 1.9 and 0.1, respectively. By quantizing Y_1 finely and Y_2 coarsely we obtain a lower average error than by quantizing X_1 and X_2 equally. In the remainder of this section we will describe general procedures for finding appropriate redundancy-removing transforms, and for optimizing related quantization schemes.

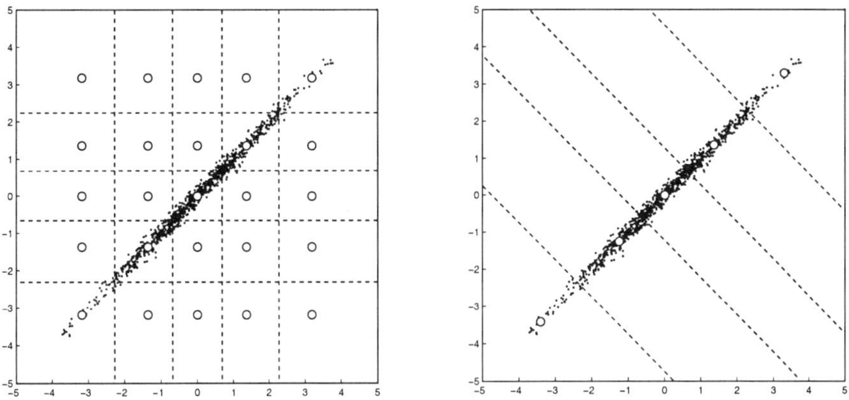

Figure 8.6 (Left) Correlated Gaussians of our image model quantized with optimal scalar quantization. Many reproduction values (shown as white dots) are wasted. (Right) Decorrelation by rotating the coordinate axes. The new axes are parallel and perpendicular to the major axis of the cloud. Scalar quantization is now much more efficient.

8.3.1 The Karhunen-Loève Transform

The previous simple example shows that removing correlations can lead to better compression. One can remove the correlation between a group of random variables using an orthogonal linear transform called the Karhunen-Loève transform (KLT), also known as the Hotelling transform.

Let $\mathbf{X}$ be a random vector that we assume has zero-mean and autocorrelation matrix $\mathbf{R}_X$. The KLT is the matrix $\mathbf{A}$ that will make the components of $\mathbf{Y} = \mathbf{AX}$ uncorrelated. It can be easily verified that such a transform matrix $\mathbf{A}$ can be constructed from the eigenvectors of $\mathbf{R}_X$, the autocorrelation matrix of $\mathbf{X}$. Without loss of generality, the rows of $\mathbf{A}$ are ordered so that $\mathbf{R_Y} = \text{diag}(\lambda_\mathbf{0}, \lambda_\mathbf{1}, \ldots, \lambda_\mathbf{N-1})$ where $\lambda_0 \geq \lambda_1 \geq \ldots \geq \lambda_{N-1} \geq 0$. This transform is optimal among all block transforms, in the sense described by the two theorems below (see [64] for proofs). The first theorem states that the KLT is optimal for mean-square approximation over a large class of random vectors.

Theorem 1 *Suppose that we truncate a transformed random vector* $\mathbf{AX}$*, keeping* m *out of the* N *coefficients and setting the rest to zero, then among all linear transforms, the KLT provides the best approximation in the mean square sense to the original vector.*

The KLT is also optimal among block transforms in the rate-distortion sense, but only when the input is a Gaussian vector and for high-resolution quantization. Optimality is achieved with a quantization strategy where the quantization noise from all transform coefficients are equal [64].

Theorem 2 *For a zero-mean, jointly Gaussian random vector, and for high-resolution quantization, among all block transforms, the KLT minimizes the distortion at a given rate.*

We emphasize that the KLT is optimal only in the context of block transforms, and partitioning an image into blocks leads to a reduction of performance. It can be shown [333] that subband transforms, which are not block-based, can provide better energy compaction properties than a block-based KLT. In the next section we motivate the use of subband transforms in coding applications using reverse waterfilling arguments.

8.3.2 Reverse Waterfilling and Subband Transforms

The limitations of block-based KLTs result from the blocking of the source. We can eliminate blocking considerations by restricting our attention to a stationary source and taking the block size to infinity. Stationary random processes have Toeplitz autocorrelation matrices. The eigenvectors of a circulant matrix are known to be complex exponentials, thus a large Toeplitz matrix with sufficiently decaying off-diagonal elements will have a diagonalizing transform close to the discrete Fourier transform (DFT). In other words, with sufficiently large block sizes, the KLT of a stationary process resembles the Fourier transform. In particular, one can make more precise statements about the KL *transform coefficients*. It has been shown [112] that in the limiting case when the block size goes to infinity, the distribution of KLT coefficients approaches that of the Fourier spectrum of the autocorrelation.

The optimality of KLT for block-based processing of Gaussian processes and the limiting results in [112] suggest that, when taking block sizes to infinity, power spectral density (psd) is the appropriate vehicle for bit allocation purposes. Similar to the case of finite-dimensional KLT, our bit allocation procedure consists of discarding very low-energy components of psd, and quantizing the remaining components such that each coefficient contributes an equal amount of distortion [64]. This concept is known as *reverse waterfilling.*

Reverse waterfilling can also be directly derived from a rate-distortion perspective. Unlike the "limiting KLT" argument described above, this explanation is not bound to high-resolution quantization and is therefore more general. Consider a Gaussian source with memory (i.e. correlated) with power spectral density $\Phi_X(\omega)$. The rate-distortion function can be expressed parametrically [22]:

$$D(\theta) = \frac{1}{2\pi^2} \int_\omega \min(\theta, \Phi_X(\omega))\, d\omega$$

$$R(\theta) = \frac{1}{4\pi^2} \int_\omega \max\left(0, \log(\frac{\Phi_X(\omega)}{\theta})\right)\, d\omega.$$

R and D are the rate and distortion pairs predicted by the Shannon limit, parameterized by θ. The goal is to design a quantization scheme that approach this theoretical rate-distortion limit. Our strategy is: at frequencies where signal power is less than θ, it is not worthwhile to spend any bits, therefore all the signal is thrown away (signal power = noise power). At frequencies where signal power is greater than θ, enough bitrate is assigned so that the noise

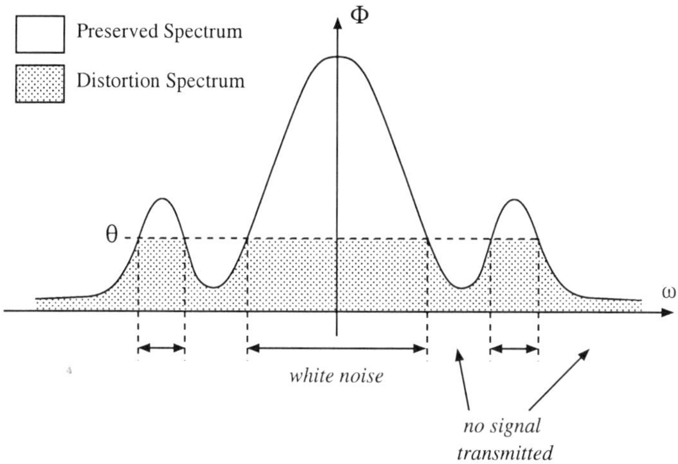

Figure 8.7 Reverse water filling of the spectrum for the rate-distortion function of a Gaussian source with memory.

power is exactly θ, and signal power over and above θ is preserved. Reverse waterfilling is illustrated in Figure 8.7.

In reverse waterfilling, each frequency component is quantized with a separate quantizer, reflecting the bit allocation appropriate for that particular component. For the Gaussian source, each frequency component is a Gaussian with variance given by the power spectrum. The process of quantizing these frequencies can be simplified by noting that frequencies with the same power density use the same quantizer. As a result, our task is simply to divide the spectrum into a partition of white segments, and to assign a quantizer to each segment. We achieve an optimal tradeoff between rate and distortion by this procedure for piecewise-constant power spectra. For other reasonably smooth power spectra, we can approach optimality by partitioning the spectrum into segments that are approximately white and quantizing each segment individually. Thus, removing blocking constraints leads to reverse waterfilling arguments which in turn motivate separation of the source into frequency bands. This separation is achieved by subband transforms, which are implemented by filter banks.

A subband transformer is a multi-rate digital signal processing system. As shown in Figure 8.8, a subband transform consists of two sets of filter banks, along with decimators and interpolators. On the left side of the figure we have the forward stage of the subband transform. The signal is sent through the input of the first set of filters, known as the *analysis filter bank*. The output of these filters is passed through decimators, which retain only one out of every M samples. The right hand side of the figure is the inverse stage of the transform. The filtered and decimated signal is first passed through a set of *interpolators*.

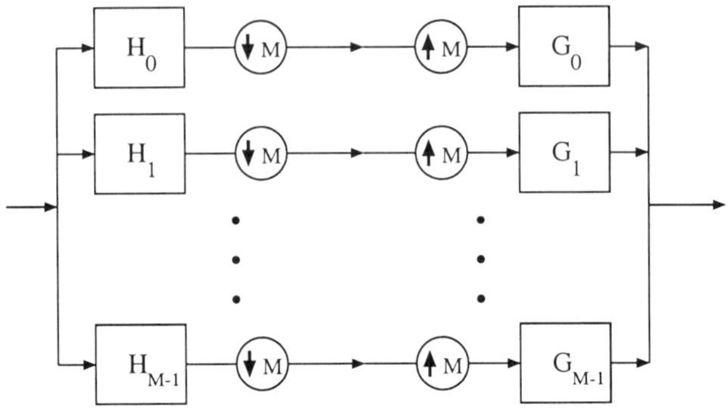

Figure 8.8 Filter bank.

Next it is passed through the *synthesis filter bank*. Finally, the components are recombined.

The combination of decimation and interpolation has the effect of zeroing out all but one out of M samples of the filtered signal. Under certain conditions, the original signal can be reconstructed exactly from this decimated M-band representation. The ideas leading to the perfect reconstruction conditions were discovered in stages by a number of investigators, including Croisier *et al.* [55], Vaidyanathan [335], Smith and Barnwell [299, 300] and Vetterli [343, 344]. For a detailed presentation of these developments, we refer the reader to the comprehensive texts by Vaidyanathan [337] and Vetterli and Kovačević [348].

8.3.3 Hierarchical Subbands, Wavelets, and Smoothness

A subset of subband transforms has been very successful in image compression applications; we refer to hierarchical subbands and in particular wavelet transforms. In this section we discuss reasons for the suitability of these transforms for image coding.

The waterfilling algorithm motivates a frequency domain approach to quantization and bit allocation. It is generally accepted that images of interest, considered as a whole, have power spectra that are stronger at lower frequencies. In particular, many use the exponentially decaying model for the tail of the power spectrum, given by

$$S_X(\omega) = e^{-\alpha|\omega|} .$$

We can now apply the waterfilling algorithm. Since the spectral model is not piecewise constant, we need to break it up in such a way that the spectrum is approximately constant in each segment. Applying a minimax criterion for the approximation yields a logarithmically distributed set of frequency bands. As we go from low frequency bands to high, the length of each successive band increases by a constant factor that is greater than 1. This in turn motivates a hierarchical structure for the subband decomposition of the signal (cf. Figure 8.9).

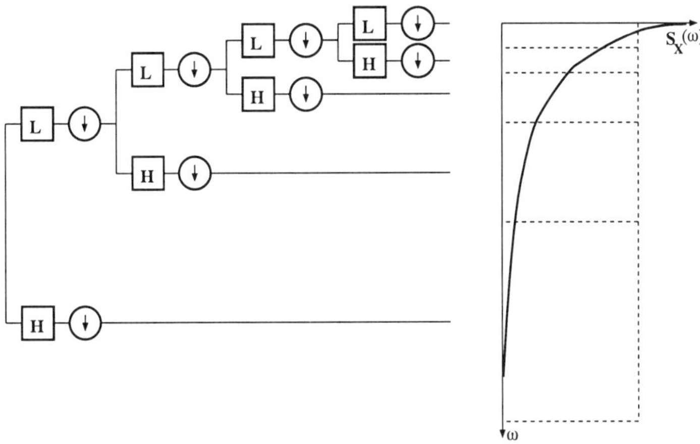

Figure 8.9 Exponential decay of power density motivates a logarithmic frequency division, leading to a hierarchical subband structure.

Hierarchical decompositions possess a number of additional attractive features. One of the most important is that they provide a measure of *scale invariance* in the transform. Consider that a shift of the location of the viewer results (roughly) in a translation and rescaling of the perceived image. We have no *a priori* reason to expect any particular viewer location; as a result, natural images possess no favored translates or scalings. Subband transforms are invariant under translates by K pixels (where K depends on the transform) since they are formed by convolution and downsampling. Hierarchical transforms add an additional degree of scale invariance. The result is a family of coding algorithms that work well with images at a wide variety of scales.

A second advantage of hierarchical subband decompositions is that they provide a convenient tree structure for the coded data. This turns out to be very important for taking advantage of remaining correlations in the signal (because image pixels, unlike our model, are not generally jointly Gaussian). We will see that zerotree coders use this structure with great efficiency.

A third advantage of hierarchical decompositions is that they leverage a considerable body of work on wavelets. The discrete wavelet transform is functionally equivalent to a hierarchical subband transform, and each framework brings to bear an important perspective on the problem of designing effective transforms. As we have seen, the subband perspective is motivated by frequency-domain arguments about optimal compression of stationary Gaussian random processes. The wavelet perspective, in contrast, emphasizes frequency as well as spatial considerations. This spatial emphasis is particularly useful for addressing nonstationary behavior in images, as we will see in the discussion of coders below.

Both the wavelet and subband perspectives yield useful design criteria for constructing filters. The subband framework emphasizes coding gain, while the wavelet framework emphasizes smoothness and polynomial reproduction. Both sets of criteria have proven useful in applications, and interesting research synthesizing these perspectives is underway.

8.4 A BASIC SUBBAND IMAGE CODER

Three basic components underly current subband coders: a decorrelating transform, a quantization procedure, and entropy coding. This structure is a legacy of traditional transform coding, and has been with subband image coding from its earliest days [365, 363]. Before discussing state-of-the-art coders (and their advanced features) in the next sections, we will describe a basic subband coder and discuss issues in the design of its components.

8.4.1 Choice of Basis

Deciding on the optimal basis to use for image coding is a difficult problem. A number of design criteria, including smoothness, accuracy of approximation, size of support, and filter frequency selectivity are known to be important. However, the best combination of these features is not known.

The simplest form of basis for images is a separable basis formed from products of one dimensional filters. The problem of basis design is much simpler in one dimension, and almost all current coders employ separable transforms. Although the two-dimensional design problem is not as well understood, recent work of Sweldens and Kovačević [184] simplifies the design of non-separable bases, and such bases may prove more efficient than separable transforms.

Unser [334] shows that spline wavelets are attractive for coding applications based on approximation theoretic considerations. Experiments by Rioul [266] for orthogonal bases indicate that smoothness is an important consideration for compression. Experiments by Antonini *et al.* [8] find that both vanishing moments and smoothness are important, and for the filters tested they found that smoothness appeared to be slightly more important than the number of vanishing moments. Nonetheless, Vetterli and Herley [347] state that "the importance of regularity for signal processing applications is still an open question." The bases most commonly used in practice have between one and two continuous derivatives. Additional smoothness does not appear to yield significant improvements in coding results.

Villasenor *et al.* [349] have examined all minimum order biorthogonal filter banks with lengths ≤ 36. In addition to the criteria already mentioned, [349] also examines measures of oscillatory behavior and of the sensitivity of the coarse-scale approximations to the translations of the signal. The best filter found in these experiments was a 7/9-tap spline variant with less dissimilar lengths from [8], and this filter is one of the most commonly used in wavelet coders.

There is one caveat with regard to the results of the filter evaluation in [349]. Villasenor *et al.* compare peak signal to noise ratios generated by a simple transform coding scheme. The bit allocation scheme they use works well for orthogonal bases, but it can be improved upon considerably in the biorthogonal case. This inefficient bit allocation causes some promising biorthogonal filter sets to be overlooked.

For biorthogonal transforms, the squared error in the transform domain is not the same as the squared error in the original image. As a result, the problem of minimizing image error is considerably more difficult than in the orthogonal case. We can reduce image-domain errors by performing bit allocation using a weighted transform-domain error measure that we discuss in Section 8.4.5. A number of other filters yield performance comparable to that of the 7/9 filter of [8] provided that we do bit allocation with a weighted error measure. One such basis is the Deslauriers-Dubuc interpolating wavelet of order 4 [73, 312], which has the advantage of having filter taps that are dyadic rationals. Others are the 10/18 filters in [329], and the 28/28 filters designed with the software in [236].

One promising new set of filters has been developed by Balasingham and Ramstad [10]. Their design procedure combines classical filter design techniques with ideas from wavelet constructions and yields filters that perform better than the popular 7/9 filter set from [8].

8.4.2 Boundaries

Careful handling of image boundaries when performing the transform is essential for effective compression algorithms. Naïve techniques for artificially extending images beyond given boundaries such as periodization or zero-padding lead to significant coding inefficiencies. For symmetrical bases, an effective strategy for handling boundaries is to extend the image via reflection [302]. Such an extension preserves continuity at the boundaries and usually leads to much smaller transform coefficients than if discontinuities were present at the boundaries. Brislawn [29] describes in detail procedures for non-expansive symmetric extensions of boundaries. An alternative approach is to modify the filter near the boundary. Boundary filters [126, 124] can be constructed that preserve filter orthogonality at boundaries. The lifting scheme [313] provides a related method for handling filtering near the boundaries.

8.4.3 Quantization

Most current subband coders employ scalar quantization for coding. There are two basic strategies for performing the scalar quantization stage. If we knew the distribution of coefficients for each subband in advance, the optimal strategy would be to use entropy-constrained Lloyd-Max quantizers for each subband. In general we do not have such knowledge, but we can provide a parametric description of coefficient distributions by sending side information. Coefficients in the high pass subbands of the transform are known *a priori* to be distributed as generalized Gaussians [203] centered around zero.

A much simpler quantizer that is commonly employed in practice is a uniform quantizer with a dead zone. The quantization bins, as shown in Figure 8.10, are of the form $[n\Delta, (n+1)\Delta)$ for $n \in \mathcal{Z}$ except for the central bin $[-\Delta, \Delta)$. Each bin is decoded to the value at the center of the bin. A slightly more

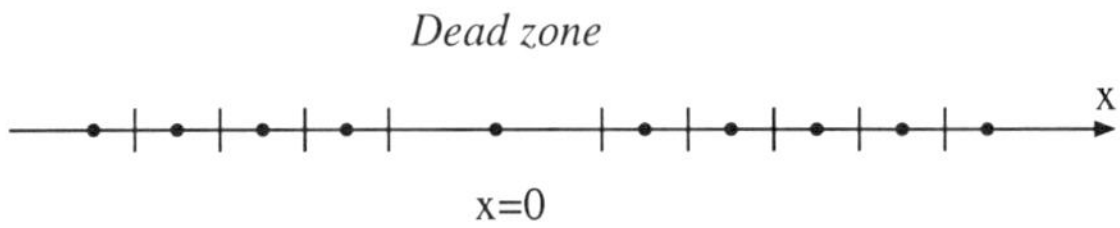

Figure 8.10 Dead-zone quantizer, with larger encoder partition around $x = 0$ (dead zone) and uniform quantization elsewhere.

sophisticated version decodes to the centroid of the bin instead of the center. Although the dead-zone quantizer does not possess the asymptotic optimality of the uniform quantizer [109], in many practical cases it works almost as well as Lloyd-Max quantizers, especially when decoded to bin centroids [90]. The main motivation for using dead-zone quantization is that the zero bin often has higher probability than other bins, thus a wider zero bin is beneficial in a Lagrangian sense, providing better rate-distortion when the quantizer is followed by entropy coding. For more details see the literature on entropy-constrained quantization [103]. Dead-zone quantization can also be nested to produce an embedded bitstream, following a procedure in [320].

8.4.4 Entropy Coding

Arithmetic coding provides a near-optimal entropy coding for the quantized coefficient values. The coder requires an estimate of the distribution of quantized coefficients. This estimate can be approximately specified by providing parameters for a generalized Gaussian or a Laplacian density. Alternatively the probabilities can be estimated online. Online adaptive estimation has the advantage of allowing coders to exploit local changes in image statistics. Efficient adaptive estimation procedures (context modeling) are discussed in [16, 79, 372, 371].

Because images are not jointly Gaussian random processes, the transform coefficients, although decorrelated, still contain considerable structure. The entropy coder can take advantage of some of this structure by conditioning the encodings on previously encoded values. Efficient context based modeling and entropy coding of wavelet coefficients can significantly improve the coding performance. In fact, several very competitive wavelet image coders are based on such techniques [320, 371, 379, 40].

8.4.5 Bit Allocation

The final question we need to address is that of how finely to quantize each subband. The general idea is to determine the number of bits b_j to devote to coding each subband j so that the total distortion $\sum_j D_j(b_j)$ is minimized subject to the constraint that $\sum_j b_j \leq B$. Here $D_j(b_j)$ is the amount of distortion incurred in coding subband j with b_j bits. When the functions $D_j(b)$ are known in closed form we can solve the problem using the Kuhn-Tucker conditions. One common practice is to approximate the functions $D_j(b)$ with the rate-distortion function for a Gaussian random variable. However, this approximation is not accurate at low bit rates. Better results may be obtained by measuring $D_j(b)$ for a range of values of b and then solving the constrained minimization problem using integer programming techniques. An algorithm of Shoham and Gersho [293] solves precisely this problem.

For biorthogonal wavelets we have the additional problem that squared error in the transform domain is not equal to squared error in the inverted image. Moulin [226] has formulated a multi-scale relaxation algorithm which provides an approximate solution to the allocation problem for this case. Moulin's algorithm yields substantially better results than the naïve approach of minimizing squared error in the transform domain.

A simpler approach is to approximate the squared error in the image by weighting the squared errors in each subband. The weight w_j for subband j is obtained as follows: we set a single coefficient in subband j to 1 and set all other wavelet coefficients to zero. We then invert this transform. The weight w_j is equal to the sum of the squares of the values in the resulting inverse transform. We allocate bits by minimizing the *weighted* sum $\sum_j w_j D_j(b_j)$ rather than the sum $\sum_j D_j(b_j)$. Further details may be found in Naveen and Woods [364]. This weighting procedure results in substantial coding improvements when using wavelets that are far from orthogonal, such as the Deslauriers-Dubuc wavelets popularized by the lifting scheme [313]. The 7/9 tap filter set of [8], on the other hand, has weights that are all nearly 1, so this weighting provides little benefit.

8.4.6 *Perceptually Weighted Error Measures*

Our goal in lossy image coding is to minimize visual discrepancies between the original and compressed images. Measuring visual discrepancy is a difficult task. There has been a great deal of research on this problem, but because of the great complexity of the human visual system, no simple, accurate, and mathematically tractable measure has been found.

Our discussion up to this point has focused on minimizing squared error distortion in compressed images primarily because this error metric is mathematically convenient. The measure suffers from a number of deficits, however. For example, consider two images that are the same everywhere except in a small region. Even if the difference in this small region is large and highly visible, the mean squared error for the whole image will be small because the discrepancy is confined to a small region. Similarly, errors that are localized in straight lines, such as the blocking artifacts produced by the discrete cosine transform, are much more visually objectionable than squared error considerations alone indicate.

There is evidence that the human visual system makes use of a multi-resolution image representation; see [351] for an overview. The eye is much more sensitive to errors in low frequencies than in high. As a result, we can improve the correspondence between the squared error metric and perceived error by weighting the errors in different subbands according to the eye's contrast sensitivity in a corresponding frequency range. Weights for the commonly used 7/9-tap filter set of [8] have been computed by Watson *et al.* in [353].

8.5 EXTENDING THE TRANSFORM CODER PARADIGM

The basic subband coder discussed in Section 8.4 is based on the traditional transform coding paradigm, namely decorrelation and scalar quantization of individual transform coefficients. The mathematical framework used in deriving the wavelet transform motivates compression algorithms that go beyond the traditional mechanisms used in transform coding. These important extensions are at the heart of the modern coding algorithms of Sections 8.6 and 8.8. We take a moment here to discuss these extensions.

Conventional transform coding relies on energy compaction in an ordered set of transform coefficients, and quantizes those coefficients with a priority according to their order. This paradigm, while quite powerful, is based on several assumptions about images that are not always completely accurate. In particular, the Gaussian assumption breaks down for the joint distributions across image discontinuities. Mallat and Falzon [204] give the following example of how the Gaussian, high-rate analysis breaks down at low rates for non-Gaussian processes:

Let $Y[n]$ be a random N-vector defined by

$$Y[n] = \begin{cases} \text{X} & \text{if } n = P \\ \text{X} & \text{if } n = P + 1(modN) \\ 0 & \text{otherwise.} \end{cases}$$

Here, P is a random integer uniformly distributed between 0 and $N - 1$ and X is a random variable that equals 1 or -1 each with probability $\frac{1}{2}$. X and P are independent. The vector Y has zero mean and a covariance matrix with entries

$$E\{Y[n]Y[m]\} = \begin{cases} \frac{2}{N} & \text{for } n = m \\ \frac{1}{N} & \text{for } |n - m| \in \{1, N - 1\} \\ 0 & \text{otherwise.} \end{cases}$$

The covariance matrix is circulant, so the KLT for this process is simply the Fourier transform, a very inefficient representation for coding Y. The energy at frequency k will be $|1 + e^{2\pi i \frac{k}{N}}|^2$ which means that the energy of Y is spread out over the entire low-frequency half of the Fourier basis with some spill-over into the high-frequency half. The KLT has "packed" the energy of the two non-zero coefficients of Y into roughly $\frac{N}{2}$ coefficients. It is obvious that Y was much more compact in its original form, and could be coded better without transformation: Only two coefficients in Y are non-zero, and we need only specify the values of these coefficients and their positions.

As suggested by the example above, the essence of the extensions to traditional transform coding is the idea of selection operators. Instead of quantizing the transform coefficients in a pre-determined order of priority, the wavelet framework lends itself to improvements, through judicious choice of which elements to code. This is made possible primarily because wavelet basis elements are spatially as well as spectrally compact. In parts of the image where the energy is spatially but not spectrally compact (like the example above) one can use selection operators to choose subsets of the transform coefficients that represent that signal efficiently. A most notable example is the Zerotree coder and its variants (Section 8.6).

More formally, the extension consists of dropping the constraint of linear image approximations, as the selection operator is nonlinear. The work of DeVore *et al.* [74] and of Mallat and Falzon [204] suggests that at low rates, the problem of image coding can be more effectively addressed as a problem in obtaining a *non-linear* image approximation. This idea leads to some important differences in coder implementation compared to the linear framework. For linear approximations, Theorems 1 and 2 in Section 8.3.1 suggest that at low rates we should approximate our images using a fixed subset of the KL basis vectors. We set a fixed set of transform coefficients to zero, namely the coefficients corresponding to the smallest eigenvalues of the covariance matrix. The non-linear approximation idea, on the other hand, uses a subset of basis functions that are selected adaptively based on the given image. Information describing the particular set of basis functions used for the approximation, called a significance map, is sent as side information. In Section 8.6 we describe zerotrees, a very important data structure used to efficiently encode significance maps.

Our example suggests that a second important assumption to relax is that our images come from a single jointly Gaussian source. We can obtain better energy compaction by optimizing our transform to the particular image at hand rather than to the global ensemble of images. Frequency-adaptive and space/frequency-adaptive coders decompose images over a large library of different bases and choose an energy-packing transform that is adapted to the image itself. We describe these adaptive coders in Section 8.7.

The selection operator that characterizes the extension to the transform coder paradigm generates information that needs to be conveyed to the decoder as "side information." This side information can be in the form of zerotrees, or more generally energy classes. Backward mixture estimation represents a different approach: it assumes that the side information is largely redundant and can be estimated from the causal data. By cutting down on the transmitted side information, these algorithms achieve a remarkable degree of performance and efficiency.

For reference, Table 8.1 provides a comparison of the peak signal to noise ratios for the coders we discuss. The test images are the 512×512 Lena image and the 512×512 Barbara image. Figure 8.11 shows the Barbara image as compressed by JPEG, a baseline wavelet transform coder, and the zerotree coder of Said and Pearlman [275]. The Barbara image is particularly difficult to code, and we have compressed the image at a low rate to emphasize coder errors. The blocking artifacts produced by the discrete cosine transform are highly visible in the image on the top right. The difference between the two wavelet coded images is more subtle but quite visible at close range. Because of the more efficient coefficient encoding (to be discussed below), the zerotree-

Table 8.1 Peak signal to noise ratios in decibels for various coders

Type of Coder	Lena (b/p)			Barbara (b/p)		
	1.0	0.5	0.25	1.0	0.5	0.25
JPEG [250]	37.9	34.9	31.6	33.1	28.3	25.2
Optimized JPEG [56]	39.6	35.9	32.3	35.9	30.6	26.7
Baseline Wavelet [65]	39.4	36.2	33.2	34.6	29.5	26.6
Zerotree (Shapiro) [290]	39.6	36.3	33.2	35.1	30.5	26.8
Zerotree (Said-Pearlman) [275]	40.5	37.2	34.1	36.9	31.7	27.8
Zerotree (R-D optimized) [375]	40.5	37.4	34.3	37.0	31.3	27.2
Frequency-adaptive [261]	39.3	36.4	33.4	36.4	31.8	28.2
Space-frequency adaptive [127]	40.1	36.9	33.8	37.0	32.3	28.7
Frequency-adaptive + Zerotrees [373]	40.6	37.4	34.4	37.7	33.1	29.3
TCQ subband [166]	41.1	37.7	34.3	–	–	–
TCQ + zerotrees [376]	41.2	37.9	34.8	–	–	–
Backward mixture estimation [199]	41.0	37.7	34.6	–	–	–
Context modeling (Chrysafis-Ortega) [40]	40.9	37.7	34.6	–	–	–
Context modeling (Wu) [371]	40.8	37.7	34.6	–	–	–

coded image has much sharper edges and better preserves the striped texture than does the baseline transform coder.

8.6 ZEROTREE CODING

The rate-distortion analysis of the previous sections showed that optimal bitrate allocation is achieved when the signal is divided into "white" subbands. It was also shown that for typical signals of interest, this leads to narrower bands in the low frequencies and wider bands in the high frequencies. Hence, wavelet transforms, with their logarithmic band structure, have very good energy compaction properties.

This energy compaction leads to efficient utilization of scalar quantizers. However, a cursory examination of the transform in Figure 8.12 shows that a significant amount of structure is present, particularly in the fine scale coefficients. Wherever there is structure, there is room for compression, and advanced wavelet compression algorithms all address this structure in the higher frequency subbands.

One of the most prevalent approaches to this problem is based on exploiting the relationships of the wavelet coefficients across bands. A direct visual inspection indicates that large areas in the high frequency bands have little or no energy, and the small areas that have significant energy are similar in shape and location, across different bands. These high-energy areas stem from poor energy compaction close to the edges of the original image. Flat and slowly varying regions in the original image are well-described by the low-frequency basis elements of the wavelet transform (hence leading to high energy compaction). At the edge locations, however, low-frequency basis elements cannot describe

Figure 8.11 Compression of the 512×512 Barbara test image at 0.25 bits per pixel. (Top left) original image. (Top right) baseline JPEG, PSNR = 24.4 dB. (Bottom left) baseline wavelet transform coder [65], PSNR = 26.6 dB. (Bottom right) Said and Pearlman zerotree coder, PSNR = 27.6 dB.

the signal adequately, and some of the energy leaks into high-frequency coefficients. This happens similarly at all scales, thus the high-energy high-frequency coefficients representing the edges in the image have the same shape.

Our *a priori* knowledge, that images of interest are formed mainly from flat areas, textures, and edges, allows us to take advantage of the resulting cross-band structure. Zerotree coders combine the idea of cross-band correlation with the notion of coding zeros jointly (which we saw previously in the case of JPEG), to generate very powerful compression algorithms.

The first instance of the implementation of zerotrees is due to Lewis and Knowles [191]. In their algorithm the image is represented by a tree-structured data construct (Figure 8.13). This data structure is implied by a dyadic discrete wavelet transform (Figure 8.9) in two dimensions. The root node of the tree

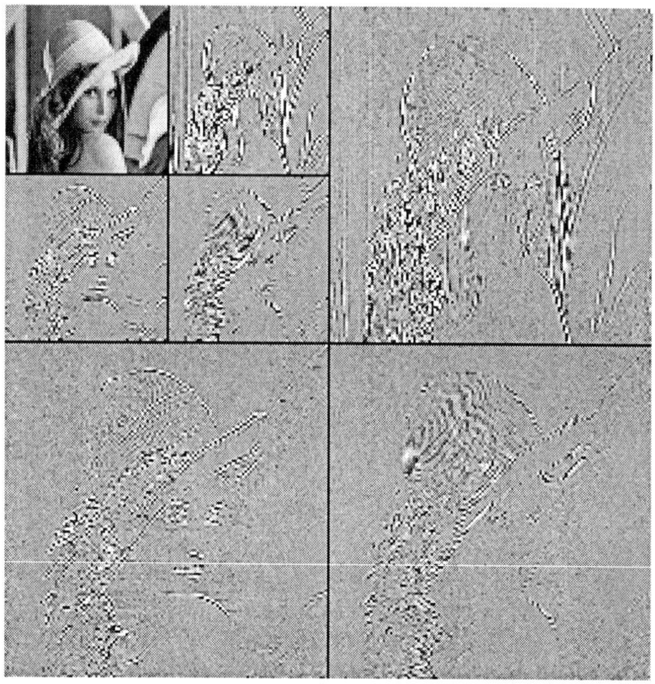

Figure 8.12 Wavelet transform of the image "Lena."

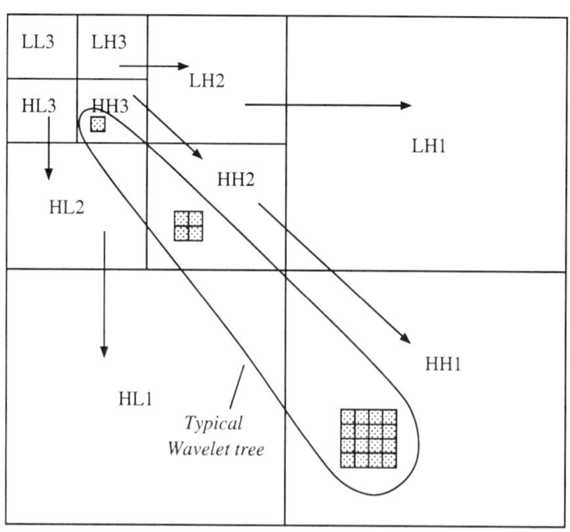

Figure 8.13 Space-frequency structure of wavelet transform.

represents the coefficient at the lowest frequency, which is the parent of three nodes. Nodes inside the tree correspond to wavelet coefficients at a frequency band determined by their height in the tree. Each of these coefficients has four children, which correspond to the wavelets at the next finer scale having the same location in space. These four coefficients represent the four phases of the higher resolution basis elements at that location. At the bottom of the data structure lie the leaf nodes, which have no children.

Note that there exist three such quadtrees for each coefficient in the low frequency band. Each of these three trees corresponds to one of three filtering orderings: there is one tree consisting entirely of coefficients arising from horizontal high-pass, vertical low-pass operation (HL); one for horizontal low-pass, vertical high-pass (LH), and one for high-pass in both directions (HH).

The zerotree quantization model used by Lewis and Knowles was arrived at by observing that often when a wavelet coefficient is small, its children on the wavelet tree are also small. This phenomenon happens because significant coefficients arise from edges and texture, which are local. It is not difficult to see that this is a form of conditioning. Lewis and Knowles took this conditioning to the limit, and assumed that insignificant parent nodes always imply insignificant child nodes. A tree or subtree that contains (or is assumed to contain) only insignificant coefficients is known as a zerotree.

The Lewis and Knowles coder achieves its compression ratios by joint coding of zeros. For efficient run-length coding, one needs to first find a conducive data structure, e.g. the zig-zag scan in JPEG. Perhaps the most significant contribution of this work was to realize that wavelet domain data provide an excellent context for run-length coding: not only are large run lengths of zeros generated, but also there is no need to transmit the length of zero runs, because they are assumed to automatically terminate at the leaf nodes of the tree. Much the same as in JPEG, this is a form of joint vector/scalar quantization. Each individual (significant) coefficient is quantized separately, but the symbols corresponding to small coefficients in fact are representing a vector consisting of that element and the zero run that follows it to the bottom of the tree.

8.6.1 The Shapiro and Said-Pearlman Coders

The Lewis and Knowles algorithm, while capturing the basic ideas inherent in many of the later coders, was incomplete. It had the key intuition that lies at the heart of more advanced zerotree coders, but did not efficiently specify significance maps, which is crucial to the performance of wavelet coders.

A significance map is a binary function whose value determines whether each coefficient is significant or not. If not significant, a coefficient is assumed to quantize to zero. Hence a decoder that knows the significance map needs no further information about that coefficient. Otherwise, the coefficient is quantized to a non-zero value. The method of Lewis and Knowles does not generate a significance map from the actual data, but uses one implicitly, based on *a priori* assumptions on the structure of the data, namely that insignificant parent nodes imply insignificant child nodes. On the infrequent occasions when this assumption does not hold, a high price is paid in terms of distortion. The methods to be discussed below make use of the fact that, by using a small number of bits to correct our mistaken assumptions about the occurrences of zerotrees, we can reduce the coded image distortion considerably.

The first algorithm of this family is due to Shapiro [290] and is known as the embedded zerotree wavelet (EZW) algorithm. Shapiro's coder was based on transmitting both the non-zero data and a significance map. The bits needed to specify a significance map can easily dominate the coder output, especially at lower bitrates. However, there is a great deal of redundancy in a general significance map for visual data, and the bitrates for its representation can be kept in check by conditioning the map values at each node of the tree on the corresponding value at the parent node. Whenever an insignificant parent node is observed, it is highly likely that the descendents are also insignificant. Therefore, most of the time, a "zerotree" significance map symbol is generated. But because p, the probability of this event, is close to 1, its information content, $-p \log p$, is very small. So most of the time, a very small amount of information is transmitted, and this keeps the average bitrate needed for the significance map relatively small.

Once in a while, one or more of the children of an insignificant node will be significant. In that case, a symbol for "isolated zero" is transmitted. The likelihood of this event is lower, and thus the bitrate for conveying this information is higher. It is essential to pay this price, however, to avoid losing significant information down the tree and therefore generating large distortions.

In summary, the Shapiro algorithm uses three symbols for significance maps: zerotree, isolated zero, or significant value. This structure, combined with conditional entropy coding of symbols, yields very good rate-distortion performance.

In addition, Shapiro's coder also generates an embedded code. Coders that generate embedded codes are said to have the *progressive transmission* or *successive refinement* property. Successive refinement consists of first approximating the image with a few bits of data, and then improving the approximation as more and more information is supplied. An embedded code has the property that for two given rates $R_1 > R_2$, the rate-R_2 code is a prefix to the rate-R_1 code. Such codes are of great practical interest for the following reasons:

- The encoder can easily achieve a precise bitrate by continuing to output bits when it reaches the desired rate.

- The decoder can cease decoding at any given point, generating an image that is the best representation possible with the decoded number of bits. This is of practical interest for broadcast applications where multiple decoders with varying computational, display, and bandwidth capabilities attempt to receive the same bitstream. With an embedded code, each receiver can decode the passing bitstream according to its particular needs and capabilities.

- Embedded codes are also very useful for indexing and browsing, where only a rough approximation is sufficient for deciding whether the image

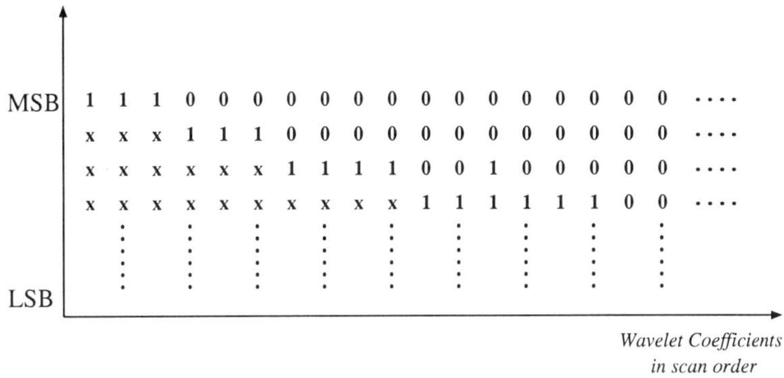

Figure 8.14 Bit plane profile for raster scan ordered wavelet coefficients.

needs to be decoded or received in full. Embedded codes speed up the process of screening images: after decoding only a small portion of the code, one knows if the target image is present. If not, decoding is aborted and the next image is requested, making it possible to screen a large number of images quickly. Once the desired image is located, the complete image is decoded.

Shapiro's method generates an embedded code by using a bit-slice approach (cf. Figure 8.14). First, the wavelet coefficients of the image are indexed into a one-dimensional array, according to their order of importance. This order places lower frequency bands before higher frequency bands, since they have more energy, and coefficients within each band appear in a raster scan order. The bit-slice code is generated by scanning this one-dimensional array, comparing each coefficient with a threshold T. This initial scan provides the decoder with sufficient information to recover the most significant bit slice. In the next pass, our information about each coefficient is refined to a resolution of $T/2$, and the pass generates another bit slice of information. This process is repeated until there are no more slices to code.

Figure 8.14 shows that the upper bit slices contain a great many zeros because there are many coefficients below the threshold. The role of zerotree coding is to avoid transmitting all these zeros. Once a zerotree symbol is transmitted, we know that all the descendent coefficients are zero, so no information is transmitted for them. In effect, zerotrees are a clever form of run-length coding, where the coefficients are ordered in a way to generate longer run lengths (more efficient) as well as making the runs self-terminating, so the length of the runs need not be transmitted.

The zerotree symbols (with high probability and small code length) can be transmitted again and again for a given coefficient, until it rises above the sinking threshold, at which point it will be tagged as a significant coefficient. After this point, no more zerotree information will be transmitted for this coefficient.

To achieve embeddedness, Shapiro uses a clever method of encoding the sign of the wavelet coefficients with the significance information. There are also further details of the priority of wavelet coefficients, the bit-slice coding, and adaptive arithmetic coding of quantized values (entropy coding), which we will not pursue further in this review. The interested reader is referred to [290] for more details.

Said and Pearlman [275] have produced an enhanced implementation of the zerotree algorithm, known as Set Partitioning in Hierarchical Trees (SPIHT). Their method is based on the same premises as the Shapiro algorithm, but with more attention to detail. The public domain version of this coder is very fast, and improves the performance of EZW by 0.3-0.6 dB. This gain is mostly due to the fact that the original zerotree algorithms allow special symbols only for single zerotrees, while in reality, there are other sets of zeros that appear with sufficient frequency to warrant special symbols of their own. In particular, the Said-Pearlman coder provides symbols for combinations of parallel zerotrees.

Davis and Chawla [66] have shown that both the Shapiro and the Said and Pearlman coders are members of a large family of tree-structured significance mapping schemes. They provide a theoretical framework that explains in more detail the performance of these coders and describe an algorithm for selecting a member of this family of significance maps that is optimized for a given image or class of images.

8.6.2 Zerotrees and Rate-Distortion Optimization

In the previous coders, zerotrees were used only when they were detected in the actual data. But consider for the moment the following hypothetical example: assume that in an image, there is a wide area of little activity, so that in the corresponding location of the wavelet coefficients there exists a large group of insignificant values. Ordinarily, this would warrant the use of a big zerotree and a low expenditure of bitrate over that area. Suppose, however, that there is a one-pixel discontinuity in the middle of the area, such that at the bottom of the would-be zerotree, there is one significant coefficient. The algorithms described so far would prohibit the use of a zerotree for the entire area.

Inaccurate representation of a single pixel will change the average distortion in the image only by a small amount. In our example we can gain significant coding efficiency by ignoring the single significant pixel, so that we can use a large zerotree. We need a way to determine the circumstances under which we should ignore significant coefficients in this manner.

The specification of a zerotree for a group of wavelet coefficients is a form of quantization. Generally, the values of the pixels we code with zerotrees are non-zero, but in using a zerotree we specify that they be decoded as zeros. Non-zerotree wavelet coefficients (significant values) are also quantized, using scalar quantizers. If we save bitrate by specifying larger zerotrees, as in the hypothetical example above, the rate that was saved can be assigned to the scalar quantizers of the remaining coefficients, thus quantizing them more accurately. Therefore, we have a choice in allocating the bitrate among two types of quantization. The question is, if we are given a unit of rate to use in coding, where should it be invested so that the corresponding reduction in distortion is maximized?

This question, in the context of zerotree wavelet coding, was addressed by Xiong *et al.* [375], using well-known bit allocation techniques [103]. The central result for optimal bit allocation states that, in the optimal state, the slope of the operational rate-distortion curves of all quantizers are equal. This result is intuitive and easy to understand. The slope of the operational rate-distortion function for each quantizer tells us how many units of distortion we add/eliminate for each unit of rate we eliminate/add. If one of the quantizers has a smaller R-D slope, meaning that it is giving us less distortion reduction for our bits spent, we can take bits away from this quantizer (i.e. we can reduce its step size) and give them to the other, more efficient quantizers. We continue to do so until all quantizers have an equal slope.

Obviously, specification of zerotrees affects the quantization levels of non-zero coefficients because total available rate is limited. Conversely, specifying quantization levels will affect the choice of zerotrees because it affects the incremental distortion between zerotree quantization and scalar quantization. Therefore, an iterative algorithm is needed for rate-distortion optimization. In phase one, the uniform scalar quantizers are fixed, and optimal zerotrees are chosen. In phase two, zerotrees are fixed and the quantization level of uniform scalar quantizers is optimized. This algorithm is guaranteed to converge to a local optimum [375].

There are further details of this algorithm involving prediction and description of zerotrees, which we leave out of the current discussion. The advantage of this method is mainly in performance, compared to both EZW and SPIHT (the latter only slightly). The main disadvantages of this method are its complexity, and perhaps more importantly, that it does not generate an embedded bitstream.

8.7 FREQUENCY, SPACE-FREQUENCY ADAPTIVE CODERS

8.7.1 Wavelet Packets

The wavelet transform does a good job of decorrelating image pixels in practice, especially when images have power spectra that decay approximately uniformly and exponentially. However, for images with non-exponential rates of spectral decay and for images which have concentrated peaks in the spectra away from DC, there exist transforms with better performance than the wavelet transform.

Our analysis of Section 8.3.2 suggests that the optimal subband decomposition for an image is one for which the spectrum in each subband is approximately flat. The octave-band decomposition produced by the wavelet transform produces nearly flat spectra for exponentially decaying spectra, but many practical examples do not fall in this category. For example, the Barbara test image shown in Figure 8.11 contains a narrow-band component at high frequencies that comes from the tablecloth and the striped clothing. Fingerprint images contain similar narrow-band high frequency components.

For such images, the best basis algorithm, developed by Coifman and Wickerhauser [48], provides an efficient way to find a fast, wavelet-like transform with good energy compaction. The new basis functions are not wavelets but rather *wavelet packets* [46].

The basic idea of wavelet packets is best seen in the frequency domain. Each stage of the wavelet transform splits the current low frequency subband into two subbands of equal width, one high-pass and one low-pass. With wavelet packets there is a new degree of freedom in the transform: again there are N stages to the transform for a signal of length 2^N, but at each stage we have the option of splitting the low-pass subband, the high-pass subband, both, or neither. The high and low pass filters used in each case are the same filters used in the wavelet transform. In fact, the wavelet transform is the special case of a wavelet packet transform in which we always split the low-pass subband. With this increased flexibility we can generate 2^N possible different transforms in 1-D. The possible transforms give rise to all possible dyadic partitions of the frequency axis. The increased flexibility does not lead to a large increase in complexity; the worst-case complexity for a wavelet packet transform is $O(N \log N)$.

8.7.2 Frequency Adaptive Coders

The *best basis algorithm* is a fast algorithm for minimizing an additive cost function over the set of all wavelet packet bases. Our analysis of transform coding for Gaussian random processes suggests that we select the basis that maximizes the transform coding gain. The approximation theoretic arguments of Mallat and Falzon [204] suggest that at low bit rates the basis that maximizes the number of coefficients below a given threshold is the best choice. The best basis paradigm can accommodate both of these choices. See [358] for an excellent introduction to wavelet packets and the best basis algorithm. Ramchandran and Vetterli [261] describe an algorithm for finding the best wavelet packet basis for coding a given image using rate-distortion criteria.

An important application of this wavelet-packet transform optimization is the FBI Wavelet/Scalar Quantization Standard for fingerprint compression. The standard uses a wavelet packet decomposition for the transform stage of the encoder [286]. The transform is fixed for all fingerprints, however, thus the FBI coder is a first-generation linear coder.

The benefits of customizing the transform on a per-image basis depend considerably on the image. For the Lena test image the improvement in peak signal to noise ratio is modest, ranging from 0.1 dB at 1 bit per pixel to 0.25 dB at 0.25 bits per pixel. This is because the octave band partitions of the spectrum of the Lena image are nearly flat. The Barbara image (cf. Figure 8.11), on the other hand, has a narrow-band peak in the spectrum at high frequencies. Consequently, the PSNR increases by roughly 2 dB over the same range of bitrates [261]. Further impressive gains are obtained by combining the adaptive transform with a zerotree structure [373].

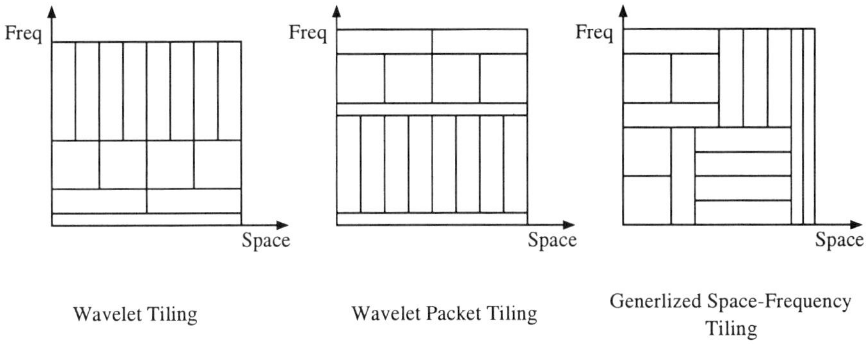

Figure 8.15 Wavelets, wavelet packets, and generalized time-frequency tiling.

8.7.3 Space-Frequency Adaptive Coders

The best basis algorithm is not limited to adaptive segmentation of the frequency domain. Related algorithms permit joint time and frequency segmentations. The simplest of these algorithms performs adapted frequency segmentations over regions of the image selected through a quadtree decomposition procedure [125, 298]. More complicated algorithms provide combinations of spatially varying frequency decompositions and frequency varying spatial decompositions [127]. These jointly adaptive algorithms work particularly well for highly nonstationary images.

The primary disadvantage of these spatially adaptive schemes are that the pre-computation requirements are much greater than for the frequency adaptive coders, and the search is also much larger. A second disadvantage is that both spatial and frequency adaptivity are limited to dyadic partitions. A limitation of this sort is necessary for keeping the complexity manageable, but dyadic partitions are not in general the best ones. Figure 8.15 shows an example of the time-frequency tiling of wavelets, wavelet packets, and space-frequency adaptive basis.

8.8 UTILIZING INTRA-BAND DEPENDENCIES

The development of the EZW coder motivated a flurry of activity in the area of zerotree wavelet algorithms. The inherent simplicity of the zerotree data structure, its computational advantages, as well as the potential for generating an embedded bitstream were all very attractive to the coding community. Zerotree algorithms were developed for a variety of applications, and many modifications and enhancements to the algorithm were devised, as described in Section 8.6.

With all the excitement incited by the discovery of EZW, it is easy to automatically assume that zerotree structures, or more generally inter-band dependencies, should be the focal point of efficient subband image compression algorithms. However, some of the best performing subband image coders known

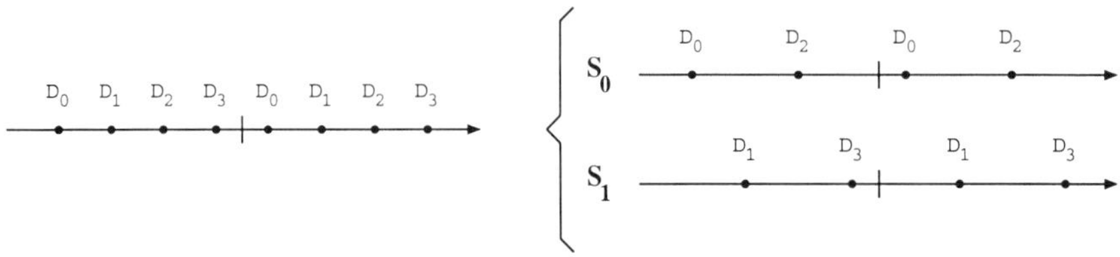

Figure 8.16 TCQ sets and supersets.

today are not based on zerotrees. In this section, we explore two methods that utilize intra-band dependencies. One of them uses the concept of Trellis Coded Quantization (TCQ). The other uses both inter- and intra-band information, and is based on a recursive estimation of the variance of the wavelet coefficients. Both of them yield excellent coding results.

8.8.1 Trellis Coded Quantization

Trellis Coded Quantization (TCQ) [210] is a fast and effective method of quantizing random variables. Trellis coding exploits correlations between variables. More interestingly, it uses non-rectangular quantizer cells that give it quantization efficiencies not attainable by scalar quantizers. TCQ grew out of the ground-breaking work of Ungerboeck [332] in trellis coded modulation.

The basic idea behind TCQ is the following: Assume that we want to quantize a stationary, memoryless uniform source at the rate of R bits per sample. Performing quantization directly on this uniform source would require an optimum scalar quantizer with 2^N reproduction levels (symbols). The idea behind TCQ is to first quantize the source more finely, with 2^{R+k} symbols. Of course this would exceed the allocated rate, so we cannot have a free choice of 2^{R+k} symbols at all times.

For example, take $k = 1$. The scalar codebook of 2^{R+1} symbols is partitioned into subsets of 2^{R-1} symbols each, generating four sets. In our example $R = 2$ (cf. Figure 8.16). The subsets are designed such that each of them represents reproduction points of a coarser, rate-$(R-1)$ quantizer. The four subsets are designated D_0, D_1, D_2, and D_3. Also, define $S_0 = D_0 \bigcup D_2$ and $S_1 = D_1 \bigcup D_3$, where S_0 and S_1 are known as *supersets.*

Obviously, the rate constraint prohibits the specification of an arbitrary symbol out of 2^{R+1} symbols. However, it is possible to exactly specify, with R bits, one element out of either S_0 or S_1. At each sample, assuming we know which one of the supersets to use, one bit can be used to determine the active subset, and $R-1$ bits to specify a codeword from the subset. The choice of superset is determined by the state of a finite state machine, described by a suitable trellis. An example of such a trellis, with eight states, is given in Figure 8.17. The subsets $\{D_0, D_1, D_2, D_3\}$ are also used to label the branches of the trellis, so the same bit that specifies the subset (at a given state) also determines the next state of the trellis.

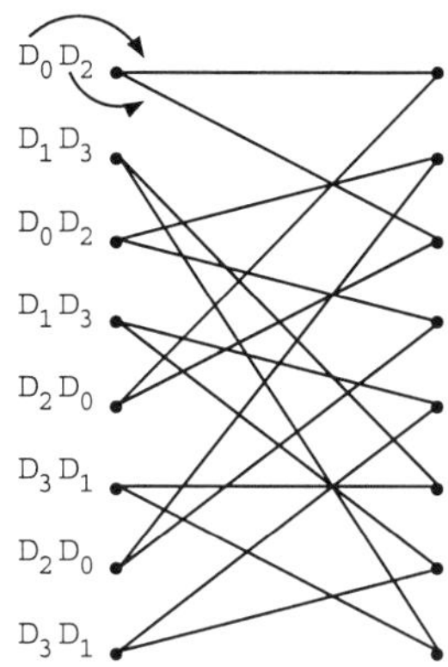

Figure 8.17 8-state TCQ trellis with subset labeling. The bits that specify the sets within the superset also dictate the path through the trellis.

Encoding is achieved by spending one bit per sample on specifying the path through the trellis, while the remaining $R-1$ bits specify a codeword out of the active subset. It may seem that we are back to a non-optimal rate-R quantizer (either S_0 or S_1). So why all this effort? The answer is that we have more codewords than a rate-R quantizer, because there is some freedom of choosing from symbols of either S_0 or S_1. Of course this choice is not completely free: the decision made at each sample is linked to decisions made at past and future sample points, through the permissible paths of the trellis. But even this limited choice improves performance. Availability of both S_0 and S_1 means that the reproduction levels of the quantizer, in effect, "slide around" and fit themselves to the data, subject to the permissible paths on the trellis.

The standard version of TCQ is not particularly suitable for image coding, because its performance degrades quickly at low rates. This is due partially to the fact that one bit per sample is used to encode the trellis alone, while interesting rates for image coding are mostly below one bit per sample. Entropy constrained TCQ (ECTCQ) improves the performance of TCQ at low rates. In particular, a version of ECTCQ due to Marcellin [209] addresses two key issues: reducing the rate used to represent the trellis (the so-called "state entropy"), and ensuring that zero can be used as an output codeword with high probability. The codebooks are designed using the algorithm and encoding rule from [38].

8.8.2 TCQ Subband Coders

Consider a subband decomposition of an image, and assume that the subbands are well represented by a *nonstationary* random process X, whose samples X_i are taken from distributions with variances σ_i^2. One can compute an "average variance" over the entire random process and perform conventional optimal quantization. But better performance is possible by sending overhead information about the variance of each sample, and quantizing it optimally according to its own pdf.

This basic idea was first proposed by Chen and Smith [33] for adaptive quantization of DCT coefficients. In their paper, Chen and Smith proposed to divide all DCT coefficients into four groups according to their "activity level," i.e. variance, and code each coefficient with an optimal quantizer designed for its group. The question of how to partition coefficients into groups was not addressed, however, and [33] arbitrarily chose to form groups with equal population.[4]

However, one can show that equally populated groups are not a always a good choice. Suppose that we want to classify the samples into J groups, and that all samples assigned to a given class $i \in \{1, ..., J\}$ are grouped into a source X_i. Let the total number of samples assigned to X_i be N_i, and the total number of samples in all groups be N. Define $p_i = N_i/N$ to be the probability of a sample belonging to the source X_i. Encoding the source X_i at rate R_i results in a mean squared error distortion of the form [157]

$$D_i(R_i) = \epsilon_i^2 \, \sigma_i^2 \, 2^{-2R_i}$$

where ϵ_i is a constant depending on the shape of the pdf. The rate allocation problem can now be solved using a Lagrange multiplier approach, much in the same way as was shown for optimal linear transforms, resulting in the following optimal rates:

$$R_i = \frac{R}{J} + \frac{1}{2} \log_2 \frac{\epsilon_i^2 \, \sigma_i^2}{\prod_{j=1}^{J} (\epsilon_j^2 \, \sigma_j^2)^{p_j}}$$

where R is the total rate and R_i are the rates assigned to each group. *Classification gain* is defined as the ratio of the quantization error of the original signal, X, divided by that of the optimally bit-allocated classified version, i.e.

$$G_c = \frac{\epsilon^2 \, \sigma^2}{\prod_{j=1}^{J} (\epsilon_j^2 \, \sigma_j^2)^{p_j}}.$$

One aims to maximize G_c over $\{p_i\}$. It is not unexpected that the optimization process can often yield non-uniform $\{p_i\}$, resulting in unequal population of the classification groups. It is noteworthy that non-uniform populations not only have better classification gain in general, but also lower overhead: Compared to a uniform $\{p_i\}$, any other distribution has smaller entropy, which implies smaller side information to specify the classes.

The classification gain is defined for X_i taken from one subband. A generalization of this result in [166] combines it with the conventional *coding gain* of the subbands. Another refinement takes into account the side information required for classification. The coding algorithm then optimizes the resulting expression to determine the classifications. ECTCQ is then used for final coding.

[4] If for a moment, we disregard the overhead information, the problem of partitioning the coefficients bears a strong resemblance to the problem of best linear transform. Both operations, namely the linear transform and partitioning, conserve energy. The goal in both is to minimize overall distortion through optimal allocation of a finite rate. Not surprisingly, the solution techniques are similar (Lagrange multipliers), and they both generate sets with maximum separation between low and high energies (maximum arithmetic to geometric mean ratio).

Practical implementation of this algorithm requires attention to a great many details, for which the interested reader is referred to [166]. For example, the classification maps determine energy levels of the signal, which are related to the location of the edges in the image, and are thus related in different subbands. A variety of methods can be used to reduce the overhead information (in fact, the coder to be discussed in the next section makes the management of side information the focus of its efforts). Other issues include alternative measures for classification and the usage of arithmetic coded TCQ. The coding results of the ECTCQ based subband coding are some of the best currently available in the literature, although the computational complexity of these algorithms is also considerably greater than the other methods presented in this paper.

Even better performance is possible (at the expense of higher complexity) by combining ideas from space-frequency quantization (SFQ) and trellis coding. Trellis coded space-frequency quantization (TCSFQ) [376] is the result of this combination. The basic idea of TCSFQ is to throw away a subset of wavelet coefficients and apply TCQ to the rest. TCSFQ can thus be thought of as taking the "best-of-the-best" from both SFQ and TCQ.

The SFQ algorithm in [375] takes advantage of the space-frequency characteristics of wavelet coefficients. It prunes a subset of spatial tree coefficients (i.e. setting the coefficients to zero) and uses scalar quantization on the rest. Optimal pruning in SFQ is achieved in a rate-distortion sense. SFQ thus uses an explicit form of subband classification, which has been shown to provide significant gain in wavelet image coding. Subband classification provides context models for both quantization and entropy coding.

SFQ only realizes the classification gain [166] with a single uniform scalar quantization applied on the non-pruned subset. Using TCQ on this set will further exploit the packing gain [87] of the trellis code, thus improving the coding performance. When combined with the conditional entropy coding scheme in [371], TCSFQ offers very good coding performance (cf. Table 8.1).

8.8.3 Context and Mixture Modeling

A common thread in successful subband and wavelet image coders is modeling of image subbands as random variables drawn from a mixture of distributions. For each sample, one needs to detect which pdf of the mixture it is drawn from, and then quantize it according to that pdf. Since the decoder needs to know which element of the mixture was used for encoding, many algorithms send side information to the decoder. This side information becomes significant, especially at low bitrates, so that efficient management of it is pivotal to the success of the image coder.

All subband and wavelet coding algorithms discussed so far use this idea in one way or another. They only differ in the constraints they put on side information so that it can be coded efficiently. For example, zerotrees are a clever way of indicating side information. The data is assumed from a mixture of very low energy (zero set) and high energy random variables, and the zero sets are assumed to have a tree structure.

The TCQ subband coders discussed in the last section also use the same idea. Different classes represent different energies in the subbands, and are transmitted as overhead. In [166], several methods are discussed to compress the side information, again based on geometrical constraints on the constituent elements of the mixture (energy classes).

A completely different approach to the problem of handling information overhead is explored in quantization via mixture modeling [199]. The version developed in [199] is named Estimation Quantization (EQ) by the authors, and is the one that we present in the following. We will refer to the the aggregate class as *backward mixture-estimation encoding* (BMEE).

BMEE models the wavelet subband coefficients as nonstationary generalized Gaussian random variables whose non-stationarity is manifested by a slowly varying variance (energy) in each band. Because the energy varies slowly, it can be predicted from causal neighboring coefficients. Therefore, unlike previous methods, BMEE does not send the bulk of mixture information as overhead, but attempts to recover it at the decoder from already transmitted data, hence the designation "backward." BMEE assumes that the causal neighborhood of a subband coefficient (including parents in a subband tree) has the same energy (variance) as the coefficient itself. The estimate of energy is found by applying a maximum likelihood method to a training set formed by the causal neighborhood.

Similar to other recursive algorithms that involve quantization, BMEE has to contend with the problem of stability and drift. Specifically, the decoder has access only to quantized coefficients, therefore the estimator of energy at the encoder can only use quantized coefficients. Otherwise, the estimates at the encoder and decoder will vary, resulting in drift problems. This presents the added difficulty of estimating variances from *quantized* causal coefficients. BMEE incorporates the quantization of the coefficients into the maximum likelihood estimation of the variance.

The quantization itself is performed with a dead-zone uniform quantizer (cf. Figure 8.10). This quantizer offers a good approximation to entropy constrained quantization of generalized Gaussian signals. The dead-zone and step sizes of the quantizers are determined through a Lagrange multiplier optimization technique, as introduced in Section 8.4.5. This optimization is performed off-line, once each for a variety of encoding rates and shape parameters, and the results are stored in a look-up table. This approach is to be credited for the speed of the algorithm, because no optimization need take place at the time of encoding the image.

Finally, the backward nature of the algorithm, combined with quantization, presents another challenge. All the elements in the causal neighborhood may sometimes quantize to zero. In that case, the current coefficient will also quantize to zero. This degenerate condition will propagate through the subband, making all coefficients on the causal side of this degeneracy equal to zero. To avoid this condition, BMEE provides for a mechanism to send side information to the receiver, whenever all neighboring elements are zero. This is accomplished by a preliminary pass through the coefficients, where the algorithm tries to "guess" which one of the coefficients will have degenerate neighborhoods, and assembles them to a set. From this set, a generalized Gaussian variance and shape parameter is computed and transmitted to the decoder. Every time a degenerate case happens, the encoder and decoder act based on this extra set of parameters, instead of using the backward estimation mode.

The BMEE coder is very fast, and especially in the low bitrate mode (less than 0.25 bits per pixel) is extremely competitive. This is likely to motivate a revisitation of the role of side information and the mechanism of its transmission in wavelet coders.

Once the quantization process is completed, another category of modeling is used for entropy coding. Context modeling in entropy coding [371, 40] attempts to estimate the probability distribution of the next symbol based on past samples. In this area, Wu's work [371] on conditional entropy coding of wavelets (CECOW) is noteworthy. CECOW utilizes a sophisticated modeling structure and seeks improvements in two directions: First, a straight forward increase in the order of models, compared to methods such as EZW and SPIHT. To avoid the problem of context dilution, CECOW determines the number of model parameters by adaptive context formation and minimum entropy quantization of contexts. Secondly, CECOW allows the shape and size of the context to vary among subbands, thus allowing more flexibility.

8.9 DISCUSSION AND SUMMARY

8.9.1 Discussion

Current research in image coding is progressing along a number of fronts on transform, quantization, and entropy coding.

At the most basic level, a new interpretation of the wavelet transform has appeared in the literature. This new theoretical framework, called the lifting scheme [312], provides a simpler and more flexible method for designing wavelets than standard Fourier-based methods. New families of non-separable wavelets constructed using lifting have the potential to improve coders. One very intriguing avenue for future research is the exploration of the nonlinear analogs of the wavelet transform that lifting makes possible. In particular, integer transforms are more easily designed with the lifting techniques, leading to efficient lossless compression, as well as computationally efficient lossy coders.

Recent developments of high performance subband/wavelet image coders (reviewed in this chapter) suggest that further improvements in performance may be possible through a better understanding of the statistical properties of subband coefficients. Subband classification is an explicit way of modeling these coefficients in quantization, while context modeling in entropy coding is aimed at the same goal. If subband classification and context modeling can be jointly optimized, improvements may be achieved over the current state-of-the-art.

With the increased performance provided by the subband coders, efforts of the community are partially channeled to other issues in image coding, such as spatial scalability, lossy to lossless coding, region-of-interest coding, and error resilience. Error resilience image coding via joint source-channel coding in particular is a promising research direction. See for example [292], [200], [264] and [100].

The adoption of wavelet based coding to video signals presents special challenges. One can apply 2-D wavelet coding in combination to temporal prediction (motion estimated prediction), which will be a direct counterpart of current DCT-based video coding methods. It is also possible to consider the video signal as a three-dimensional array of data and attempt to compress it with 3-D wavelet analysis. 3-D wavelet video coding has been explored by a number of researchers (see the collection of papers in this area in [327]). This 3-D wavelet based approach presents difficulties that arise from the fundamental properties of the discrete wavelet transform. The discrete wavelet transform (as well as any subband decomposition) is a space-varying operator, due to the presence of decimation and interpolation. This space variance is not conducive to compact representation of video signals, as described below.

Video signals are best modeled by 2-D projections whose position in consecutive frames of the video signal varies by unknown amounts. Because vast amounts of information are repeated in this way, one can achieve considerable gain by representing the repeated information only once. This is the basis of motion compensated coding. However, since the wavelet representation of the same 2-D signal will vary once it is shifted[5], this redundancy is difficult to reproduce in the wavelet domain. A frequency domain study of the difficulties of 3-D wavelet coding of video is presented in [239], and leads to the same insights. Some attempts have also been made on applying 3-D wavelet coding on the residual 3-D data after motion compensation, but have met with indifferent success.

8.9.2 Summary

Image compression is governed by the general laws of information theory and specifically rate-distortion theory. However, these general laws are nonconstructive and more specific techniques of quantization theory are needed for the actual development of compression algorithms.

Vector quantization can theoretically attain the maximum achievable coding efficiency. However, VQ has three main impediments: computational complexity, delay, and the curse of dimensionality. Transform coding techniques, in conjunction with entropy coding, capture important gains of VQ, while avoiding most of its difficulties.

Theoretically, the KLT is optimal for Gaussian processes, among block transforms. Approximations to the KLT, such as the DCT, have led to very successful image coding algorithms such as JPEG. However, even if one argues that image pixels can be individually Gaussian, they cannot be assumed to be jointly Gaussian, at least not across the image discontinuities. Image discontinuities are the place where traditional coders spend the most rate, and suffer the most distortion. This happens because traditional Fourier-type transforms

[5]Unless the shift is exactly by a correct multiple of M samples, where M is the downsampling rate

(e.g., DCT) disperse the energy of discontinuous signals across many coefficients, while the compaction of energy in the transform domain is essential for good coding performance.

Smooth subband bases of compact support, in particular the wavelet transform, provide an elegant framework for signal representation in which both smooth areas and discontinuities can be represented compactly in the transform domain.

State of the art wavelet coders assume that image data comes from a source with fluctuating variance. Each of these coders provides a mechanism to express the local variance of the wavelet coefficients, and quantizes the coefficients optimally or near-optimally according to that variance. The individual wavelet coders vary in the way they estimate and transmit this variances to the decoder, as well as the strategies for quantizing according to that variance.

Zerotree coders assume a two-state structure for the variances: either negligible (zero) or otherwise. They send side information to the decoder to indicate the positions of the non-zero coefficients. This process yields a non-linear image approximation, in contrast to the linear truncated KLT-based approximation motivated by the Gaussian model. The set of zero coefficients are expressed in terms of wavelet trees (Lewis & Knowles, Shapiro, others) or combinations thereof (Said & Pearlman). The zero sets are transmitted to the receiver as overhead, as well as the rest of the quantized data. Zerotree coders rely strongly on the dependency of data across scales of the wavelet transform.

Frequency-adaptive coders improve upon basic wavelet coders by adapting transforms according to the local inter-pixel correlation structure within an image. Local fluctuations in the correlation structure and in the variance can be addressed by spatially adapting the transform and by augmenting the optimized transforms with a zerotree structure.

Other wavelet coders use dependency of data within the bands (and sometimes across the bands as well). Coders based on Trellis Coded Quantization (TCQ) partition coefficients into a number of groups, according to their energy. For each coefficient, they estimate and/or transmit the group information as well as coding the value of the coefficient with TCQ, according to the nominal variance of the group. Another newly developed class of coders transmit only minimal variance information while achieving impressive coding results, indicating that perhaps the variance information is more redundant than previously thought.

Subband transforms and the ideas arising from wavelet analysis have had an indelible effect on the theory and practice of image compression, and are likely to continue their dominant presence in image coding research in the near future.

9 SCALABLE PICTURE CODING FOR MULTIMEDIA APPLICATIONS

Iraj Sodagar and Ya-Qin Zhang

Sarnoff Research Center
Princeton, NJ
[isodagar,yzhang]@sarnoff.com

9.1 INTRODUCTION

Digital image and video coding has received considerable attention lately in academia and industry in terms of both coding algorithms and standards activities. The development of digital compression standards such as JPEG, JBIG, MPEG-1, MPEG-2, H.261 and H.263, and their extensive use in communication and broadcasting applications, indicates the importance of digital image and video compression in today's world. Each of the above standards targets the efficient compression of a single media (still image or video) for a narrow group of applications. JPEG addresses compression of still images. MPEG-1 and MPEG-2 were developed to compress natural video for broadcasting applications. ITU H.261 and H.263 were developed for real-time coding of video at low bit rates and used for conversational services. With the increasing technological convergence of broadcasting, telecommunication and computers, and explosive growth of the Internet and the World Wide Web (WWW), the means of communication is undergoing a dramatic transformation. Many applications are moving from single-media environments to multimedia environments in which different types of contents such as natural and synthetic images, video, audio, graphics and text are created, transmitted and used at the same time. In many of these multimedia applications, features such as content-based access, analysis, and manipulation are important requirements in addition to the efficient coding of multimedia contents. Currently, MPEG-4 and JPEG2000 are working towards standardization for coding image and video, providing enhanced functionalities such as content-based access, content-based manipulation, content-based editing, combined natural and synthetic data coding, robustness, content-based scalability, in addition to improved coding efficiency

[147, 149]. This chapter describes scalable image and video coding techniques using zerotree wavelet aglorithms adopted in the MPEG4 visual texture coding of still images and computer texture maps.

The discrete wavelet transform (DWT) has recently emerged as a powerful tool in image and video compression due to its flexibility in representing nonstationary image signals and its ability in adapting to human visual characteristics. The wavelet representation provides a multiresolution/multifrequency expression of a signal with localization in both time and frequency. This property is very desirable in image and video coding applications. First, real-world image and video signals are nonstationary in nature. A wavelet transform decomposes a nonstationary signal into a set of multiscaled wavelets where each component becomes relatively more stationary and hence easier to code. Also, coding schemes and parameters can be adapted to the statistical properties of each wavelet, and hence coding each stationary component is more efficient than coding the whole nonstationary signal. In addition, the wavelet representation matches to the spatially tuned frequency-modulated properties experienced in early human vision as indicated by the research results in psychophysics and physiology. Compared with block-based transform coding, the wavelet approach avoids the "blocking artifacts" due to the nature of its global decomposition.

Zerotree wavelet coding techniques use the combination of the wavelet transform and the zerotree structure to provide very high coding efficiency as well as spatial and quality scalability features. The described algorithms not only outperform discrete cosine transform (DCT) based algorithms in terms of coding efficiency, they also provide spatial and quality scalabilities needed for multimedia applications. The zerotree wavelet coding technique has been adopted in the MPEG-4 standard as the default visual texture coding tool [151].

The organization of the chapter is as follows. We begin with a general overview of scalable image coding and its required features, followed by a brief description of the DWT. We will then describe three variations of wavelet-based zerotree coding algorithms including the embedded zerotree wavelet (EZW), the zerotree entropy (ZTE) coding, and the multiscale zerotree entropy (MZTE) coding. The chapter concludes with the application of zerotree wavelet coding to very low bit rate video coding.

9.2 SCALABLE IMAGE CODING

In conventional compression techniques, an input image is compressed for a given target bit rate. The result is a single bitstream that has to be fully decoded to reconstruct the image. "Lossy" compression introduces some distortion in the reconstructed image with the amount of distortion increasing at higher compression ratios. In such techniques, any given bitstream represents only a single compressed image. Common compression techniques such as JPEG and JBIG are designed with such property and are widely used in many communication systems where "point-to-point" communications is the main application.

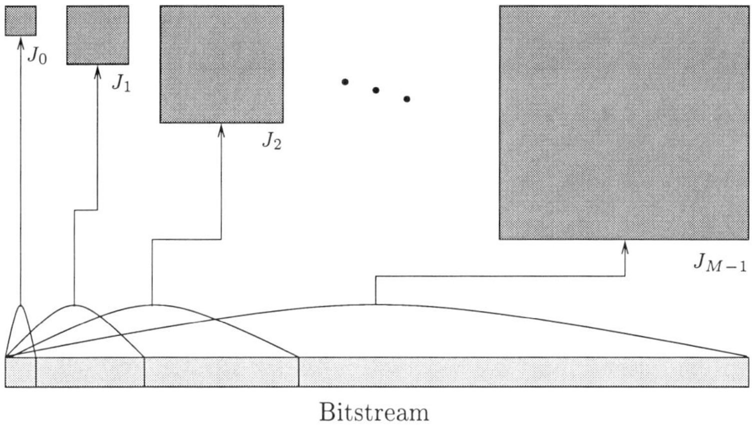

Figure 9.1 Decoded frame in M different spatial layers. J_m denotes m^{th} layer, $m \in \{0, 1, \ldots, M-1\}$.

In an interactive multimedia environment with variable bandwidth (such as Internet), scalabe representation with embedded bitstreams is desirable. In scalable compression, the bitstream can be progressively decoded to provide different versions of the image or parsed into many different bitstreams, each with different rates. Two important scalability features are spatial and quality scalabilities.

Figures 9.1 and 9.2 show two examples of such scalabilities. In Figure 9.1, the bitstream has M layers of spatial scalability. In this case, the bitstream consists of M different segments. By decoding the first segment, the user can access a preview version of the decoded image at a lower resolution. Decoding the second segment results in a larger reconstructed image. By progressively decoding the additional segments, the viewer can increase the spatial resolution of the image. Figure 9.2 shows an example in which the compressed bitstream includes N layers of quality scalability. In this figure, the bitstream consists of N different segments. Decoding the first segment provides an early view of the reconstructed image. Further decoding of successive segments results in an increase in the quality of the reconstructed image at N steps. Figure 9.3 shows a more complex case of combined spatial-quality scalabilities. In this example, the bitstream consists of M spatial layers, each including N layers of quality scalability. As in the previous cases, both the spatial resolution and the quality of the decoded image are improved by progressively decoding the bitstream.

Figures 9.4 and 9.5 demonstrate the progressive decoding application using the spatial and quality scalabilities. In Figure 9.4, the bitstream is spatially scalable. A decoder can retrieve the first part of a bitstream and obtain a low-resolution version of the decoded image. By progressively decoding the second part of the bitstream, an image with medium resolution can be obtained. Full resolution is achieved by fully decoding the bitstream. In Figure 9.5, the

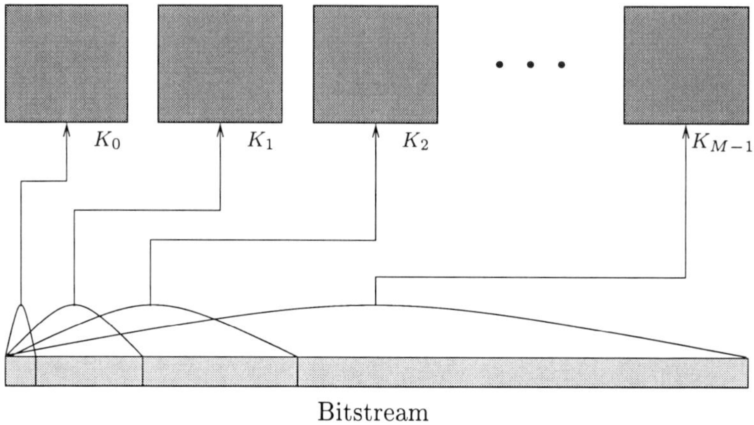

Figure 9.2 Decoded frame in N different quality layers. K_n denotes n^{th} layer, $n \in \{0, 1, \ldots, N-1\}$.

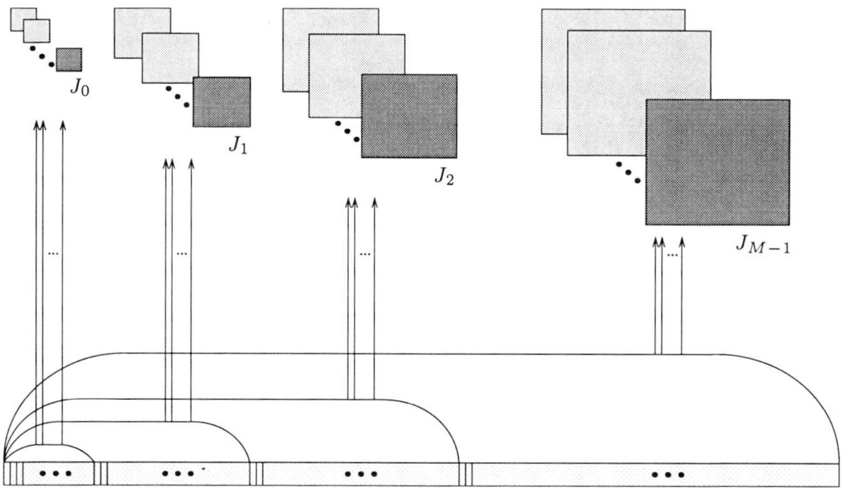

Figure 9.3 Decoded frame in $N \times M$ spatial/quality layers.

bitstream is quality scalable. Therefore, a decoder can decode the first part of the bitstream and obtain a lower-quality version of the decoded image. By progressive decoding of the second part of the bitstream, a decoded image with a medium-quality can be obtained. Finally, full decoding the bitstream the decoded image can be displayed with the highest quality possible.

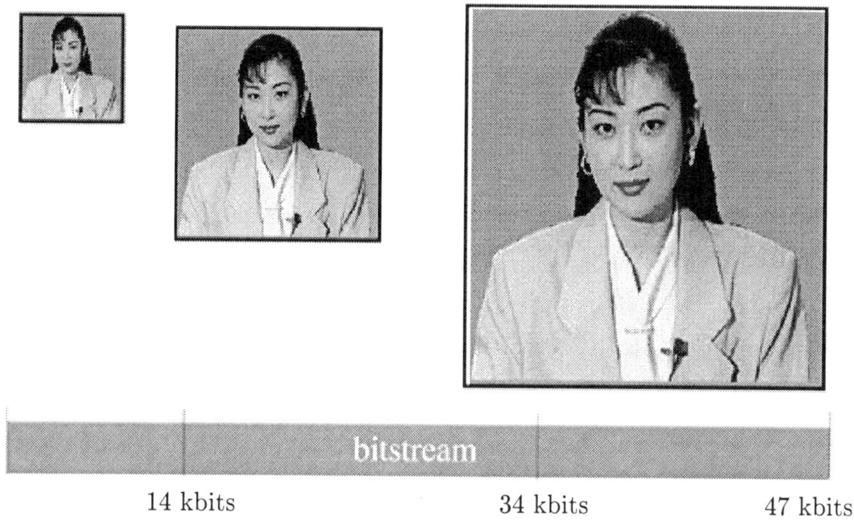

Figure 9.4 An example of progressively decoding an image by resolution.

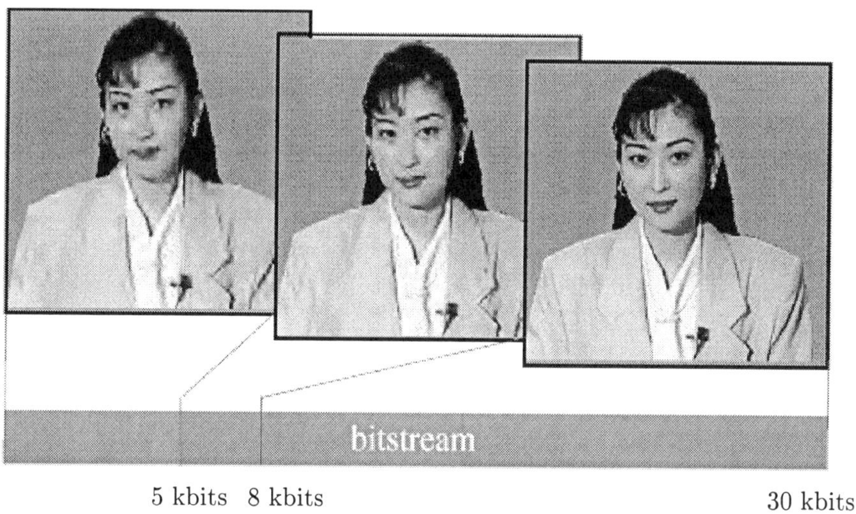

Figure 9.5 An example of progressively decoding an image by quality.

9.3 IMAGE COMPRESSION USING ZEROTREE WAVELETS

Zerotree wavelet coding is a proven technique for coding wavelet transform coefficients [290, 348, 2, 276, 320, 374, 82, 379, 285]. Besides superior compression performance, the advantages of zerotree wavelet coding include simplicity, an embedded bitstream, scalability, and precise bit rate control. Zerotree wavelet coding is based on three key ideas: (1) using wavelet transforms for decorrelation, (2) exploiting the self-similarity inherent in the wavelet transform to predict the location of significant information across scales, and (3) universal lossless data compression using adaptive arithmetic coding. In this section, we first give a brief description of the wavelet transform. Then, the EZW technique [290] which generates a bitstream with the maximum possible number of quality scalability layers is discussed. A description of the ZTE coding technique [211, 212], which only provides spatial scalability, is then given. The most general technique, MZTE [151], which provides a flexible framework in which to encode images with an arbitrary number of spatial or quality scalabilities is also described.

9.3.1 Discrete Wavelet Transform

The idea of representing a signal with various resolutions and analyzing it in this representation has emerged independently in the mathematics, physics and signal processing communities. In the mid-1980s, the wavelet transform was proposed by the mathematics community for the multiresolution representation of continuous signals. The mathematical aspects of wavelet transforms were introduced by Grossman and Morlet [113] and developed by Meyer [219], Daubechies [60], Mallat [203, 205] and Strang [307]. It was the discovery of the relationship between wavelets and filter bank that initiated major research activity in the use of wavelets in image and video applications. Mallat [203] originally demonstrated that the computation of the discrete wavelet representation can be accomplished with a hierarchical structure of filter banks. Daubechies [60] used discrete filters to generate compactly supported wavelets. Vetterli and Herley [346] have also studied the connection between the discrete wavelet transform and perfect-reconstruction filter banks. It was discovered that there is a one-to-one correspondence between the discrete wavelet transform and the octave-band filter bank. Before the wavelet transform was formally introduced, tree-structured filter banks were already employed for multiresolution representations of discrete signals in the speech processing literature [282, 231].

The discrete wavelet transform has been extensively covered in the literature [47, 41, 42, 61, 135, 86, 336, 305, 325, 268]. A key result from the signal processing perspective is that the structure of computations in the DWT and octave-band filter bank is identical. In fact, the DWT basis functions are often generated by starting with discrete-time filter banks with certain conditions. From the coding perspective, the octave-band filter bank can provide a linear time-frequency transform that significantly decorrelates the signal. In this multiresolution representation, the coarsest subband is a lowpass approximation of the original signal and each higher resolution further improves the accuracy of this representation.

The one-dimensional DWT is usually implemented using an octave-band tree structure [60, 203, 205]. Each stage of the tree structure is an identical two-band decomposition and only the low-frequency channel output of this two-band system is further divided. The inverse transform is performed using a similar

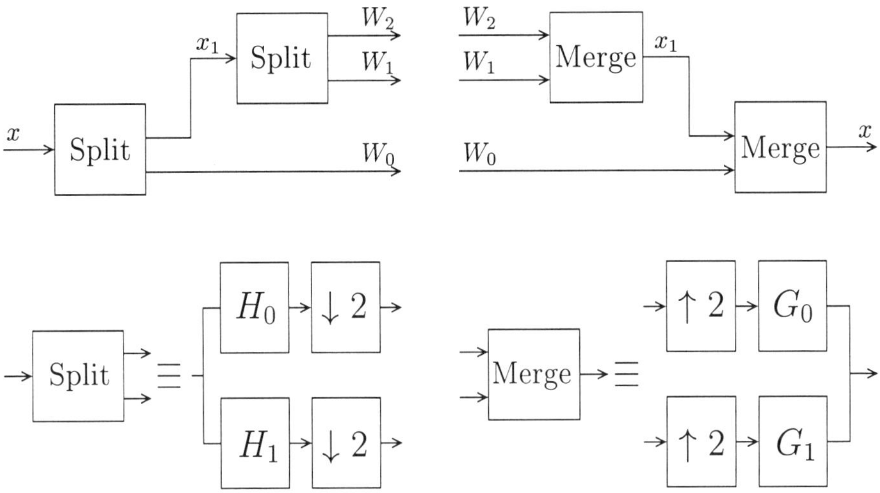

Figure 9.6 Block diagram of the two-stage tree structure used for the DWT and its inverse.

structure with the corresponding two-band synthesis filters as shown in Figure 9.6. In this figure, H_0 and H_1 represent the lowpass and highpass analysis filters, respectively, while G_0 and G_1 represent the corresponding lowpass and highpass synthesis filters. The first subband decomposition is obtained by filtering the input signal by H_0 and H_1 and then down-sampling the output by a factor of 2. Therefore, each subband signal has half the resolution of the original signal. Subband X_1 is further decomposed into two subbands, W_2 and W_1, using a second analysis filter bank. Subbands W_2 and W_1 have resolutions equal to one quarter of the original signal. Therefore, using this two-stage DWT, the input signal is decomposed into three subbands. The lowpass subband, W_2, represents a coarse approximation of the original signal while subbands W_1 and W_0 represent the refining "details." To decorrelate the input signal, the filter bank must be designed such that a significant amount of the input signal energy is concentrated in the low-frequency subbands.

The inverse DWT (IDWT) is also shown in Figure 9.6. Similar to the DWT, the IDWT consists of a 2-stage tree structure based on 2-band synthesis filter banks. A DWT is invertible, i.e. the IDWT perfectly reconstructs the original signal, only if the synthesis filter bank corresponds to the analysis filter bank of the DWT and this two-band analysis-synthesis filter bank system is perfect reconstructing. A perfect-reconstruction filter bank is one in which the synthesis filter bank output is exactly identical to the input of the analysis filter bank.

Two-Dimensional DWT. Similar to the 1-D case, multidimensional DWTs are usually implemented in the form of hierarchical tree structures of filter banks. Each stage of the tree is a multidimensional filter bank. In the 2-D

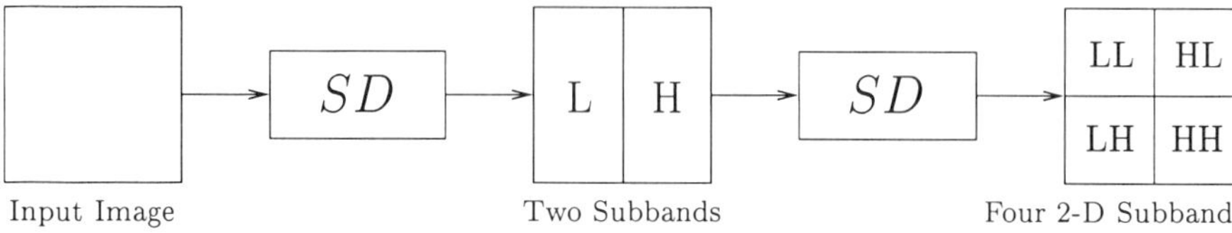

Figure 9.7 Decomposition of an image using a separable 2-D filter bank. SD denotes subband decomposition.

case, the 2-D DWT can be implemented using an octave-band tree structure of 2-D filter banks. A 2-D analysis filter bank consists of 2-D filters and 2-D downsamplers. In the most general case, the filters are non-separable and downsampling varies with the direction. Design and implementation of such general 2-D perfect-reconstruction filter banks are quite complex. One special case is the separable 2-D filter bank which uses both separable filters and separable downsampling-upsampling functions. In this case, one can decompose the image by first applying an 1-D analysis filter bank to each row of the image and then applying another 1-D analysis filter bank to each column of the subbands' output. The final four-subband output is shown in Figure 9.7. To reconstruct the image back from the subbands, the inverse procedure must be applied using the corresponding 1-D synthesis filter banks. Separable 2-D filter banks are very efficient in terms of implementation complexity due to the fact that the decomposition is applied in each dimension separately. Although separable filter banks do not necessarily maximize the coding gain of the transform for a given image, their performance in wavelet structures is close to optimal for natural images, and therefore, they are widely used in image compression.

Finite-Length DWT. The DWT is a non-expansive transform for infinite-length signals, meaning that the transform output has the same size as the transform input. This property requires the use of critically sampled filter banks in the corresponding subband decomposition. It is clear that the non-expansive property is highly desired for compression applications.

As mentioned above, the DWT can be implemented as a tree structure of subband decompositions. In each stage of the subband decomposition, the input signal is filtered using FIR filters. Clearly, the output of the linear convolution of a finite-size signal with an FIR filter is longer than the original length. This means that the wavelet transform of a finite-length signal is usually longer than the input signal. For example, if a finite-size image is decomposed with a wavelet transform, the number of wavelet coefficients is more than the number of pixels in the original image. This expansion clearly has a reverse effect on the signal compression. Two different schemes have been suggested previously, circular convolution and symmetric extension [302]. In the first method, the input signal is replicated in all dimensions and then filtered using circular convolution. Although this method avoids signal expansion, it results in noise migration from one boundary of the image to the opposite boundary. The second approach extends the input signal by mirroring it at each boundary. If the linear phase analysis/synthesis filters are used, it can be shown that the output signal is also symmetric around the boundaries and, therefore, expansion of the signal can be avoided. It has also been shown that the symmetric extension has superior performance compared to circular convolution for compression applications. Therefore, in most image/video compression applications, analysis/synthesis filter bank linear phase filters are used in conjunction with symmetric extension.

Wavelet Packets. In conventional time-frequency transforms such a wavelets, the underlying basis functions are fixed in time/space and define a specific tiling of the time-frequency plane. Consequently, they are fundamentally mismatched to applications in which signal characteristics are changing in time. A more flexible approach is obtained if the basis functions of the transform adapt to the signal properties. In this scenario, the time-frequency tiling of the transform can be changed from a set with good frequency localization to one with good time localization and vice versa [47]. This general concept has been introduced in the context of analysis as well as adaptive nonoverlapping block transforms such as Karhunen-Loève transform. However, the concept of time-varying overlapping block transforms such as wavelet packets is quite new [304, 303, 230, 261].

Wavelet packets are also implemented by a tree structure of filter banks. In this case, however, both the low-frequency and high-frequency bands can be decomposed in each stage of the tree. This flexibility provides a wide range of decompositions, each of which yield different time-frequency tiling and thus different time-frequency localization. One trivial case is the decomposition of both bands in each stage of tree structure, which is equivalent to an uniform filter bank. Figure 9.8 shows a few examples of possible wavelet packets.

9.3.2 Embedded Zerotree Wavelet Coding

Embedded zerotree wavelet coding is applied to coefficients resulting from a 2-D DWT. The 2-D DWT decomposes the input image into a set of subbands of varying resolutions. The coarsest subband is a lowpass approximation of the original image, and the other subbands are finer-scale refinements.

In the hierarchical subband system such as that of the wavelet transform, with the exception of the highest frequency subbands, every coefficient at a given scale can be related to a set of coefficients of similar orientation at the next finer scale. Coefficients at the coarse scale are called parents whereas all coefficients at the same spatial location, and similar orientation at the next finer scale, are called children. As an example, Figure 9.9 shows a wavelet tree descending from a coefficient in the subband HH3. For the lowest frequency subband, LL3 in the example, the parent-child relationship is defined such that each parent node has three children, one in each subband at the same scale and spatial location but different orientation.

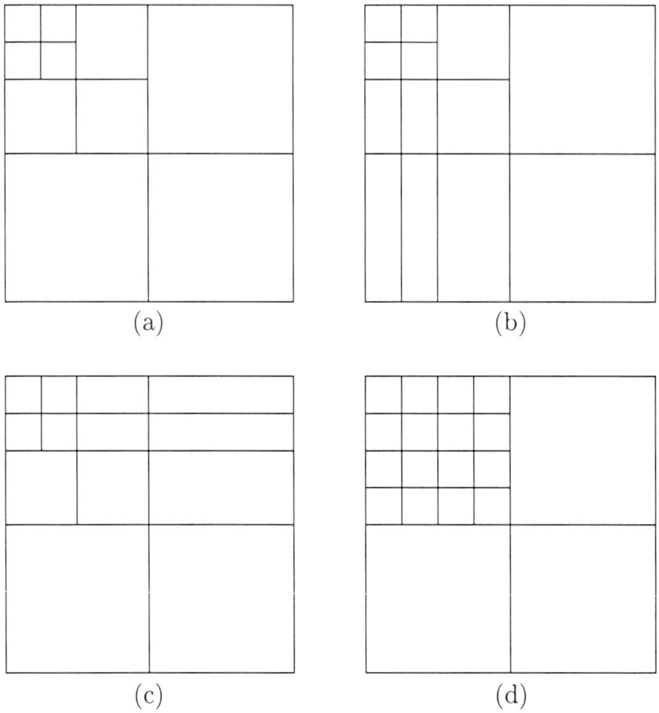

Figure 9.8 Examples of wavelet packet decompositions.

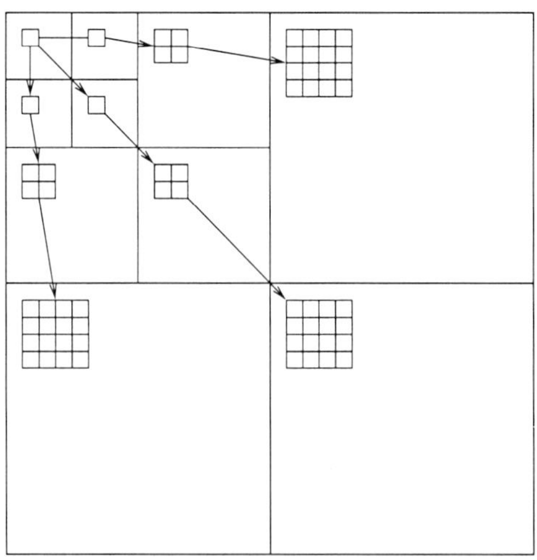

Figure 9.9 The parent-child relationship of wavelet coefficients.

EZW introduces a data structure called a *zerotree*, built on the parent-child relationship. The zerotree structure takes advantage of the principle that if a wavelet coefficient at a coarse scale is insignificant (quantized to zero) with respect to a given threshold, T, then all wavelet coefficients of the same orientation at the same spatial location at finer wavelet scales are also likely to be insignificant with respect to that T. The zerotree structure is similar to the zigzag scanning and end-of-block symbol commonly used in coding DCT coefficients.

EZW scans wavelet coefficients subband by subband. Parents are scanned before any of their children, but only after all neighboring parents have been scanned. Each coefficient is compared against the current threshold, T. A coefficient is significant if its amplitude is greater than T. Such a coefficient is then encoded using one of the symbols *negative significant* (NS) or *positive significant* (PS). The *zerotree root* (*ZTR*) symbol is used to signify a coefficient below T, with all its children in the zerotree data structure also below T. The *isolated zero* (*IZ*) symbol signifies a coefficient below T, but with at least one child in excess of T. For significant coefficients, EZW further encodes coefficient values using a successive approximation quantization (SAQ) scheme. Coding is done bit-plane by bit-plane. The successive approximation approach to quantization of the wavelet coefficient leads to the embedded nature of an EZW coded bitstream.

9.3.3 Implicit Quantization in EZW

The quantization that is implemented by EZW can be characterized as a family of quantizers, each of which is a mid-rise uniform quantizer with a dead-zone around zero. An example of such a quantizer is plotted in Figure 9.10. After each iteration of the EZW algorithm, all coefficients are effectively quantized by one of these quantizers. As the algorithm proceeds through the next iteration, the effective quantization for each coefficient becomes that of the next finer quantizer in the family of quantizers until, at the end of the iteration, the quantization of all coefficients is that of this new quantizer. If the bit allocation is depleted at the end of one iteration, all coefficients will have been quantized according to the same quantizer. It is more likely, however, that the bit allocation is exhausted before an iteration is complete, in which case the final effective quantization for each coefficient is according to one of two quantizers depending upon where in the scan of coefficients the algorithm stops.

Each iteration of the EZW algorithm is characterized by a threshold and the quantizer effectively implemented is a function of that threshold. As the iterations proceed, the threshold decreases and the quantization becomes finer. All thresholds are powers of two. The initial threshold T_d is set at a power of two such that the magnitude of at least one coefficient lies between T_d and $2T_d$ and no coefficient has a magnitude greater than $2T_d$. As each coefficient with magnitude c is processed by this iteration, its quantized magnitude c_q becomes

$$c_q = \begin{cases} T_d \text{int}[c/T_d] + T_d/2 & \text{if } c \geq T_d \\ 0 & \text{otherwise} \end{cases}$$

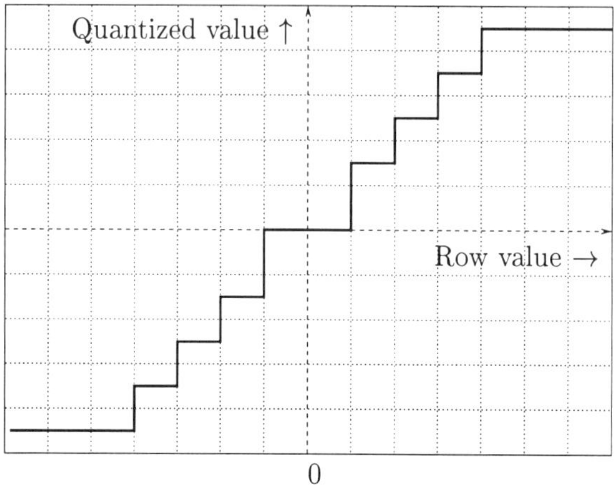

Figure 9.10 An example of a mid-rise quantizer.

where int[·] means "take the integer part of." After this iteration, a new threshold T_s is used, where $T_s = T_d/2$. Coefficients are scanned again, and become quantized according to

$$c_q = \begin{cases} T_s \text{int}[c/T_s] + T_s/2 & \text{if } c \geq 2T_s \\ 0 & \text{otherwise.} \end{cases}$$

Notice that the only difference between this quantizer and the first one is that the dead-zone around zero extends to the threshold in the first quantizer but to twice the threshold in the second.

After this iteration, the threshold T_d is again used, where here T_d is set to $T_d = T_s$. Coefficients are scanned again and are quantized according to the first rule above. After that iteration, T_s is set and the second rule applies. This changing of quantizers continues until the bit allocation for the image is exhausted.

9.3.4 Zerotree Entropy Coding

Zerotree entropy coding is an efficient technique for entropy coding of quantized wavelet coefficients. This technique is based on, but differs significantly from, the EZW algorithm. Like EZW, the ZTE algorithm exploits the self-similarity inherent in the wavelet transform of images and video residuals to predict the location of information across wavelet scales. ZTE coding organizes quantized wavelet coefficients into wavelet trees and then uses zerotrees to reduce the number of bits required to represent those trees. ZTE differs from EZW in four major ways: (1) quantization is explicit instead of implicit and can be performed distinct from the zerotree growing process or can be incorporated into the process thereby making it possible to adjust the quantization according to where the transform coefficient lies and what it represents in the frame, (2)

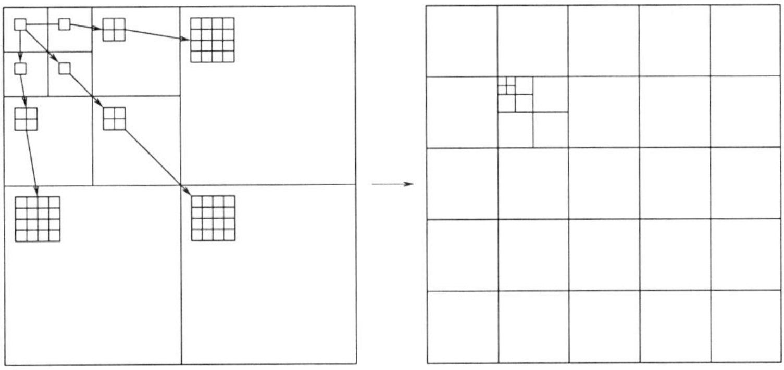

Figure 9.11 Building wavelet blocks after taking 2-D DWT.

coefficient scanning, tree growing, and coding is performed in one pass instead of bit-plane by bit-plane, (3) coefficient scanning can be changed from subband to subband through a depth-first traversal of each tree, and (4) the alphabet of symbols for classifying the tree nodes is changed to one that performs significantly better for very low bit rate encoding of video. The ZTE algorithm does not produce an fully embedded bitstream as EZW does, but by sacrificing the embedding property this scheme gains flexibility and other advantages over EZW coding, including substantial improvement in coding efficiency and speed.

Wavelet Blocks and Explicit Quantization. In ZTE coding, the coefficients of each wavelet tree are reorganized to form a wavelet block as shown in Figure 9.11. Each wavelet block comprises those coefficients at all scales and orientations that correspond to the image at the spatial location of that block. The concept of the wavelet block provides an association between wavelet coefficients and what they represent spatially in the image.

The ZTE improves upon EZW for very low bit rate video coding in several significant ways. In EZW, quantization of the wavelet coefficients is done implicitly using successive approximation. When using ZTE, the quantization is explicit and can be made adaptive to scene content. Quantization can be done entirely before ZTE or it can be integrated into ZTE and performed as the wavelet trees are traversed and the coefficients encoded.

If coefficient quantization is performed as the trees are built, then it is possible to dynamically specify a global quantizer step size for each wavelet block as well as an individual quantizer step size for each coefficient of a block. These quantizers can then be adjusted according to what spatially the coefficients of a particular block represent (scene content), to what frequency band the coefficients represents, or both. The advantages of incorporating quantization into ZTE are: (1) the status of the encoding process and bit usage are available to the quantizer for adaptation purposes, and (2) by quantizing coefficients as the wavelet trees are traversed, information such as spatial location and frequency band is available to the quantizer for it to adapt accordingly and thus provide content-based coding.

To use ZTE, a symbol is assigned to each node in a wavelet tree describing the wavelet coefficient corresponding to that node. Quantization of the wavelet transform coefficients can be done prior to the construction of the wavelet tree, as a separate task, or quantization can be incorporated into the wavelet tree construction. In the second case, as a wavelet tree is traversed for coding, the wavelet coefficients can be quantized in an adaptive fashion according to spatial location and/or frequency content.

Significant Symbols in ZTE. The extreme quantization required to achieve a very low bit rate produces many zero coefficients. Zerotrees exist at any tree node where the coefficient is zero and all the node's children are zerotrees. The wavelet trees are efficiently represented and coded by scanning each tree depth-first from the root in the low-low band through the children, and assigning one of four symbols, *zerotree root*, *valued zerotree root* (*VZTR*), *value* (*VAL*) or *isolated zero* to each node encountered. A *ZTR* denotes a coefficient that is the root of a zerotree. Zerotrees do not need to be scanned further because it is known that all coefficients in such a tree have amplitude zero. A *VZTR* is a node where the coefficient has nonzero amplitude and all four children are zerotree roots. The scan of this tree can stop at this symbol. A *VAL* symbol identifies a coefficient with amplitude nonzero and also with some nonzero descendant. Finally, an *IZ* symbol identifies a coefficient with amplitude zero but with some nonzero descendant. The symbols and quantized coefficients are then losslessly encoded using an adaptive arithmetic coder.

Encoding of Wavelet Coefficients. In ZTE, the wavelet coefficients of the low-low subband are encoded independently from the other bands. After explicit quantization of the low-low subband, the magnitude of the minimum and maximum values of the differential quantization indices, `band_offset` and `band_max_value`, respectively, are encoded into the bitstream. The parameter `band_offset` is a negative or a zero integer and the parameter `band_max_value` is a positive integer, so only the magnitudes of these parameters are encoded. The differential quantization indices are then encoded using the arithmetic encoder in raster scan order, starting from the upper left index and ending at the lowest right one. The adaptive probability model of the arithmetic encoder is updated with encoding of each bit of the predicted quantization index to adopt the probability model to the statistics of the low-low subband. The value `band_offset` is subtracted from all the values and a forward predictive scheme is applied. Each of the current coefficients, w_X, is predicted from three other quantized coefficients in its neighborhood, namely w_A, w_B, and w_C (cf. Figure 9.12) and the predicted value, pw_X is subtracted from the current coefficient,

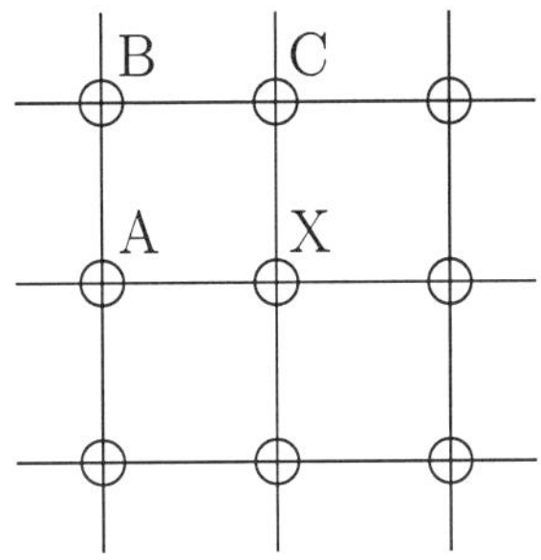

Figure 9.12 DPCM encoding of DC band coefficients.

that is

```
if (|w_A − w_B|) < (|w_B − w_C|)
    pw_X = w_C
else
    pw_X = w_A
    w_X = w_X − pw_X.
```

If any of the nodes A, B or C is not in the image, its value is set to zero for the purpose of the forward prediction.

In ZTE coding, the quantized wavelet coefficients are scanned either in the tree-depth fashion or in band-by-band fashion. In tree-depth scanning, all the coefficients of each tree are encoded before encoding the next tree. In band-by-band scanning, all coefficients are encoded from the lowest to the highest frequency subbands. Table 9.1 shows the tree-depth scanning order for a 16×16 image, with three levels of decomposition. In this figure, the indices 0, 1, 2 and 3 represent the DC band coefficients which are encoded separately. The remaining coefficients are encoded in the order shown in the figure. As an example, indices $4, 5, \ldots, 24$ represent one tree. At first, coefficients in this tree are encoded starting from index 4 and ending at index 24. Then, the coefficients in the second tree are encoded starting from index 25 and ending at 45. The third tree is encoded starting from index 46 and ending at index 66 and so on. Table 9.2 shows the wavelet coefficients scanned in the subband-by-subband fashion, from the lowest to the highest frequency subbands. This figure shows an example for a 16×16 image with three levels of decomposition. The DC band is located at upper left corner (with indices 0, 1, 2, 3) and is encoded separately as described in DC band decoding. The remaining coefficients are encoded in the order which is shown in the figure, starting from index 4 and ending at index 255.

Zerotree symbols and quantized coefficients are losslessly encoded using an adaptive arithmetic coder with a given symbol alphabet. The arithmetic encoder adaptively tracks the statistics of the zerotree symbols and encoded values using three models: (1) type to encode the zerotree symbols, (2) magnitude to

Table 9.1 Tree-depth scanning order in ZTE encoding.

0	1	4	67	5	6	68	69	9	10	13	14	72	73	76	77
2	3	130	193	7	8	70	71	11	12	15	16	74	75	78	79
25	88	46	109	131	132	194	195	17	18	21	22	80	81	84	85
151	214	172	235	133	134	196	197	19	20	23	24	82	83	86	87
26	27	89	90	47	48	110	111	135	136	139	140	198	199	202	203
28	29	91	92	49	50	112	113	137	138	141	142	200	201	204	205
152	153	215	216	173	174	236	237	143	144	147	148	206	207	210	211
154	155	217	218	175	176	238	239	145	146	149	150	208	209	121	213
30	31	34	35	93	94	97	98	51	52	55	56	114	115	118	119
32	33	36	37	95	96	99	100	53	54	57	58	116	117	120	121
38	39	42	43	101	102	105	106	59	60	63	64	122	123	126	127
40	41	44	45	103	104	107	108	61	62	65	66	124	125	128	129
156	157	160	161	219	220	223	224	177	178	181	182	240	241	244	245
158	159	162	163	221	222	225	226	179	180	183	184	242	243	246	247
164	165	168	169	227	228	231	232	185	186	189	190	248	249	252	253
166	167	170	171	229	230	233	234	187	188	191	192	250	251	254	255

Table 9.2 Band-by-band scanning order in ZTE encoding.

0	1	4	7	16	17	28	29	64	65	68	69	112	113	116	117
2	3	10	13	18	19	30	31	66	67	70	71	114	115	118	119
5	8	6	9	40	41	52	53	72	73	76	77	120	121	124	125
11	14	12	15	42	43	54	55	74	75	78	79	122	123	126	127
20	21	32	33	24	25	36	37	160	161	164	165	208	209	212	213
22	23	34	35	26	27	38	39	162	163	166	167	210	211	214	215
44	45	56	57	48	49	60	61	168	169	172	173	216	217	220	221
46	47	58	59	50	51	62	63	170	171	174	175	218	219	222	223
80	81	84	85	128	129	132	133	96	97	100	101	144	145	148	149
82	83	86	87	130	131	134	135	98	99	102	103	146	147	150	151
88	89	92	93	136	137	140	141	104	105	108	109	152	153	156	157
90	91	94	95	138	139	142	143	106	107	110	111	154	155	158	159
176	177	180	181	224	225	228	229	192	193	196	197	240	241	244	245
178	179	182	183	226	227	230	231	194	195	198	199	242	243	246	247
184	185	188	189	232	233	236	237	200	201	204	205	248	249	252	253
186	187	190	191	234	235	238	239	202	203	206	207	250	251	254	255

encode the values in a bit-plane fashion, and (3) sign to encode the sign of the value. Each coefficient's zerotree symbol is encoded first and, if necessary, then its value is encoded. The value is encoded in two steps. First, its absolute value is encoded in a bit-plane fashion using the appropriate probability model and then the sign is encoded using a binary probability model with '0' and '1' meaning a positive and negative sign, respectively. The sign model is initialized to the uniform probability distribution.

9.3.5 Multi-Scale Zerotree Entropy Coding

The multi-scale zerotree entropy coding technique is based on ZTE [211, 212] but utilizes a new framework to improve and extend the ZTE method to achieve a fully scalable yet very efficient coding technique. In this scheme, the low-low band is separately encoded. In order to achieve a wide range of scalability levels efficiently, the other bands are encoded using the MZTE coding scheme. This multiscale scheme provides a very flexible approach to support the right trade-off between layers and types of scalability, complexity and coding efficiency for any multimedia application. Figure 9.13 ilustrates a block diagram implementation of this technique.

The wavelet coefficients of the first spatial (and/or quality) layer are first quantized with the quantizer Q_0. These quantized coefficients are scanned using the zerotree concept and the significant maps and quantized coefficients are entropy coded. The output of entropy coder at this level, BS_0, is the first portion of the bitstream. The quantized wavelet coefficients of the first layer are also reconstructed and subtracted from the original wavelet coefficients. These residual wavelet coefficients are fed into the second stage of the coder in which the wavelet coefficients are quantized with Q_1, zerotree scanned and entropy coded. The output of this stage, BS_1, is the second portion of the output bitstream. The quantized coefficients of the second stage are also reconstructed and subtracted from the original coefficients. As is shown in Figure 9.13, $N+1$ stages of the scheme provides $N+1$ layers of scalability. Each level presents one layer of quality or spatial (or both) scalability.

In MZTE, the wavelet coefficients are quantized by quantizers which approximate a uniform and mid-rise quantizer with a dead-zone of double size as closely as possible at each scalability layer. Each quality layer of each spatial layer has a quantization *value* (Q-value) associated with it. Each spatial layer has a corresponding sequence of these Q-values. The quantization of coefficients is performed in three steps: (1) construction of initial quantization value sequence from input parameters, (2) revision of the quantization sequence, and (3) quantization of the coefficients.

Let n be the total number of spatial layers and $k(i)$ be the number of quality layers associated with spatial layer i. We define the total number of scalability layers associated with spatial layer i, $L(i)$, as the sum of all the quality layers from that spatial layer and all higher spatial layers, that is,

$$L(i) = k(i) + k(i+1) + \ldots + k(n).$$

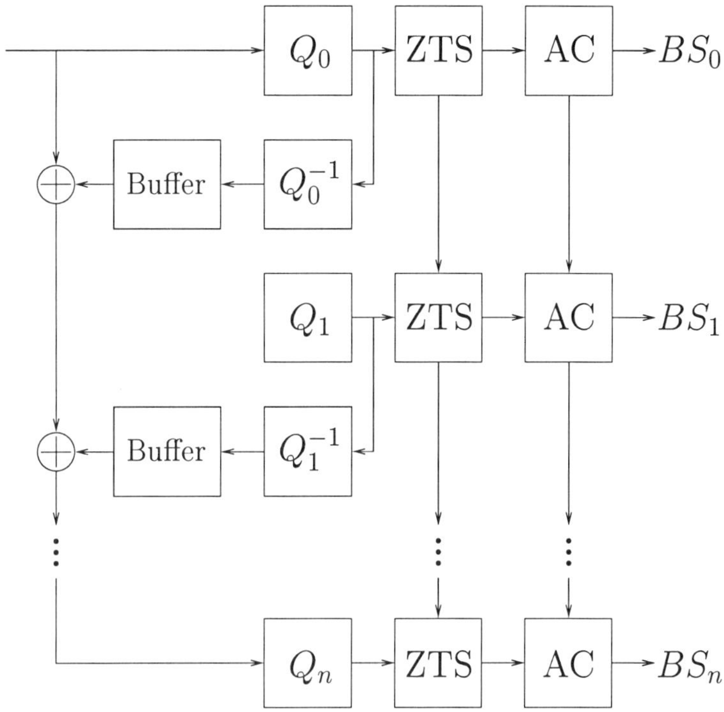

Figure 9.13 Multi-scale ZTE encoding structure (BS - Bitstream).

Let $Q(m, n)$ be the Q-value corresponding to spatial layer m and quality layer n. The quantization sequence (or Q-sequence) associated with spatial layer i is defined as the sequence of the Q-values from all the quality layers from the ith spatial layer and all higher spatial layers ordered by increasing quality layer then increasing spatial layer, namely

$$\begin{aligned} Q_i &= [Q_i(0), Q_i(1), \dots, Q_i(m)] \\ &= [Q(i,1), Q(i,2), \dots, Q(i,k(i)), Q(i+1,1), Q(i+1,2), \\ &\quad \dots, Q(i+1,k(i+1)), \dots, Q(n,1), Q(n,2), \dots, Q(n,k(n))]. \end{aligned}$$

The sequence Q_i represents the procedure for successive refinement of the wavelet coefficients which are first quantized in the spatial layer i. In order to make this successive refinement efficient, the sequence Q_i is revised before starting the quantization. Let $Q_i(j)$ denote the jth value of the quantization sequence Q_i. Consider the case when $Q_i(j) = pQ_i(j+1)$. If p is an integer greater than one, each quantized coefficient of layer j is efficiently refined at layer $(j = 1)$ as each quantization step size $Q_i(j)$ is further divided into p equal partitions in layer $(j+1)$. If p is greater than one, but not an integer, the partitioning of the $j+1$ layer will not be uniform. This is due to the fact that $Q_i(j)$

Table 9.3 Revision of the quantization sequence.

Condition on $p = Q_i(j)/Q_i(j+1)$	Revision procedure
$p < 1.5$	$QR_i(j+1) = Q_i(j)$ (no quantization at layer $j+1$)
$p \geq 1.5$ (p is integer)	$QR_i(j+1) = Q_i(j+1)$ (no revision)
$p \geq 1.5$ (p is non-integer)	$QR_i(j+1) = \text{ceil}(Q_i(j)/q)$ where $q = \text{round}(Q_i(j)/Q_i(j+1))$

corresponds to quantization levels which cover $Q_i(j)$ possible coefficient values that cannot be evenly divided into $Q_i(j+1)$ partitions. In this case, $Q_i(j+1)$ is revised such as to be as close to an integer factor of $Q_i(j)$ as possible. The last case is when $Q_i(j+1) \geq Q_i(j)$. In this case, no further refinement can be obtained at the $(j+1)^{th}$ scalability layer over the jth layer so we simply revised $Q_i(j+1)$ to be $Q_i(j)$. The revised quantization sequence is referred as QR_i. Table 9.3 summarizes the revision procedure.

Categorizing the coefficients in the image according to which spatial layer wherein they first appear, we make the following definitions:

$$S(i) = \{\text{all coefficients which first appear in spatial layer } i\}$$
$$T(i) = \{\text{all coefficients which appear in spatial layer } i\}.$$

Since once a coefficient appears in a spatial layer it appears in all higher spatial layers we have the relationship

$$T(1) \subset T(2) \subset \cdots \subset T(n-1) \subset T(n). \tag{9.1}$$

To quantize each coefficient in $S(i)$ we use the Q-values in the revised quantization sequence, QR_i. These Q-values are positive integers and they represent the range of values the quantization level spans at that scalability layer. For the initial quantization we simply divide the value by the Q-value for the first scalability layer. This gives us our initial quantization level (note that it also gives us a double-sized dead-zone). For successive scalability layers we need only send the information that represents the refinement of the quantizer. The refinement information values are called residuals and are the index of the new quantization level within the old level wherein the original coefficient value lies.

We then partition the inverse range of the quantized value from the previous scalability layer in such a way that makes the partitions as uniform as possible based on the previously calculated number of refinement levels, m. This partitioning always leaves a discrepancy of zero between the partition sizes if the previous Q-value is evenly divisible by the current Q-value (e.g. previous $Q = 25$ and current $Q = 5$). If the previous Q-value is not evenly divisible by the current Q-value (e.g. previous $Q = 25$ and current $Q = 10$) then we have a maximum discrepancy of 1 between partitions. The larger partitions are always the ones closer to zero.

We then number the partitions. The residual index is simply the number of the partition wherein the original (which is not quantized) value actually lies. We have the following two cases for this numbering:

I: If the previous quality level is quantized to zero (that is the value was in the dead-zone) then the residual has to be one of the $2m-1$ values in $\{-m, \ldots, 0, \ldots, m\}$.

II: If the previous quality level is quantized to a nonzero value the residual has to be one of the m values in $\{0, \ldots, m-1\}$ since the sign is already know at the inverse quantizer.

The restriction of the possible values of the residuals is based solely on the relationship between successive quantization values and whether the value was quantized to zero in the last scalability pass (both of these facts are known at the decoder). This is one reason why using two probability models (one for the first case and one for the second case) increases coding efficiency.

For inverse quantization, we map the quantization level at the current quality layer to the midpoint of its inverse range. Thus we get a maximum quantization error of one-half the inverse range of the quantization level to which we dequantize. One can reconstruct the quantization levels given the list of Q-values associated with each quality layer, the initial quantization value, and the residuals.

Zerotree Significant Maps in MZTE At the first scalability layer, the zerotree symbols and the corresponding values are encoded for the wavelet coefficients of that scalability layer. The zerotree symbols are generated in the same way as in the ZTE method. For subsequent scalability layers, the zerotree map is updated along with corresponding value refinements. In each scalability layer, a new zerotree symbol is encoded for a coefficient only if it was encoded as *ZTR*, *VZTR* or *IZ* in the previous scalability layer. If the coefficient was decoded as *VAL* in the previous layer, a *VAL* symbol is also assigned at the current layer. Only its refinement value is encoded in the bitstream.

Figure 9.14 shows the relation between the symbols of one layer to the next layer. If the node is not coded before (shown as "x" in the figure), it can be encoded using any of four symbols. If it is *ZTR* in one layer, it can remain *ZTR* or any of other 3 symbols in the next layer. If it is detected as *IZ*, it can remain *IZ* or only become *VAL*. If it is *VZTR*, it can remain *VZTR* or become *VAL*. Once it is assigned *VAL* it stays *VAL* and no symbol is transmitted.

9.3.6 *Entropy Coding in EZW, ZTE and MZTE*

Symbols and quantized coefficient values generated by the zerotree stage are all encoded using an adaptive arithmetic coder, as presented in [359]. The arithmetic coder is run over several data sets simultaneously. A separate model with an associated alphabet is used for each. The arithmetic coder uses adaptive models to track the statistics of each set of input data then encodes each set close to its entropy. The symbols encoded differ based upon whether EZW,

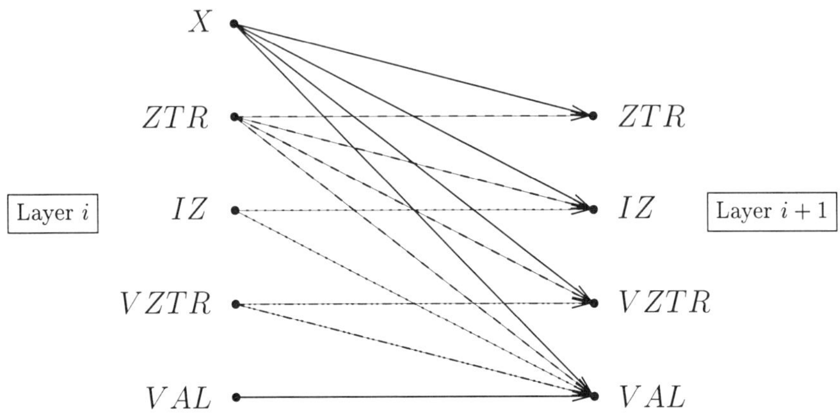

Figure 9.14 Zerotree mapping from one scalability level to the next.

ZTE or MZTE coding is used. For EZW, a four-symbol alphabet is used for the significance map and a different two-symbol alphabet is used for the SAQ information. The arithmetic coder is restarted every time a new significance map is encoded or a new bit-plane is encoded by SAQ. For ZTE, symbols describing node type (zerotree root, valued zerotree root, value or isolated zero) are encoded. The list of nonzero quantized coefficients that correspond one-to-one with the valued zerotree root or value symbols are encoded using an alphabet that does not include zero. The remaining coefficients, which correspond one-to-one to the value symbols, are encoded using an alphabet that does include zero. For any node reached in a scan that is a leaf with no children, neither root symbol can apply. Therefore, bits are saved by not encoding any symbol for this node and encoding the coefficient along with those corresponding to the value symbol using the alphabet that includes zero.

In MZTE, one additional probability model, *residual*, is used for decoding the refinements of the coefficients that were decoded with a *VAL* or *VZTR* symbol in any previous scalability layers. If in the previous layer, a *VAL* symbol was assigned to a node, the same symbol is kept for the current pass and no zerotree symbol is encoded. If in the previous layer, a *VZTR* symbol was assigned, a new symbol is decoded for the current layer, but it can only be *VAL* or *VZTR*. The residual model, same as the other probability models, is also initialized to the uniform probability distribution at the beginning of each scalability layer. The number of bins for the residual model is calculated based on the ratio of the quantization step sizes of the current and previous scalability. The values of the new *VAL* or *VZTR* coefficients are encoded in the same way as in ZTE.

When a residual model is used, only the magnitude of the refinement are encoded as these values are always zero or positive integers. Furthermore, to utilize the high correlation of zerotree symbols between scalability layers, a context modeling based on the zerotree symbol of the coefficient in the previous

Table 9.4 Context models for non-leaf coefficients.

Context model	Possible values
INIT	$ZTR(2), IZ(0), VZTR(3)$, or $VAL(1)$
ZTR	$ZTR(2), IZ(0), VZTR(3)$, or $VAL(1)$
ZTR DESCENDANT	$ZTR(2)$
IZ	$IZ(0)$ or $VAL(1)$

Table 9.5 Context models for leaf coefficients.

Context model	Possible values
LEAF_INIT	$ZTR(0)$ or $VZTR(1)$
LEAF_ZTR	$ZTR(0)$ or $VZTR(1)$
LEAF_ZTR DESCENDANT	$ZTR(0)$ or $VZTR(1)$

Table 9.6 PSNR comparison between the image decoded by MZTE and JPEG.

Compression scheme	PSNR-Y	PSNR-U	PSNR-Y
DCT based JPEG	28.6	34.74	34.98
Wavelet based MZTE	30.98	41.68	40.14

scalabilty layer in MZTE is used to better estimate the distribution of zerotree symbols. In MZTE, only INIT and LEAF_INIT are used for the first scalability layer for the non-leaf subbands and leaf subbands, respectively. Subsequent scalability layers in the MZTE use the context associated with the symbols. The different zerotree symbol models and their possible values are summarized in Tables 9.4 and 9.5. If a spatial layer is added then the contexts of all previous leaf subband coefficients are switched into the corresponding non-leaf contexts. The coefficients in the newly added subbands use the LEAF_INIT context initially.

9.3.7 Quality Comparison of MZTE versus JPEG

To compare the quality of MZTE and JPEG compression, a composite synthetic image is used. The images in Figures 9.15 and 9.16 are generated by JPEG and MZTE compression schemes, respectively, at the same compression ratio of 45:1. The results show that the MZTE scheme generates much better image quality with more detail information without any blocking effects as compared to the JPEG scheme. The PSNR values for both reconstructed images are tabulated in Table 9.6.

Figure 9.15 A synthetic image compressed and decompressed by JPEG.

9.4 ZEROTREE WAVELET CODING FOR VERY LOW BIT RATE VIDEO COMPRESSION

Very low bit rate video coding has lately become a very important topic due to its use in video-conferencing and video streaming applications. In this section, we briefly describe an approach to very low bit rate video coding using the zerotree wavelet.

Figure 9.17 shows the block diagram of a video encoder based on motion compensation and zerotree wavelet coding. The structure of this encoder is similar to other motion-compensated, block-based, DCT transform video coders such as MPEG-1 and H.263, but it uses a DWT, overlapping motion estimation/compensation(to better match the DWT), and the zerotree concept for

Figure 9.16 A synthetic image compressed and decompressed by MZTE.

coding the wavelet coefficients. The wavelet transform reduces the blocking artifacts seen at very low bit rates while providing a better way to address the scalability functionalities. The 5 specific components of the coder are: (1) a global motion estimation to track camera zooming, panning and other global motion activities; (2) overlapping block motion compensation to remove temporal redundancy; (3) an adaptive DWT of the residual to remove spatial correlation; (4) quantization of the wavelet coefficients to remove irrelevancy; and (5) the use of zerotrees and an arithmetic coder to losslessly encode the quantized coefficients with a minimum number of bits.

Overlapping in the block motion compensation significantly reduces the artifacts that would otherwise arise from the mismatch between the block nature

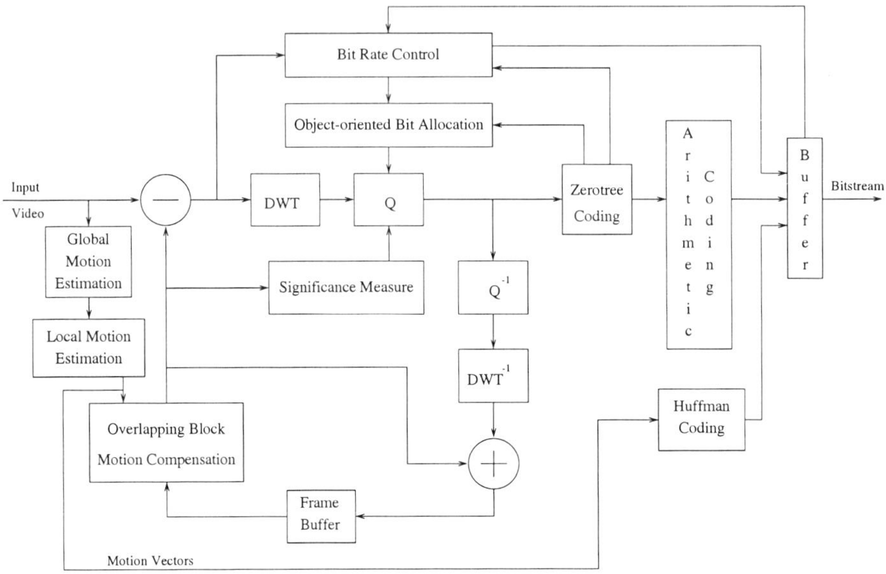

Figure 9.17 Block diagram of the video codec.

of motion compensation and the global nature of the DWT [242, 227]. At high compression ratios, ringing is the major artifact of the wavelet transform. To reduce the ringing artifacts the codec uses different filter lengths at each stage of the subband decomposition. Quantization and coding of the wavelet coefficients are done using the zerotree coding concept. The coder incorporates the ZTE coding algorithm specifically tailored to give the best performance for encoding wavelet coefficients of very low bit rate video as well as spatial and quality scalability.

This coder also includes several advanced options that can improve performance and add greater functionality. These options are: (1) global motion estimation using an affine transform to compensate for large motion, camera movement, zoom, or pan; (2) variable block size motion estimation requiring potentially fewer bits to encode motion vectors than fixed block size motion estimation; (3) object-oriented quantizer step size adjustment; and (4) a rate control scheme to control the quantization factor to achieve the desired mean bit rate.

In this encoder, the input frames of video are encoded either as *Intra* or *Inter*, where Intra is used for the first frame and Inter is used for the remaining frames of the test sequences. The first Intra frame can be coded using either the EZW or ZTE/MZTE algorithm. All succeeding frames of the video sequence are encoded as forward predicted "P" frames from the preceding frame. Bidirectionally-predicted "B" frames are also possible. First, global motion is detected and compensated by means of an affine transform. Then a block-based motion estimation scheme is used to detect local motion. The prediction is done using the fixed block size motion estimation scheme of H.263 [152] where estimation is done to half-pixel accuracy on blocks of 16×16 or 8×8. There is an option for a new variable block size motion estimation scheme that allows a greater range of block sizes.

Each block is predicted using the overlapping block motion compensation scheme of H.263. Overlapping block motion compensation is used with both the fixed block size and the variable block size motion estimation schemes. After all blocks have been predicted, the residuals are pieced together to form a complete residual frame for subsequent processing by the wavelet transform. The overlapping in the motion compensation ensures that a coherent residual is presented to the wavelet transform without any artificial block discontinuities. For those blocks where prediction fails, the Intra mode is selected and the original image block is coded. To turn this block into a residual similar to the other predicted blocks, the mean is subtracted from the Intra blocks and sent as overhead. In this way, all blocks now contain a prediction error, constituting either the difference between the current block and its overlapping motion-compensated prediction or the difference of the block pixels from their mean.

The wavelet transform is flexible, allowing the use of different filters at each level of the decomposition. Shorter filters are used at the later levels of the decomposition to reduce their effective length and thereby reduce the extent of ringing after quantization of the coefficients. At the start of the decomposition, longer filters are needed to avoid any artificial blockiness that short filters would cause.

The wavelet coefficients are organized into wavelet trees, each of which is rooted in the low-low band of the decomposition and extending into the higher frequency bands at the same spatial location. Each wavelet tree provides a correspondence between the wavelet coefficients and the spatial region they represent in the frame. The coefficients of each wavelet tree can be rearranged to form a wavelet block. Each wavelet block is a fixed size block of the frame and comprises those wavelet coefficients at all scales and orientations that correspond to that block.

The wavelet trees are scanned and the wavelet coefficients are quantized. Because each tree relates to a distinct block of the frame, quantization can be varied according to what content of the image is covered by each block. It is also possible to vary the quantization within the block in a frequency-dependent fashion. The quantizer is a mid-rise uniform quantizer with dead-zone around zero. There are two advanced schemes for modifying the quantization process based on content. First, using gradient-based significance analysis of the residuals, the quantizer dead-zone can be varied to enhance quality where most visible at the expense of committing greater errors where they are not visible. Second, the quantizer step size can be adjusted in a content-dependent manner to improve the quality of important objects in the scene.

9.5 CONCLUSIONS

In this chapter, scalable texture coding is discussed for multimedia application. Spatial and quality scalabilities are two important features desired in many multimedia applications. We have presented three zerotree wavelet techniques which provide high compression efficiency as well as scalability of the bitstream. Shapiro's EZW is a zerotree wavelet algorithm that provides high granularity quality scalability. ZTE coding demonstrated its effectiveness for very high coding efficiency while providing spatial scalability. ZTE also uses the zerotree concept but differs from EZW in ways that allow it to deliver greater compression performance. In the ZTE algorithm, quantization is explicit, coefficient scanning is performed in one pass, and the encoded symbols are tailored for very low bit rate coding of image, video I-frames and motion-compensated residuals. MZTE coding technique is based on ZTE coding but utilizes a new framework to improve and extend the ZTE method to achieve a fully scalable yet very efficient coding technique. MZTE provides a very flexible approach to support the right trade-off between layers and types of scalability, complexity and coding efficiency for multimedia applications. MZTE is adopted in MPEG-4 for encoding still pictures of either images or synthetized textures.

10 MULTIRESOLUTION AND OBJECT-BASED VIDEO WATERMARKING USING PERCEPTUAL MODELS

Mitchell D. Swanson[1], Bin Zhu[1] and Ahmed H. Tewfik[1,2]

[1] Cognicity Inc.
Minneapolis, MN
[swanson,zhu]@cognicity.com

[2] University of Minnesota
Minneapolis, MN*
tewfik@ee.umn.edu

10.1 INTRODUCTION

In today's communications environment, digital media such as images, audio and video, are readily manipulated, reproduced, and distributed over information networks. Such prodigious data-sharing leads to problems regarding copyright protection. As a result, creators and distributors of digital data are hesitant to provide access to their digital intellectual property. Technical solutions for copyright protection of multimedia data are actively being pursued.

Digital watermarking has been proposed as a means to identify the owner and distribution path of digital data. Watermarking is the process of encoding hidden copyright information into digital data by making small modifications to the data samples, e.g., pixels. Many watermark algorithms have been proposed. Some techniques modify spatial/temporal data samples (e.g., [339, 20, 360, 252, 319]) while others modify transform coefficients (e.g., [20, 176, 52, 24, 326, 116]). Unlike encryption, watermarking does not restrict access to the data. Furthermore, a watermark is designed to *permanently* reside in the host data whereas once encrypted data is decrypted intellectual property rights are no

*This work was supported by AFOSR under grant AF/F49620-94-1-0461.

longer protected. When the ownership of data is in question, this information can be extracted to completely characterize the owner or distribution path.

In this chapter, we present a pair of novel video watermarking schemes. One technique employs temporal multiresolution techniques to construct a watermark [310]. The second algorithm breaks each video frame into objects that are individually watermarked [309]. Both watermarking procedures explicitly exploit the human visual system (HVS) to guarantee that the embedded watermark is imperceptible. Similar to our image and audio watermarking procedures based on perceptual models [311, 25], the video watermarks adapt to and are highly dependent on the video being watermarked. This guarantees an invisible and robust watermark.

Watermarking digital video introduces some concerns that generally do not have a counterpart in images and audio. Due to large amounts of data and inherent redundancy between frames, video signals are highly susceptible to pirate attacks, including frame averaging, frame dropping, frame swapping, collusion, statistical analysis, et cetera. Many of these attacks may be accomplished with little or no damage to the video signal. However, the watermark may be adversely affected. Scenes must be embedded with a consistent and reliable watermark which survives such pirate attacks. Applying an identical watermark to each frame in the video leads to problems of maintaining statistical invisibility. Furthermore, such an approach is necessarily video *independent*, as the watermark is fixed. Applying independent watermarks to each frame is also a problem since regions in each video frame with little or no motion remain the same frame after frame. Motionless regions in successive video frames may be statistically compared or averaged to remove independent watermarks. Two approaches to address these issues are:

1. **Multiresolution watermarking algorithm.** We employ a watermark which consists of fixed and varying components. The components are generated from a temporal wavelet transform representation of each video scene. A wavelet transform applied along the temporal axis of the video results in a *multiresolution temporal representation* of the video. In particular, the representation consists of temporal lowpass frames and highpass frames. The lowpass frames consist of the static components in the video scene. The highpass frames capture the motion components and changing nature of the video sequence. Our watermark is designed and embedded in each of these components. The watermarks embedded in the lowpass frames exist throughout the *entire* video scene. The watermarks embedded in the motion frames are highly localized in time and change rapidly from frame to frame. Thus, the watermark is a composite of *static* and *dynamic* components. The combined representation overcomes the aforementioned drawbacks associated with a fixed or independent watermarking procedure. Although averaging frames simply damages the dynamic watermark components, the static components survive such attacks and are easily recovered for copyright verification.

To generate a watermark, the visual masking properties of the wavelet coefficient frames are computed and used to filter (i.e., shape) a pseudo-random sequence which represents the author or distribution path of the video. Based on pseudo-random sequences, the noise-like watermark is statistically undetectable, thereby helping thwart pirate attacks. Furthermore, due to the combined static and dynamic watermark representation, the watermark is readily extracted from a *single frame* of the video *without knowledge of the location* of that particular frame in the video, even after printing and rescanning.

2. **Object-based watermarking algorithm.** We also propose an object-based watermarking procedure to address these issues. We employ a segmentation algorithm [30] to extract objects from the video. Each segmented object is embedded with a unique watermark according to its perceptual characteristics. In particular, each object in the video has an associated watermark. As the object experiences translations and transformations over time, the watermark remains embedded with it. An interframe transformation of the object is estimated and used to modify the watermark accordingly. If the object is modified too much, or if the watermark exceeds the tolerable error level (see Section 10.3) of the object pixels, a new object and new watermark are defined. Objects defined in the video are collected into an *object database*. As new frames are processed, segmented objects may be compared with previously defined objects for similarity. Objects which appear visually similar use the same watermark (subject to small modifications according to affine transformations). As a result, the watermark for each frame changes according to the perceptual characteristics while simultaneously protecting objects against statistical analysis and averaging.

In the next section, we introduce our author representation. Our frequency and spatial masking models are reviewed in Section 10.3. The wavelet transform is reviewed in Section 10.4. Our watermarking design algorithms are introduced in Sections 10.5 and 10.6. The detection algorithms are introduced in Section 10.7. Finally, experimental results for the two algorithms are presented in Sections 10.8 and 10.9. The robustness of our watermarking procedures are illustrated for an assortment of signal processing operations and distortions including colored noise, MPEG coding, multiple watermarks, frame dropping, and printing and scanning. We present our conclusions in Section 10.10.

10.2 AUTHOR REPRESENTATION AND THE DEADLOCK PROBLEM

The main function of a video watermarking algorithm is to unambiguously establish and protect ownership of video data. Unfortunately, most current watermarking schemes are unable to resolve rightful ownership of digital data when multiple ownership claims are made, i.e. when a deadlock problem arises [53]. The inability to deal with deadlock is independent of how the watermark is inserted in the video data or how robust it is to various types of modifications.

Currently, watermarking techniques which do *not* require the original (non-watermarked) signal during detection are the most vulnerable to ownership deadlocks. A pirate simply adds his or her watermark to the watermarked data. The data now has two watermarks. Current watermarking schemes are unable to establish who watermarked the data first.

Watermarking procedures that require the original data set for watermark detection also suffer from deadlocks. In such schemes, a party other than the owner may counterfeit a watermark by "subtracting off" a second watermark from the publicly available data and claim the result to be his or her original. This second watermark allows the pirate to claim copyright ownership since he or she can show that both the publicly available data and the original of the rightful owner contain a copy of their counterfeit watermark.

To understand how our procedure solves the deadlock problem, let us assume that two parties claim ownership of a video. To determine the rightful owner of the video, an arbitrator examines only the video in question, the originals of both parties and the key used by each party to generate their watermark.

We use a two step approach to resolve deadlock: dual watermarks and a video dependent watermarking scheme. Our dual watermark employs a *pair* of watermarks. One watermarking procedure requires the original data set for watermark detection. This chapter provides a detailed description of that procedure and of its robustness. The second watermarking procedure does *not* require the original data set and hence, is a simple data hiding procedure. Any watermarking technique which does not require the original for watermark detection and satisfies the requirements outlined in [54] can be used to insert the second watermark.

In case of deadlock, the arbitrator first checks for the watermark that requires the original for watermark detection. If the pirate is clever and has used the attack suggested in [53] and outlined above, the arbitrator would be unable to resolve the deadlock with this first test. The arbitrator then checks for the watermark that *does not* require the original video sequence in the video segments that each ownership contender claims to be his *original*. Since the original video sequence of a pirate is derived from the watermarked copy produced by the rightful owner, it will contain the watermark of the rightful owner. On the other hand, the true original of the rightful owner will not contain the watermark of the pirate since the pirate has no access to that original and the watermark does not require subtraction of another data set for its detection.

Further protection against deadlock is provided by the technique that we use to select the pseudo-random sequence that represents the author. This technique is similar to an approach developed independently in [54]. Both techniques solve the shortcomings of the solution proposed in [53] for solving the deadlock problem. Specifically, the author has two random keys, or seeds, x_1 and x_2, from which a pseudo-random sequence y can be generated using a suitable pseudo-random sequence generator [270]. Popular generators include Rivest/Shamir/Adelman (RSA), Rabin, Blum/Micali, and Blum/Blum/Shub [111]. With the two proper keys, the watermark may be extracted. Without the two keys, the data hidden in the video is statistically undetectable and practically impossible to recover. Note that we do *not* use the classical maximal length pseudo-noise sequence (m-sequence) generated by linear feedback shift registers to generate a watermark. Sequences generated by shift registers are cryptographically insecure: one can solve for the feedback pattern, or, equivalently, the keys, given a small number of output bits y.

The noise-like sequence y, after some processing, is the actual watermark hidden in the video stream. Key x_1 is *author* dependent, key x_2 is *signal* dependent and key x_1 is the secret key assigned to (or chosen by) the author. Key x_2 is *computed from the video signal* which the author wishes to watermark. It is computed from the video using a one-way hash function. In particular, the tolerable error levels supplied by the masking models are hashed to key x_2. Any one of a number of well-known secure one-way hash functions may be used to compute x_2, including RSA, Message Digest - Version 4 (MD4), and the Secure Hash Algorithm (SHA) [270, 111]. For example, the Blum/Blum/Shub pseudo-random generator uses the one way function $y = g_n(x) = x^2 \bmod n$ where $n = pq$ for primes p and q so that $p = q = 3 \bmod 4$. It can be shown that generating x or y from partial knowledge of y is *computationally infeasible* for the Blum/Blum/Shub generator.

The signal dependent key x_2 makes counterfeiting very difficult. The pirate can only provide key x_1 to the arbitrator. Key x_2 is automatically computed by the watermarking algorithm from the original signal. The pirate generates a counterfeit original by subtracting off a watermark. However, the watermark (partially generated from the signal dependent key) *depends* on the counterfeit original. Thus, the pirate must generate a watermark which creates a counterfeit original which, in turn, generates the watermark! As it is computationally infeasible to invert the one-way hash function, the pirate is unable to fabricate a counterfeit original which generates the desired watermark.

10.3 VISUAL MASKING

We use image masking models based on the HVS to ensure that the watermark embedded into each video frame is perceptually invisible and robust. Visual masking refers to a situation wherein a signal raises the visual threshold for other signals around it. Masking characteristics are used in high quality low bit rate coding algorithms to further reduce bit rates [156]. The masking models reviewed here are based on image models. A detailed discussion of the models may be found in [380]. We are currently developing a watermarking algorithm that takes temporal masking [108] into account.

10.3.1 Frequency Masking

Our frequency masking model is based on the knowledge that a masking grating raises the visual threshold for signal gratings around the masking frequency [189]. The model we use [380], based on the discrete cosine transform (DCT), expresses the contrast threshold at frequency f as a function of f, the masking

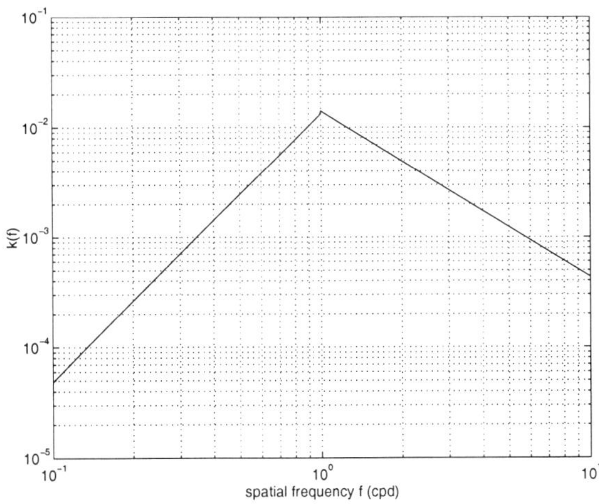

Figure 10.1 The masking characteristic function $k(f)$.

frequency f_m and the masking contrast c_m

$$c(f, f_m) = c_0(f) \cdot \max\{1, [k(f/f_m)c_m]^\alpha\},$$

where the detection threshold at frequency f, $c_0(f)$, and $\alpha = 0.62$ are determined by psycho-visual tests [189]. The mask weighting function $k(f)$ is shown in Figure 10.1.

To find the contrast threshold $c(f)$ at a frequency f in an image, we first use the DCT to transform the image into the frequency domain and find the contrast at each frequency. Then, we use a summation rule of the form

$$c(f) = \left[\sum_{f_m} c(f, f_m)^2\right]^{1/2},$$

to sum the masking effects from all the masking signals near f. If the contrast error at f is less than $c(f)$, the model predicts that the error is invisible to human eyes.

10.3.2 Spatial Masking

Our spatial masking model is based on the threshold vision model proposed by Girod [108]. The model accurately predicts the masking effects near edges and in uniform background. Assuming that the modifications to the image are small, the upper channel of Girod's model can be linearized [380] to obtain the tolerable error level for each coefficient. This is a reasonable assumption for transparent watermarking.

Under certain simplifying assumptions [380], the tolerable error level for a pixel $p(x, y)$ can be obtained by first computing the contrast saturation at

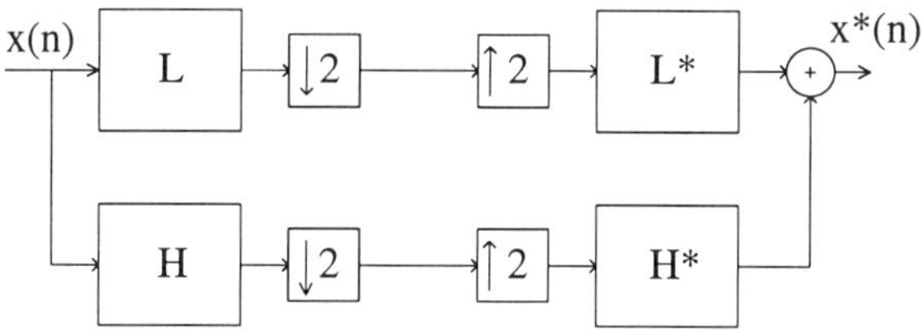

Figure 10.2 Diagram of a two-band filter bank.

(x, y), i.e.

$$dc_{sat}(x, y) = dc_{sat} = \sqrt{\frac{T}{\sum_{x',y'} w_4(0, 0, x', y')}},$$

where the weight $w_4(x, y, x', y')$ is Gaussian distributed and centered at the point (x, y). Parameter T is a threshold based on a series of psycho-visual tests. Once $dc_{sat}(x, y)$ is computed, the luminance on the retina, dl_{ret}, is obtained from the equation

$$dc_{sat}(x, y) = w_2(x, y) \cdot dl_{ret}(x, y).$$

From dl_{ret}, the tolerable error level $ds(x, y)$ for the pixel $p(x, y)$ is computed from

$$dl_{ret}(x, y) = w_1(x, y) \cdot ds(x, y).$$

The weights $w_1(x, y)$ and $w_2(x, y)$ are based on Girod's model. The masking model predicts that changes to pixel $p(x, y)$ less than $ds(x, y)$ introduce no perceptible distortion.

10.4 TEMPORAL WAVELET TRANSFORM

A wavelet transform [268, 337] is a powerful tool employed to represent signals at multiple resolutions. The multiresolution nature of a wavelet decomposition provides signal specific information localized in time, space, or frequency which can be exploited for signal analysis and processing.

We employ the wavelet transform along the *temporal axis* of the video sequence. The wavelet transform is used to provide a compact multiresolution temporal representation of a video, leading to static and dynamic video components. A wavelet transform can be computed using a 2-band perfect reconstruction filter bank as shown in Figure 10.2. The video signal is simultaneously passed through lowpass, L, and highpass, H, filters and then decimated by a factor of 2 to give *static* (no motion) and *dynamic* (motion) components of the original signal. The two decimated signals may be upsampled and passed through complementary filters and summed to reconstruct the original signal. Wavelet filters are widely available [337].

Figure 10.3 Example of temporal wavelet transform.

In Figure 10.3, we show an example of the temporal wavelet transform. The top row consists of four consecutive frames from a sample "Football" video. The bottom row consists of the four temporal wavelet coefficient frames computed from the original Football sequence. The two temporal lowpass frames (bottom left) represent the static components of the Football frames. The detail, or dynamic, components are represented by the two temporal highpass frames (bottom right) in Figure 10.3. Note that a filter bank may be cascaded with additional filter banks to provide further temporal resolutions of the input signal. The output of the cascaded filter banks consists of multiple temporal resolutions of the input.

10.5 MULTIRESOLUTION WATERMARK DESIGN

The first video watermarking algorithm we describe is based on the temporal wavelet transform. The first step in our watermarking algorithm consists of breaking the video sequence into *scenes* [228]. Recall from the Introduction that segmentation into scenes allows the watermarking procedure to take into account temporal redundancy. Visually similar regions in the video sequence, frames from the same scene, must be embedded with a consistent watermark. To address assorted pirate attacks on the watermark, we perform a temporal wavelet transform on the video scenes. The multiresolution nature of the wavelet transform allows the watermark to exist across *multiple temporal scales*, resolving the above mentioned pirate attacks. For example, the embedded watermark in the lowest frequency (DC) wavelet frame exists in *all frames* in the scene.

We denote indexed temporal variables by capital letters with subscripts, for example the ith frame F_i in a video scene. Frames are ordered sequentially according to time. The tilde representation is used to denote a wavelet representation as in $\tilde{F}_i$ is the ith wavelet coefficient frame. Without loss of gen-

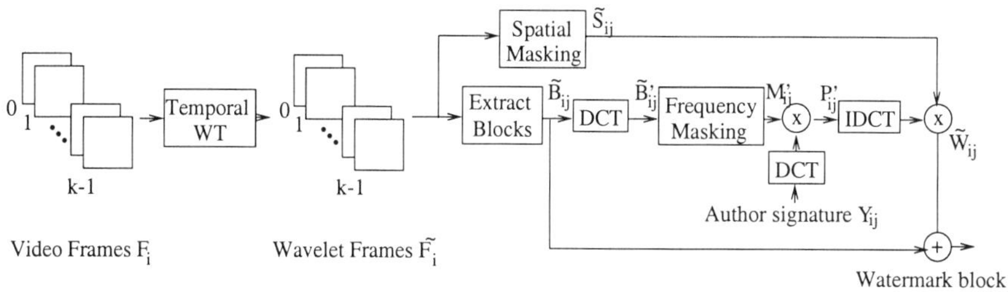

Figure 10.4 Diagram of video watermarking procedure.

erality, wavelet frames are ordered from lowest frequency to highest frequency; $\tilde{F}_0$ is a DC frame. Finally, primed capital letters, such as F_i', denote the DCT representation of an indexed variable.

In Figure 10.4, we show our video watermarking procedure. Consider a scene of k frames from the video sequence. Let each frame from the scene be of size $n \times m$. The video may be gray-scale (8 bits/pixel) or color (24 bits/pixel). Let F_i denote the frames in the scene, where $i = 0, \ldots, k-1$. Initially, we compute the wavelet transform of the k frames F_i to obtain k *wavelet coefficient frames* $\tilde{F}_i$, $i = 0, \ldots, k-1$. The watermark is constructed and added to the video using the following steps:

1. Segment each wavelet frame $\tilde{F}_i$ into 8×8 blocks $\tilde{B}_{ij}$, $i = 0, 1, \ldots, \lfloor \frac{n}{8} \rfloor$ and $j = 0, 1, \ldots, \lfloor \frac{m}{8} \rfloor$

2. For each block $\tilde{B}_{ij}$,

   (a) Compute the DCT, $\tilde{B}'_{ij}$, of the frame block, $\tilde{B}_{ij}$

   (b) Compute the frequency mask, M'_{ij}, of the DCT block, $\tilde{B}'_{ij}$

   (c) Use the mask M'_{ij} to weight the noise-like author, Y'_{ij}, for that frame block, creating the frequency-shaped author signature, $P'_{ij} = M'_{ij}Y_{ij}$;

   (d) Create the wavelet coefficient watermark block, $\tilde{W}_{ij}$, by computing the inverse DCT of P'_{ij} and locally increase the watermark to the maximum tolerable error level provided by the spatial mask $\tilde{S}_{ij}$

   (e) Add the watermark $\tilde{W}_{ij}$ to the block, $\tilde{B}_{ij}$, creating the watermarked block

3. Repeat for each wavelet coefficient frame $\tilde{F}_i$

The watermark for each wavelet coefficient frame is the block concatenation of all the 8×8 watermark blocks, $\tilde{W}_{ij}$, for that frame. The wavelet coefficient frames with the embedded watermarks are then converted back to the temporal domain using the inverse wavelet transform. As the watermark is designed and embedded in the wavelet domain, the individual watermarks for each wavelet coefficient frame are spread out to varying levels of support in the temporal domain. For example, watermarks embedded in highpass wavelet frames are localized temporally. Conversely, watermarks embedded in lowpass wavelet frames are generally located throughout the scene in the temporal domain.

10.6 OBJECT-BASED WATERMARK DESIGN

The second watermarking algorithm is based on video objects. As described in Section 10.1, a segmentation algorithm [30] is employed to extract objects from the video. Each segmented object is embedded with a unique watermark according its perceptual characteristics. As the object experiences translations and transformations over time, the watermark remains embedded with it. An interframe transformation of the object is estimated and used to modify the watermark accordingly. If the object is modified too much, or if the watermark exceeds the tolerable error level (TEL) of the object pixels, a new object and new watermark are defined.

Objects defined in the video are collected into an *object database.* As new frames are processed, segmented objects may be compared with previously defined objects for similarity. Objects which appear visually similar use the same watermark (subject to small modifications according to affine transformations). As a result, the watermark for each frame changes according to the perceptual characteristics while simultaneously protecting objects against statistical analysis and averaging. Our object-based video watermarking algorithm has several other advantages. As it is object based, the algorithm may be easily incorporated into the MPEG-4 object-based coding framework. In addition, the detection algorithm does not require information regarding the location (i.e., index) of the test frames in the video. We simply identify the objects in the test frames. Once objects are identified, their watermarks may be retrieved from the database and used to determine ownership.

In this discussion, we implement a simplified block-based (MPEG) approach to our object watermarking algorithm. Rather than watermarking true objects with irregular boundaries, we watermark rectangular blocks using a modified form of MPEG motion tracking. Specifically, we perform frame-by-frame block tracking in terms of translation, rotation, and scaling between the current reference block and candidate blocks in the next frame. Given a block in the current frame, an affine transformation vector is obtained by minimizing a cost function measuring the mismatch between the block and each predictor candidate. The range of predictor candidates are limited by scale, rotation, and translation. The error corresponding to the best matching candidate is compared to a similarity threshold. Candidate blocks with mismatches less than the threshold are signed with identical watermarks.

As shown in Figure 10.5, the author representation, visual masking, and block tracking are combined to form our video watermarking algorithm. The watermark is computed frame-by-frame. Initially, the spatial (S) and frequency (M) masking values for the current frame are computed. The frequency masking values are obtained from the discrete cosine transform (DCT) coefficients

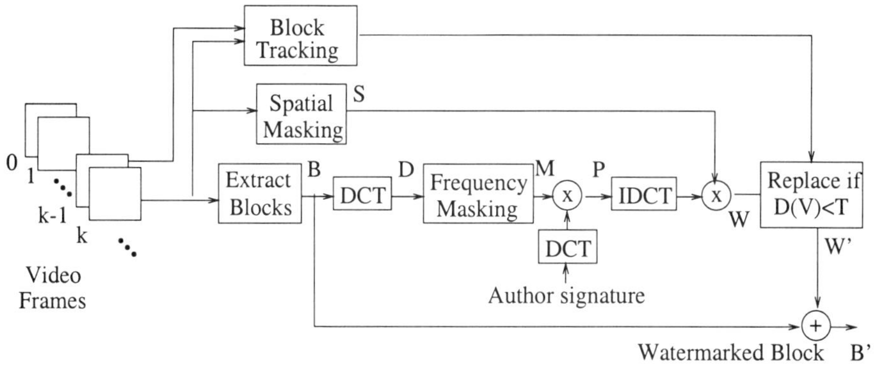

Figure 10.5 Diagram of video watermarking technique.

(D) of 8×8 blocks (B) in the frame. Segmenting the frame into blocks ensures that the frequency masking estimates are localized. Each block of frequency masking values is then multiplied by part of the pseudo-random author representation. The inverse DCT of the product (P) is computed. The result is multiplied by the spatial masking values for the frame, creating the perceptually shaped pseudo-noise, W.

Finally, MPEG block tracking is included. The watermark for a macroblock in the current frame is replaced with the watermark for the macroblock from the previous frame (according to the offset by the motion vector) if the distortion $D(V)$ is less than a threshold T.

10.7 WATERMARK DETECTION

Each watermark is designed to be easily extracted by the owner, even when signal processing operations are applied to the host video. As the embedded watermark is noise-like, a pirate has insufficient knowledge to directly remove the watermark. Therefore, any destruction attempts are done blindly. Unlike other users, the owner has a copy of the original video and the noise-like author signature which was embedded into the video. Typically, the owner is presented with one or more video frames to which he or she wishes to prove ownership rights. Two methods have been developed to extract the potential watermark from a test video or test video frame. Both employ hypothesis testing [340]. One test employs index knowledge during detection, i.e., we know the placement of the test video frame(s) relative to the original video. The second detection method does *not* require knowledge of the location of the test frame(s). This is extremely useful in a video setting, where thousands of frames may be similar, and we are uncertain where the test frames reside.

Detection I: Watermark Detection With Index Knowledge. When the location of the test frame is known, a straightforward hypothesis test may be applied. Let K denote the number of frames to be tested. For each frame

R_k, $1 \leq k \leq K$, in the test video, we perform a hypothesis test

$$\begin{array}{lll} \mathbf{H}_0: & X_k = R_k - F_k = N_k & \text{(No watermark)} \\ \mathbf{H}_1: & X_k = R_k - F_k = W_k^* + N_k & \text{(Watermark)} \end{array} \tag{10.1}$$

where F_k is the original frame, W_k^* is the (potentially modified) watermark recovered from the frame, and N_k is noise. Detection for the object-based watermark proceeds in the same manner, only R_k, F_k, and W_k denote the test object, original object, and corresponding watermark, respectively. The hypothesis decision is obtained by computing the scalar similarity between each extracted signal, X_k, and original watermark, W_k, denoted as

$$S_k = Sim_k(X_k, W_k) = \frac{X_k \cdot W_k}{W_k \cdot W_k}.$$

The overall similarity between the extracted and original watermark is computed as the mean of S_k for all k: S=mean(S_k). The overall similarity is compared with a threshold to determine whether the test video is watermarked. As shown in the experimental results, our experimental threshold is chosen around 0.1, i.e., a similarity value ≥ 0.1 indicates the presence of the owner's copyright. In such a case, the video is deemed the property of the author, and a copyright claim is valid. A similarity value < 0.1 indicates the absence of a watermark.

When the length (in terms of frames) of the test video is the same as the length of the original video, we perform the hypothesis test in the *wavelet domain* for the multiresolution watermarking algorithm. A temporal wavelet transform of the test video is computed to obtain its wavelet coefficient frames, $\tilde{R}_k$. Substituting wavelet transform values in (10.1)

$$\begin{array}{lll} \mathbf{H}_0: & \tilde{X}_k = \tilde{R}_k - \tilde{F}_k = N_k & \text{(No watermark)} \\ \mathbf{H}_1: & \tilde{X}_k = \tilde{R}_k - \tilde{F}_k = \tilde{W}_k^* + N_k & \text{(Watermark)} \end{array} \tag{10.2}$$

where $\tilde{F}_k$ are the wavelet coefficient frames from the original video, $\tilde{W}_k^*$ is the potentially modified watermarks from each frame, and N_k is noise. This test is performed for each wavelet frame to obtain $\tilde{X}_k$ for all k. Similarity values are computed as before: $S_k = Sim_k(\tilde{X}_k, \tilde{W}_k)$.

Using the original video signal to detect the presence of a watermark, we can handle virtually *all* types of distortions, including cropping, rotation, rescaling, et cetera, by employing a generalized likelihood ratio test [340]. We have also developed a second detection scheme which is capable of recovering a watermark after many distortions *without* a generalized likelihood ratio test. The procedure is fast and simple, particularly when confronted with the large amount of data associated with video.

Detection II: Watermark Detection Without Index Knowledge. In many cases, we may have no knowledge of the indices of the test frames. Pirate tampering may lead to many types of derived videos which are often difficult to process. For example, a pirate may steal *one* frame from a video. A pirate may also create a video which is *not* the same length as the original video.

Temporal cropping, frame dropping, and frame interpolation are all examples. A pirate may also swap the order of the frames. Most of the better watermarking schemes currently available use different watermarks for different images. As such, they generally require knowledge of *which* frame was stolen. If they are unable to ascertain which frame was stolen, they are unable to determine which watermark was used.

Our second method can extract the watermark *without* knowledge of where a frame or object belongs in the video sequence. No information regarding cropping, frame order, interpolated frames, et cetera, is required! As a result, no searching and correlation computations are required to locate the test frame index. For the multiresolution video watermarking algorithm, the hypothesis test is formed by removing the *low temporal wavelet frame* from the test frame and computing the similarity with the watermark for the low temporal wavelet frame. The hypothesis test is formed as

$$\begin{array}{llr} \mathbf{H}_0: & X_k = R_k - \tilde{F}_0 = N_k & \text{(No watermark)} \\ \mathbf{H}_1: & X_k = R_k - \tilde{F}_0 = \tilde{W}_k^* + N_k & \text{(Watermark)} \end{array} \qquad (10.3)$$

where R_k is the test frame in the *spatial* domain and $\tilde{F}_0$ is the lowest temporal *wavelet* frame. The hypothesis decision is made by computing the scalar similarity between each extracted signal, X_k, and original watermark for the low temporal wavelet frame, $\tilde{W}_0$: $Sim_k(X_k, \tilde{W}_0)$. This simple yet powerful approach exploits the wavelet property of varying temporal support.

For the object-based watermarking case, the term R_k represents the test object. The term $\tilde{F}_0$ is replaced with O_k, the *original object from the object database* corresponding to R_k. In this case, W_k denotes the corresponding watermark for O_k. Again, the frame index is not required during detection.

10.8 RESULTS FOR THE MULTIRESOLUTION WATERMARKING ALGORITHM

In this section, we present the visual and robustness results for the multiresolution watermarking procedure. Results for the object-based watermarking algorithm are presented in the following section.

10.8.1 Visual Results

We illustrate the invisibility and robustness of our watermarking scheme on two gray-scale (8 bpp) videos: "Ping-Pong" and "Football." Each frame is of size 240×352. An original frame from each video is shown in Figures 10.6(a) and 10.7(a). The corresponding watermarked frame for each is shown in Figures 10.6(b) and 10.7(b). In both cases, the watermarked frame appears visually identical to the original. In Figures 10.6(c) and 10.7(c), the watermark for each frame, scaled to gray-levels for display, are shown. Although the watermarks are computed on the *wavelet* frames, we display them in the *spatial* domain for visual convenience. The watermark for each frame is the same size as the host frame, namely 240×352. For each frame, the watermark values corresponding

Table 10.1 Statistical properties of the video watermark.

Video	Maximum	Minimum	Variance	PSNR (dB)
Ping-Pong	42	-44	11.20	37.64
Football	43	-47	14.44	36.54

Table 10.2 Blind testing of watermarked videos.

Video	Preferred original to watermarked
Ping-Pong	48.5 %
Football	50.5 %

to smoother background regions are generally smaller than watermark values near motion and edge regions. This is to be expected, as motion and edge regions have more favorable masking characteristics.

Some statistical properties for each of the watermarks are shown in Table 10.1. The values are computed for the frames presented in Figures 10.6 and 10.7, which are representative of the watermarks for the other frames in each of the videos. The maximum and minimum values are in terms of the watermark values over the 240×352 watermark. Peak signal-to-noise ratio (PSNR), a common image quality metric, is defined as $20 \log_{10}(255/\sqrt{SNR})$. The signal-to-noise ratio (SNR) is computed between the original and watermarked frame.

To determine the quality of the watermarked videos, we performed a series of informal visual tests. For each test video, we displayed the original to the viewer. Then two randomly selected videos "A" and "B" were sequentially displayed to the viewer. The ordered pair was randomly selected as (original, watermarked) or (watermarked, original). The viewer was asked to select the video A or B which was visually more pleasing. This test was performed ten times for each video. A total of 10 viewers (not including the authors) took part in the blind test. The results of this test are displayed in Table 10.2. As predicted by the visual masking models, the original and watermarked videos appeared visually similar and each was preferred approximately 50% of the time. We conclude that the watermark causes no degradations to the host video.

10.8.2 Robustness Results

To be effective, the watermark must be robust to incidental and intentional signal distortions incurred by the host video. Clearly, any lossy signal operations performed on the host video affect the embedded watermark.

The robustness of our watermarking approach is measured by the ability to *detect a watermark when one is present in the video.* Robustness is further based on the ability of the algorithm *to reject a video when a watermark is not*

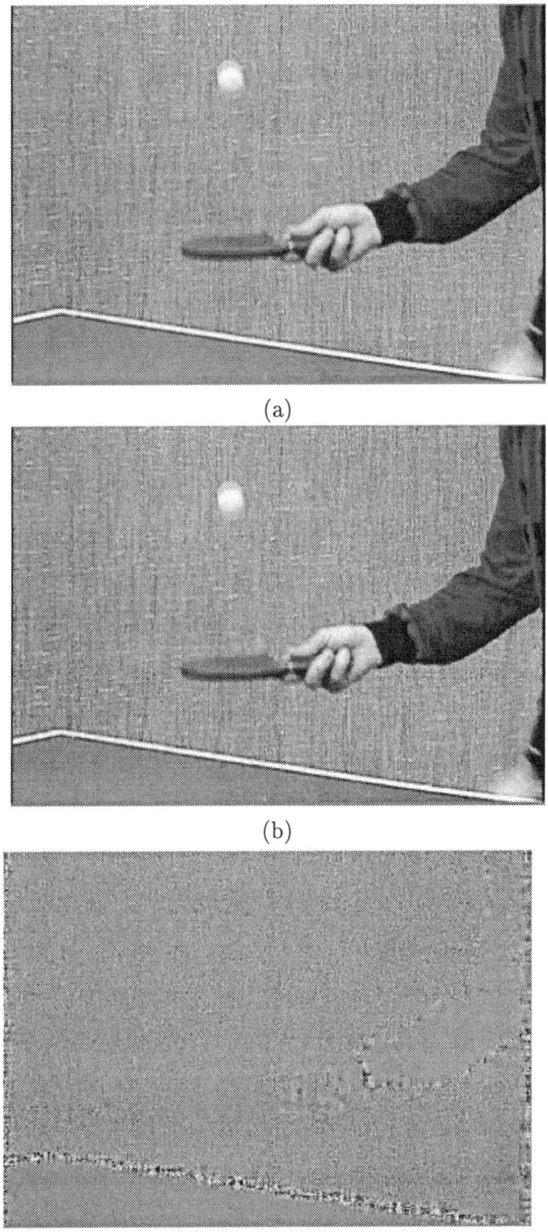

(a) (b) (c)

Figure 10.6 Frame from Ping-Pong video: (a) original, (b) watermarked, and (c) watermark.

(a)

(b)

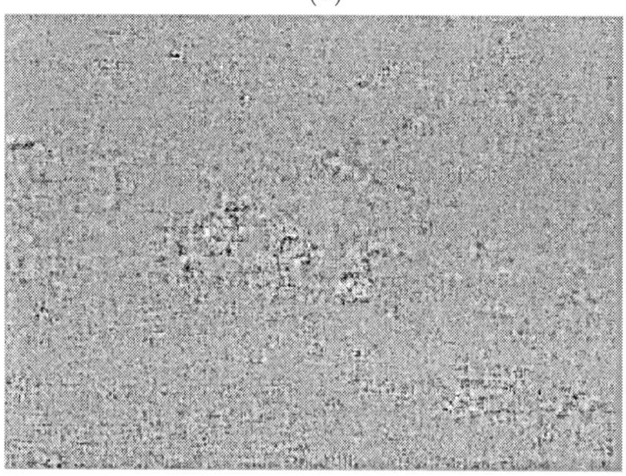

(c)

Figure 10.7 Frame from Football video: (a) original, (b) watermarked, and (c) watermark.

present. For a given distortion, the overall performance may be ascertained by the relative difference between the similarity when a watermark is present (hypothesis $\mathbf{H}_1$) and the similarity when a watermark is not present (hypothesis $\mathbf{H}_0$). In each robustness experiment, similarity results were obtained for both hypotheses. In particular, the degradation was applied to the video when a watermark was present. It was also applied to the video when a watermark was *not* present. The similarity was computed between the original watermark and the recovered signal (which may or may not have a watermark). A large similarity indicates the presence of a watermark ($\mathbf{H}_1$), while a low similarity suggests the lack of a watermark ($\mathbf{H}_0$). As shown in our tests, no overlap between the hypotheses occurred during the degradations and distortions. This indicates a high probability of detection and a low probability of false alarm.

We use the first 32 frames from each video for our tests. Both detection approaches were performed during each experiment. Specifically, we performed detection when the entire test sequence was available and the indices were known (Detection I, (10.2)). We also performed detection on a frame-by-frame basis without knowledge of the frame index (Detection II, (10.3)). In this case, we assume that the index of the frame is *unknown*, so we do not know the location of the frame in the video.

10.8.3 Colored Noise

To model perceptual coding techniques, we corrupted the watermark with *worst case colored noise which follows the visual masks.* Colored noise was generated by shaping (multiplying) white noise with the frequency and spatial masks for the video. As the colored noise is generated in the same fashion as the watermark, it acts like another *interfering* watermark. We generated colored noise and added it to the video with and without the watermark. The variance of the noise for each test sequence was chosen 9 times greater than the watermark embedded in the video. For example, the average variance of the watermark over all frames from the Football sequence is 14.0. The colored noise sequence was constructed with a variance of approximately 126.0 (PSNR = 27.1 dB). Noisy frames from each of the watermarked videos are shown in Figures 10.8(a) and (b). These frames correspond to those shown in Figures 10.6 and 10.7.

For each video, this testing process was repeated 100 times with a new noise sequence for each run. In the first test, we use all of the frames in the video for detection (Detection I). The similarity values for each video sequence with and without the watermark are shown in Table 10.3. The maximum, mean, and minimum similarity values are computed over all 100 noise runs. It is important to note that the minimum similarity values with watermark are much larger than the maximum similarity values without watermark; an overlap between the two indicates possible errors in detection. In this case, the minimum similarity value of the Ping-Pong sequence with watermark is 0.91, which is much larger than the maximum value of 0.03 without watermark. Similar results hold for the Football sequence. As a result, one may readily decide whether a watermark exists in the video. Selecting a decision threshold, T, somewhere in the range

(a)

(b)

Figure 10.8 Frame from videos with colored noise (PSNR $= 25.1$ dB): (a) Ping-Pong and (b) Football.

Table 10.3 Similarity results with colored noise.

Video	PSNR (dB)	With watermark			No watermark		
		Max	Mean	Min	Max	Mean	Min
Ping-Pong	27.8	1.00	0.96	0.91	0.03	0.00	-0.02
Football	27.1	1.00	0.97	0.93	0.04	0.00	-0.03

of approximately $0.1 \leq T \leq 0.9$ essentially guarantees a correct hypothesis decision for these test videos in colored noise.

We also performed testing on a frame-by-frame basis without knowledge of the frame index (Detection II). Detection was performed by removing the low-pass temporal frame, $\tilde{F}_0$, from the test frame, and correlating the result with the watermark $\tilde{W}_0$ corresponding to $\tilde{F}_0$. The similarity values obtained during testing indicate easy discrimination between the two hypotheses as shown in Figure 10.9. The upper similarity values in each plot correspond to each frame with a watermark. The lower similarity curve corresponds to each frame without a watermark. The error bars around each similarity value indicate the *maximum* and *minimum* similarity values over the 100 runs. The x-axis corresponds to frame number and runs from 0 to 31. Observe that the upper value is widely separated from the lower value for each frame. Consequently, an error free hypothesis decision is easy to obtain without knowledge of the position of the frame in the video scene.

In all of the following distortion experiments, we add colored noise to each video *prior* to distortion, e.g. coding, printing and scanning. The colored noise is used to simulate additional attacks on the watermark, including masking-based coders and *other watermarks*. The strength of the colored noise is approximately the same as that of the watermark and is not visible.

10.8.4 Coding

In most applications involving storage and transmission of digital video, a lossy coding operation is performed on the video to reduce bit rates and increase efficiency. We tested the ability of the watermark to survive MPEG-1 coding [188] at very low quality. In our experiment, we set the MPEG tables at the coarsest possible quantization levels to maximize compression.

A watermarked Ping-Pong video frame coded at 0.08 bpp is shown in Figure 10.10(a). The corresponding compression ratio (CR) is 100:1. The original (non-coded) frame is shown in Figure 10.6(b). Note that a large amount of distortion is present in the frame. Using the same quantization tables, a frame from the Football video at 0.18 bpp (CR 44:1) is shown in Figure 10.10(b). Note that the two videos used the same quantization tables. However, the Football sequence has more motion present than the Ping-Pong sequence. As a result, it requires additional bits/pixel to encode the video.

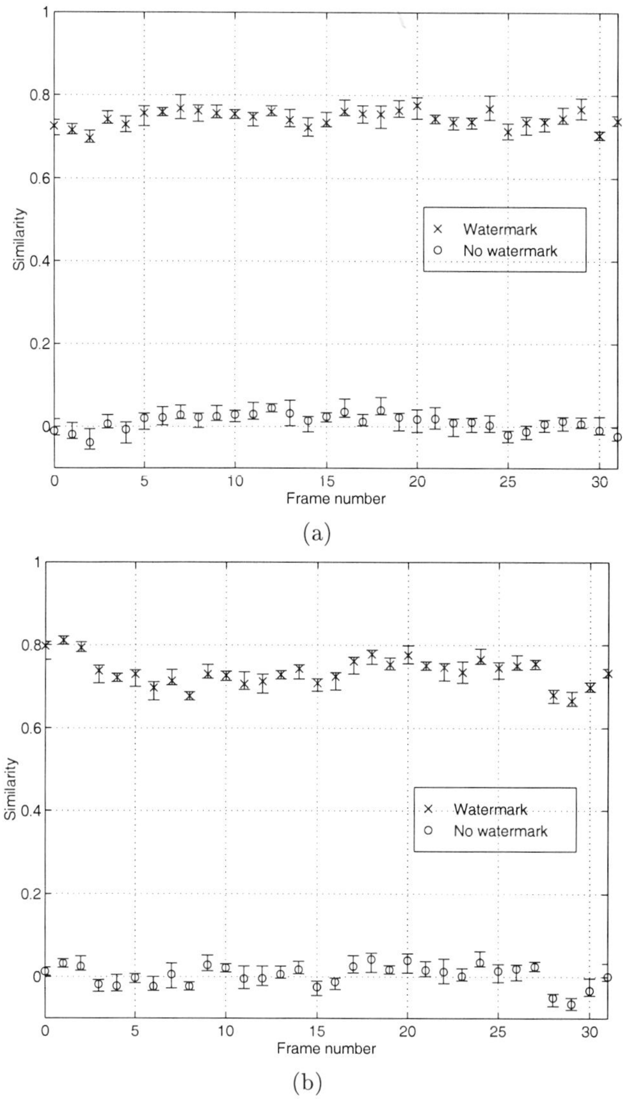

Figure 10.9 Similarity values versus frame number in colored noise: (a) Ping-Pong and (b) Football. The error bars around each similarity value indicate the maximum and minimum similarity values over 100 runs.

(a)

(b)

Figure 10.10 MPEG coded frame: (a) Ping-Pong (0.08 bits/pixel, CR 100:1) and (b) Football (0.18 bits/pixel, CR 44:1).

Table 10.4 Similarity results after MPEG coding.

Video	CR	PSNR (dB)	With watermark			No watermark		
			Max	Mean	Min	Max	Mean	Min
Ping-Pong	100:1	26.8	0.41	0.35	0.28	0.06	0.00	-0.08
Football	44:1	24.4	0.37	0.32	0.27	0.07	0.01	-0.05

To simulate additional attacks on the watermark, we added colored noise to each video *prior* to MPEG coding. Each video was tested 100 times, with a different colored noise sequence used during each run. In the first test, we use all of the frames in the video for detection (Detection I). The maximum, mean, and minimum similarity values for each video sequence with and without the watermark are shown in Table 10.4. Again, observe that the minimum similarity values with watermark are much larger than the maximum values without watermark. Even at very low coding quality, the similarity values are widely separated, allowing the existence of a watermark to be easily ascertained.

We also performed detection on single frames from the video (Detection II) without index knowledge. The plots are shown in Figure 10.11(a) and (b). The error bars indicate no overlap between the two similarity curves. Even at very low bit rate, the presence of a watermark is easily observed.

10.8.5 Multiple Watermarks

We also tested the ability to detect watermarks in the presence of other watermarks. This distortion seems likely to occur, as watermarks may be embedded sequentially to track legitimate multimedia distribution. Furthermore, a pirate may use additional watermarks to attack a valid watermark. We embedded three *consecutive* watermarks into each test video, one after another. All three use the original (non-watermarked) video as their original during detection. We then added colored noise to the videos and MPEG coded the result. The Ping-Pong sequence was coded at 0.28 bpp (CR 29:1, PSNR 27.45 dB). Using the same MPEG parameters, the Football sequence was coded at 0.51 bpp (CR 16:1, PSNR 25.43 dB). The test was performed 100 times by generating a new colored noise sequence each time. Similarity values associated with the three watermarks without index knowledge are plotted in Figures 10.12(a) and (b). The presence of the three watermarks is easily determined.

10.8.6 Frame Averaging

Distortions of particular interest in video watermarking are those associated with temporal processing, of which temporal cropping, frame dropping, and frame interpolation are all examples. As we have shown, temporal cropping is handled with our Detection II approach which does not require information regarding frame indices. To test frame dropping and interpolation, we dropped

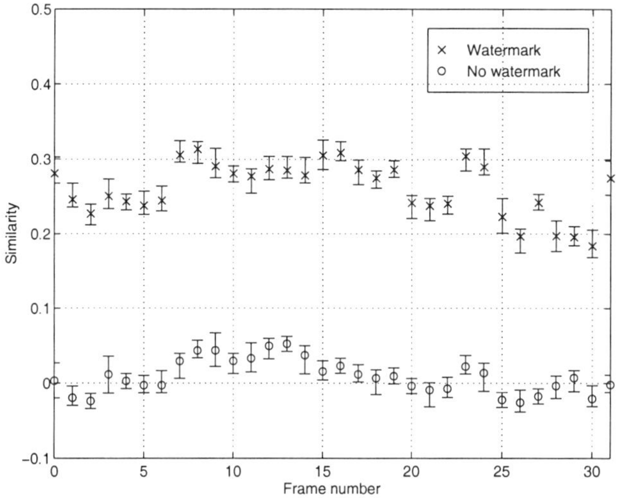

(a)

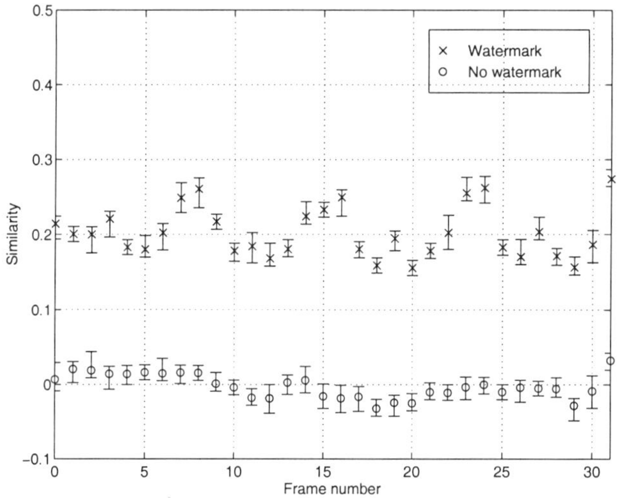

(b)

Figure 10.11 Similarity values versus frame number after MPEG coding: (a) Ping-Pong and (b) Football. The error bars around each similarity value indicate the maximum and minimum similarity values over 100 runs.

the odd index frames from the test sequences. The missing frames were replaced with the average of the two neighboring frames, $F_{2n+1} = (F_{2n} + F_{2n+2})/2$. Again, we applied Detection II. The resulting similarity values are shown in Figures 10.13(a) and (b). The plots with and without watermark are widely separated.

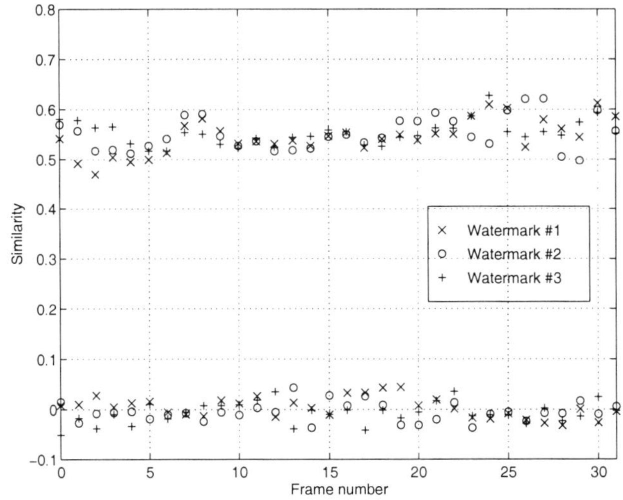

(a)

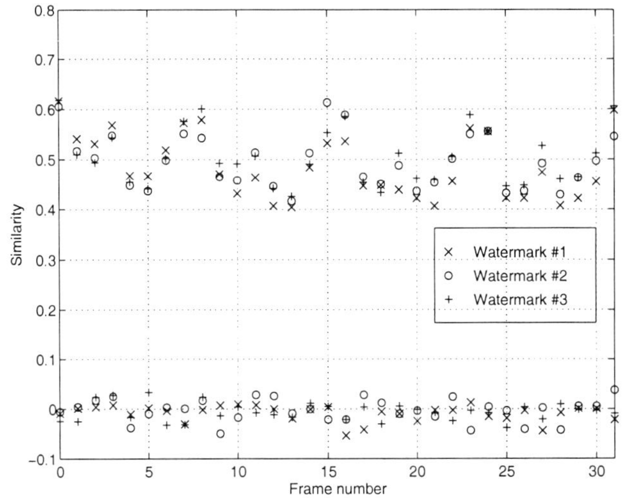

(b)

Figure 10.12 Similarity values for three watermarks after MPEG coding: (a) Ping-Pong and (b) Football.

10.8.7 Printing and Scanning

One important copyright issue is that of protecting individual video frames from being duplicated in print, such as magazines and technical documents. For this test, we created a hardcopy of the original and watermarked frames shown in Figures 10.6 and 10.7 and used a flatbed scanner to re-digitize them. The similarity results obtained from printing and scanning are shown in Table 10.5.

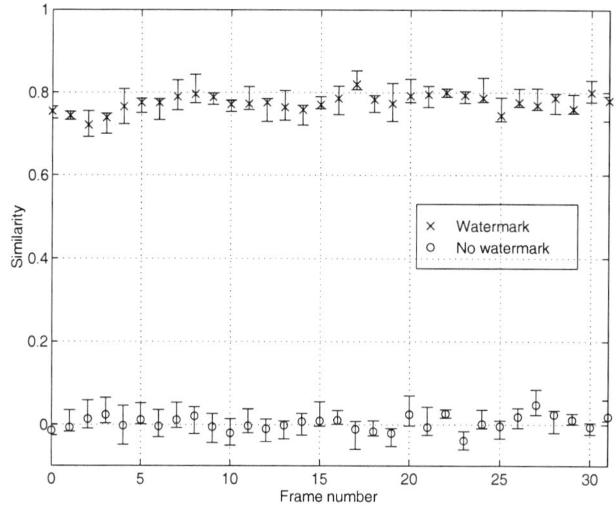

(a)

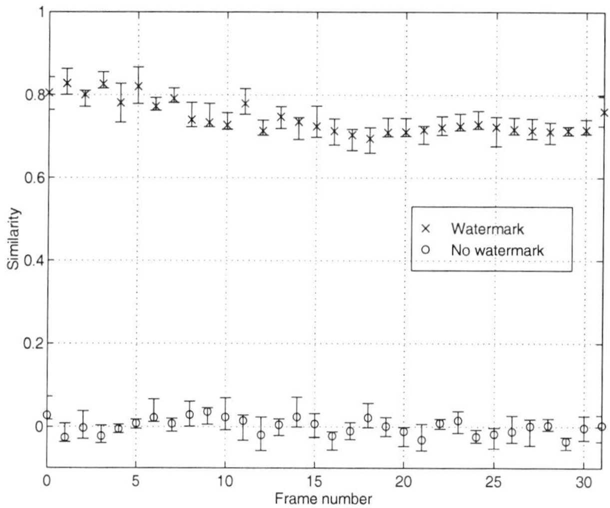

(b)

Figure 10.13 Similarity values after frame dropping and averaging: (a) Ping-Pong and (b) Football. The error bars around each similarity value indicate the maximum and minimum similarity values over 100 runs.

Detection was performed *without* knowledge of frame location (Detection II). The resulting similarity values offer easy discrimination between watermarked and non-watermarked printed frames even without knowledge of frame position.

Table 10.5 Similarity results after printing and scanning.

Video Frame	Similarity	
	With watermark	Without watermark
Ping-Pong	0.734	0.011
Football	0.611	0.052

Figure 10.14 Original, watermarked, and rescaled watermark frames: (a) Garden and (b) Football.

10.9 RESULTS FOR THE OBJECT-BASED WATERMARKING ALGORITHM

We illustrate our watermarking scheme on the two gray-scale video sequences: "Garden" and Football. Each frame is of size 240×352. Original and watermarked frames from each video sequence are shown in Figures 10.14(a) and (b). The watermark for each frame, rescaled for display, is also shown in Figure 10.14. The values for the watermark over each frame typically range from ± 2 (smooth regions) to ± 50 (edges and textures).

In an attempt to defeat the watermark, we introduce several large distortions to the video sequences. Additive noise with standard deviation of 100 was added to each video sequence. One of the frames from the noisy Garden sequence is shown in Figure 10.15. The maximum, mean, and minimum similarity values for each video sequence with and without the watermark are shown in the first two rows of Table 10.6. Even in extremely noisy environments, the minimum similarity of the sequence with the watermark is much greater than the maximum similarity of the sequence without the watermark. Robustness to MPEG coding at very high compression ratios (CR) was tested. The MPEG quantization tables were set at the *coarsest possible*. Although the resulting

Table 10.6 Similarity results in noise and MPEG coding.

Video	CR	PSNR (dB)	With watermark Max	Mean	Min	No watermark Max	Mean	Min
Garden	N/A	10.25	0.84	0.76	0.71	0.09	0.01	0.00
Football	N/A	10.06	0.96	0.80	0.66	0.10	0.00	0.00
Garden	68:1	22.8	0.66	0.24	0.19	0.05	0.01	0.00
Football	166:1	24.6	0.62	0.27	0.18	0.07	0.00	0.00

video sequences are highly distorted (cf. Figure 10.16), the results in rows 3 and 4 in Table 10.6 show that the watermark is easily detected.

We also tested the robustness of our procedure to printing and scanning. One of the frames from the Garden sequence (cf. Figure 10.17) was printed and scanned back to a digital form. When the watermark was present in the frame, the similarity measure was 0.73. When the watermark was not present, the similarity measure was 0.03. We also tested cropping on the Football video. A 120×100 region was cropped and MPEG coded at 36:1. This amounts to only 14.2% of the original frame size. With watermark, the maximum, minimum and mean were 0.83, 0.51, and 0.22, respectively. Without watermark, they were 0.09, 0.00, and 0.00. Again, the watermark was easily detected.

10.10 CONCLUSION

We presented a pair of watermarking procedures to embed copyright protection into digital video by directly modifying the video samples. One watermark technique was based on a multiresolution temporal representation of the video. A second watermarking technique embedded a watermark into each object in the video. Each watermarking technique directly exploits the masking phenomena of the human visual system to guarantee that the embedded watermark is imperceptible. The owner of the digital video piece is represented by a pseudo-random sequence defined in terms of two secret keys. One key is the owner's personal identification. The other key is calculated directly from the original video signal. The signal dependent watermarking procedure shapes the noise-like author representation according to the masking effects of the host signal. The embedded watermark is perceptually and statistically undetectable. We illustrated the robustness of the watermarking procedure to several video degradations, including colored noise, MPEG coding, multiple watermarks, frame dropping, and printing and scanning. The watermark was readily detected with and without index knowledge in all of these distortions.

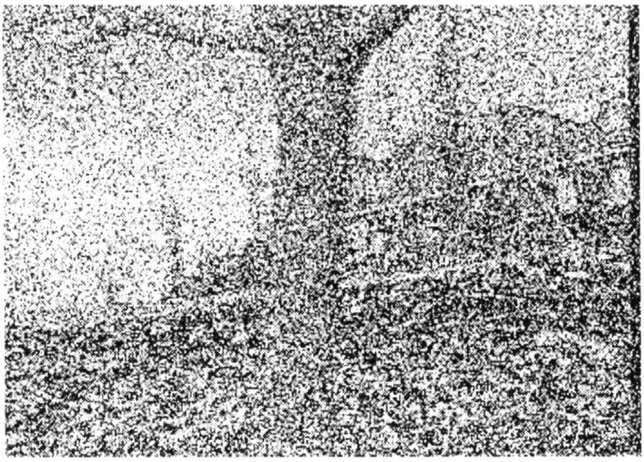

Figure 10.15 Noisy frame.

Figure 10.16 MPEG coded.

Figure 10.17 Printed frame.

11 TRANSFORMS IN TELEMEDICINE APPLICATIONS

Mark J. T. Smith[1] and Alen Docef[2]

[1]Georgia Institute of Technology
Atlanta, GA
mark.smith@carnegie.gatech.edu

[2]University of British Columbia
Vancouver, BC
alen@ece.ubc.ca

Digital representations and transform technology have played a prominent role in the medical field. However, the full potential of this technology in the medical area is far from being realized. Most medical files at this time are recorded and archived in non-electronic form, a notable example being x-ray images which are stored primarily on film. This general practice is changing. Medical equipment manufacturers are moving toward fully digital medical record management systems, often called PACS (Picture Archiving and Communication Systems) in the medical equipment industry. With the industry converging toward digital formats comes a new freedom — the freedom to transmit and exchange patient data electronically. Such capabilities have important implications in the current health care environment, where medical specialists are concentrated primarily in large urban hospitals. One can now envision having a virtual hospital with specialists distributed geographically throughout the country and the world. With cable connections, wireless links, telephone and ISDN lines, and ATM and internet connections now available, one can entertain the notion of a seamless network of on-line practitioners, specialists, and medical center staff personnel on call. In this scenario, physicians could work from home, from the office, or on the road. Such an arrangement could have dramatic impact in terms of medical resource accessibility, quality, response time, and cost. This vision and variations of it fall under the umbrella of *telemedicine.*

One of the major emerging challenges facing the telemedicine community is the size of medical record data, and the current limitations of the national and global telecommunications infrastructure. Digitizing medical records consumes a tremendous amount of memory. For instance, it is not uncommon for digital x-rays to be of size 2,000 × 2,000 pixels or larger with a depth of 10 or 12 bits per pixel. In addition to the size, a simple patient study may contain several images, which only exacerbates the data representation problem. For transmission purposes, raw data sets of this size would demand high capacity lines to facilitate rapid transmission. Transmission lines of this type are not available everywhere. Instead, we have common telephone lines that form the backbone of the present national and international telecommunications infrastructure. Telephones are everywhere, providing data rates of about 28.8 kilobits per second under reasonably favorable conditions. With this limited channel capacity, transmission of a 2000 × 2000 12 bit/pixel x-ray would take approximately one-half hour. If 64 kbit/second ISDN lines are considered instead, the time drops to 12 minutes, which is still a long time for a single image. For the more expensive T1 lines (with 1.5 Mbit/second channel capacity) the transmission time reduces to about one-half minute.

Ideally, we would like the convenience of being able to exchange medical images within a few seconds over a common telephone or even cellular phone and store and retrieve digitized images from a terminal. How can this be done given the inherent bandlimited telecommunications infrastructure of today? The answer lies in compression. By compressing medical images first, data and hence transmission time can be reduced significantly. At the receiver, the compressed bit stream is decoded and reconstructed, thereby enabling records to be exchanged rapidly.

The notion of compression for real-time transmission is obviously not limited to the medical area. In fact, most of the compression research presently in progress is being driven by teleconferencing and multimedia applications. Interestingly, the application of compression to the medical area is somewhat unique, specifically in regard to reconstruction quality issues, the volume and precision of data, and the diversity of images. In addition, we would like to take full advantage of digital technology to enhance the formation and clarity of medical images. Discrete-time transforms play a central role in both compression and image formation aspects of telemedicine. The emphasis in all of our discussions will focus on the role of transforms in telemedicine. Such transformations may be based on Fourier transforms, block transforms, or subband decompositions. In the first part of this chapter, we will explore several of the common medical image modalities, and discuss their characteristics and their method of formation. In the second part, we will discuss some of the attractive techniques for medical image compression and transmission, and the issues involved in deploying these techniques in practice.

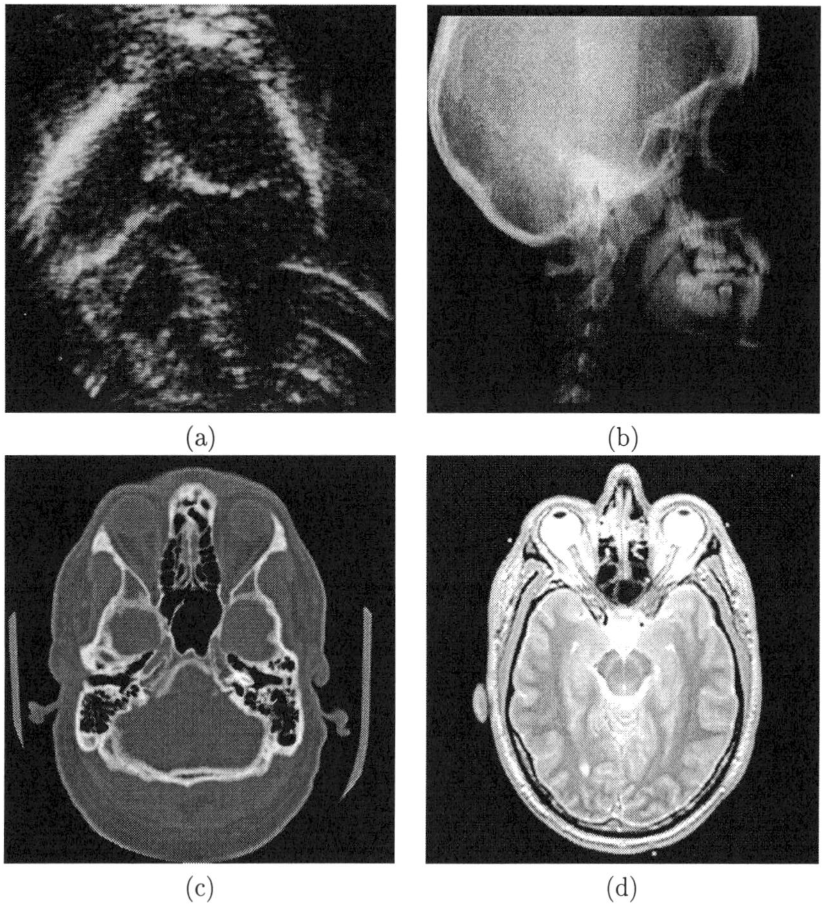

Figure 11.1 A few typical medical images generated via: (a) B-mode ultrasound; (b) head x-ray; (c) brain x-ray computed tomography; (d) brain magnetic resonance imaging.

11.1 MEDICAL IMAGES AND THEIR CHARACTERISTICS

In order to more fully appreciate transform-based processing in the area of telemedicine, familiarity with the formation, usage, and statistical properties of the various image modalities is key, and motivates the discussion that follows. Medical imaging technologies exploit the interaction between the human anatomy and the output of emissive materials or emissions devices. These emissions are then used to obtain pictures of the human anatomy. There are many medical modalities of this type, some of the most popular being ultrasound, x-rays, x-ray computed tomography, and magnetic resonance imaging. Images generated from these technologies are markedly different from natural images we acquire from optical cameras. The nature of these medical image modalities is shown in Figure 11.1 for ultrasound, x-ray, x-ray computed tomography, and magnetic resonance images. These images provide important anatomical

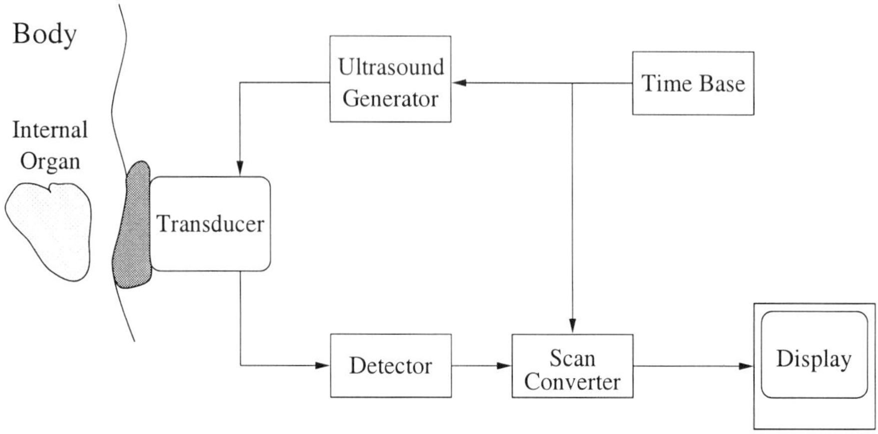

Figure 11.2 Generic ultrasound system.

information to physicians and specialists upon which diagnoses can be made. In the remainder of this section, we examine image formation issues associated with these modalities, and provide a conceptual view of the underlying physics.

11.1.1 Ultrasound Images

Ultrasound is well-suited for the imaging of soft tissues, its major areas of application being obstetrics, cardiology, and the imaging of the abdominal region. According to some reports, more money is spent presently on ultrasound equipment than on any other medical imaging technology. Perhaps this should not be surprising, since ultrasound is one of the few painless low-cost, non-invasive, and safe procedures available for viewing internal organs in the body.

Ultrasound imaging is based on emitting a short pulse of ultrasound in the examined tissue and measuring the time of flight and intensity of reflected sound waves [139]. A generic ultrasound imaging system is sketched in Figure 11.2. The time of flight is proportional to the depth where the reflection occurred, and the intensity is proportional to the gradient of the tissue density at the specific depth. Diagnostic ultrasound operates in the range of 1 to 15 MHz, which translates to wavelengths of 1.5 mm to 0.1 mm, assuming a soft tissue acoustic propagation speed of 1500 meters/second.

According to the way the tissue is scanned by the ultrasonic transducer, a number of ultrasound imaging modes can be created. The most popular imaging mode is the B mode or *brightness* mode. Here, the ultrasound beam is steered electronically in a plane to form an image of the body density gradient. Acoustically reflective components within the body show up as bright spots in the ultrasound image. Two properties of acoustic wave-tissue interaction have a particularly significant effect on the images we see, attenuation and scattering. *Attenuation* is characterized by the reduction in wave intensity, I, resulting from absorption and scattering in the tissue. The wave intensity may

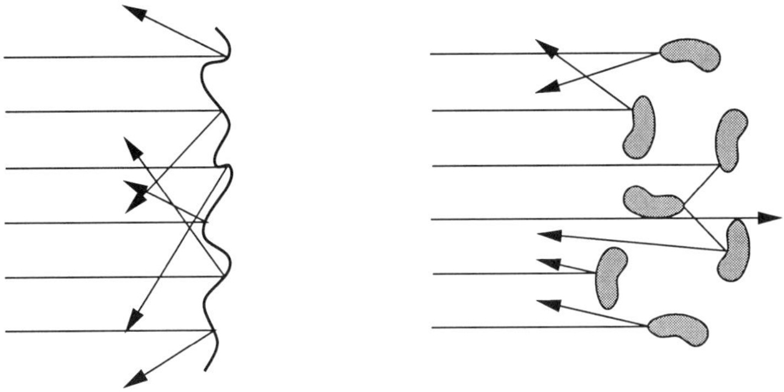

Figure 11.3 Scattering of ultrasound waves.

be described by an exponential decay of the initial intensity as a function of distance, z, traveled by the wave. That is,

$$I = I_0 e^{-\alpha z},$$

where I_0 is the initial intensity and α is the attenuation index. The index α increases with frequency, f, and is governed by the equation

$$\alpha = \text{constant} \cdot f^{1.1}.$$

Scattering is a phenomenon caused by the nonhomogeneous nature of the tissues. If the boundary between two media has surface irregularities or small reflecting inhomogeneities in the medium of the order of magnitude of the wavelength, the reflected wave is fragmented into waves traveling in random directions. This is shown in Figure 11.3. The interference of these waves at the receiver leads to the presence of speckle.

The spatial resolution in an ultrasound image is determined by the wavelength, the width of the beam, and the duration of the ultrasound pulse. Resolution is not isotropic and it varies with depth, but the theoretical limit is given by the wavelength. For typical ultrasound frequencies, the smallest details that can be seen are on the order of one millimeter. In the way ultrasound images are generally acquired, the region of interest in a digitized NTSC image like the one shown in Figure 11.1 is about 256 by 256 pixels.

The temporal resolution of ultrasound video is typically 30 frames/sec, but higher frame rates such as 60 frames/sec are often required for fast motion studies, such as those in pediatric cardiology. The maximum useful dynamic range, limited by noise and power limitations, is generally 30 dB. This implies that five bits are sufficient for the representation of ultrasound images.

For telemedicine purposes (more specifically teleultrasound purposes) the ultrasonic operating frequency can be important. Low-frequency and high-frequency ultrasound have different uses and properties. *Low-frequency ultrasound* has a frequency of around 2MHz, and can penetrate deep into the human body. This makes it useful for examining a variety of internal organs, like the heart. Cardiology ultrasound video is used to obtain qualitative information, such as the shape of the organ and tissue type as well as quantitative information, such as thickness and motion of the interior walls and leaflets within the heart. *High-frequency ultrasound* has a frequency of around 8MHz. It can be used for tissues closer to the skin, such as the liver, the carotid, and the uterus. Since the wavelength is smaller, speckle is less apparent and smaller details can be visualized. The quality of texture is important here to be able to differentiate between healthy and malignant tissue. On the other hand, the amount of motion in this type of video is typically significantly less than in cardiology ultrasound. Owing to these diverse characteristics, teleultrasound is one of the more challenging subareas in telemedicine. One often has to contend with the high quality transmission of still images and sequences of images, each with a different mix of spatial and temporal resolution.

11.1.2 Magnetic Resonance Images

Magnetic resonance imaging is fast becoming the preferred modality in a number of clinical applications because it is minimally invasive and can be applied to both hard structures and soft tissues. A magnetic resonance image (MRI) is based on measuring magnetic properties of hydrogen nuclei. The size of a typical MRI of the head is 256 by 256 pixels. Considering a field of view of about 24 cm, the pixel density is about one pixel per mm [370]. The pixel depth is often 12 bits. As can be seen in Figure 11.1(d), MRIs provide a high quality view of the internal tissue over a wide range in density. In spite of the quality, there is a level of noise present in the MRI. The source of this noise is thermal noise induced in the probe by the body itself. It is modeled as zero-mean, additive, white Gaussian noise. It has been shown [370] that the signal-to-noise ratio can be improved only by decreasing the spatial resolution when the acquisition time is fixed. Given the rising popularity of MRIs, techniques for efficient and effective transmission for this class of images is of particular interest. MRIs are usually generated in sets, each of which consists of a collection of images corresponding to pictures of axial slices of the body. Since there is some level of redundancy among the slices, it is interesting to consider how that can be exploited for telemedicine.

11.1.3 X-ray Images and X-ray Computed Tomography

X-ray photography is the oldest and simplest medical imaging technology. It is based on measuring x-ray absorption in the human body, absorption that is proportional to tissue density. X-ray imaging works best on hard structures, such as bones. A newer x-ray technology is computed tomography (CT), which results in images of cross-sectional slices of the body, like we saw with the MRIs. Caution must be exercised with x-rays, as repeated exposure can produce damage to the tissue. This is a serious problem that limits x-ray use in sensitive situations, such as pregnancies.

The resolution of x-ray images, either acquired digitally or by scanning an analog film, is 6 to 8 lines/mm, which translates into digital images of size

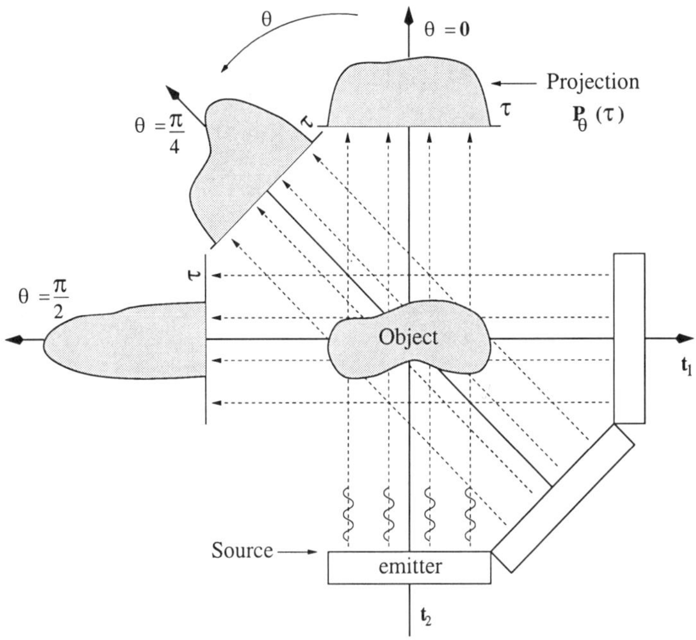

Figure 11.4 Illustration of a computed tomography system.

larger than 2048 by 2048 pixels. X-ray CT images have a lower resolution, typically about 2 lines/mm, or 512 by 512 pixels in digital format. Both types of images have a pixel depth of 12 bits. Like MRIs, a noise component can be associated with x-ray CTs. The quantum noise is dominant and comes from the quantization of energy into photons. It is Poisson distributed and independent of the measurement noise. The measurement noise is additive Gaussian noise and is usually negligible relative to the quantum noise.

Both MRIs and x-ray CTs share the same basic formation algorithm to create the images we see. What is acquired from the physical device in both cases is a set of 1-D signals, each with an associated angular rotation. Signal processing must be performed in order to convert these 1-D signals into 2-D cross-sectional slices. The theory underlying this processing is based on transforms, and is called *reconstruction from projections.*

11.1.4 Reconstruction from Projections

Transform-based signal processing is involved in the formation of cross-sectional images from a set of projections. To appreciate how this works, consider the general framework shown in Figure 11.4, where the emitter irradiates the object of interest, $x(t_1, t_2)$, resulting in a projection signal at the detector. A set of these projection signals is collected, which we denote $p_\theta(\tau)$. The variable θ, where $0 \leq \theta \leq \pi$, is discrete in practice and corresponds to the angle associated with a given projection. The projection $p_\theta(\tau)$ is obtained by the integral

$$p_\theta(\tau) = \int_{-\infty}^{\infty} x(\tau\cos\theta - \xi\sin\theta, \tau\sin\theta + \xi\cos\theta)\, d\xi. \tag{11.1}$$

For the various CT and MRI modalities, the detector outputs can be related to a projection of the form given in equation (11.1). For example, the appropriate relationship for x-ray CTs (with $\theta = 0$) would be

$$p_0(\tau) = -\log\left(\frac{I(\tau)}{I_0(\tau)}\right) = \int_{-\infty}^{\infty} x(\tau, \xi)\, d\xi,$$

where $I_0(\tau)$ is the input intensity in photons per second, and

$$I(\tau) = I_0(\tau) e^{\int x(\tau,\xi)\, d\xi}.$$

Since each projection is a 1-D signal, each projection has associated with it a 1-D Fourier transform $P_\theta(\varpi)$, given by

$$\begin{aligned} P_\theta(\varpi) &= \int_{-\infty}^{\infty} p_\theta(\tau) e^{-j\varpi\tau}\, d\tau \\ &= \int_{-\infty}^{\infty}\int_{-\infty}^{\infty} x(\tau\cos\theta - \xi\sin\theta, \tau\sin\theta + \xi\cos\theta) e^{-j\varpi\tau}\, d\xi\, d\tau. \end{aligned} \tag{11.2}$$

The set of projections in the frequency domain can be related to the Fourier transform of the object $x(t_1, t_2)$ by way of the *Projection-Slice Theorem.* This theorem states that the Fourier transform of a projection corresponds to a radial slice of the 2-D Fourier transform of $x(t_1, t_2)$ with angular orientation θ. That is,

$$\begin{aligned} P_\theta(\varpi) &= \int\int x(t_1, t_2) e^{-j(\overbrace{\varpi\cos\theta}^{\Omega_1} t_1 + \overbrace{\varpi\sin\theta}^{\Omega_2} t_2)}\, dt_1\, dt_2 \\ &= X(\varpi\cos\theta, \varpi\sin\theta) = X(\Omega_1, \Omega_2), \end{aligned} \tag{11.3}$$

where $X(\Omega_1, \Omega_2)$ is the 2-D Fourier transform of $x(t_1, t_2)$. Thus, if we had an infinite set of projections (i.e. if θ were continuous), we would have the 2-D Fourier transform of $x(t_1, t_2)$. Practical equipment constraints only allow a finite set of projections to be acquired. Moreover, since the signal processing is performed digitally on computer, the projections obtained are sampled. This results in a discrete sampled representation of the projection set in the variables θ and τ. By taking the discrete Fourier transform (DFT) of each sampled projection, we can obtain a radially sampled representation of the 2-D DFT of $x(t_1, t_2)$. This radial sampling is shown by the pattern of open circles in Figure 11.5. It is now a simple matter to see that the 2-D DFT of the sampled object can be obtained by interpolating from the radially sampled grid to the rectangular 2-D lattice shown by the solid black circles in Figure 11.5. To complete the reconstruction after interpolation, one merely takes the inverse 2-D DFT to

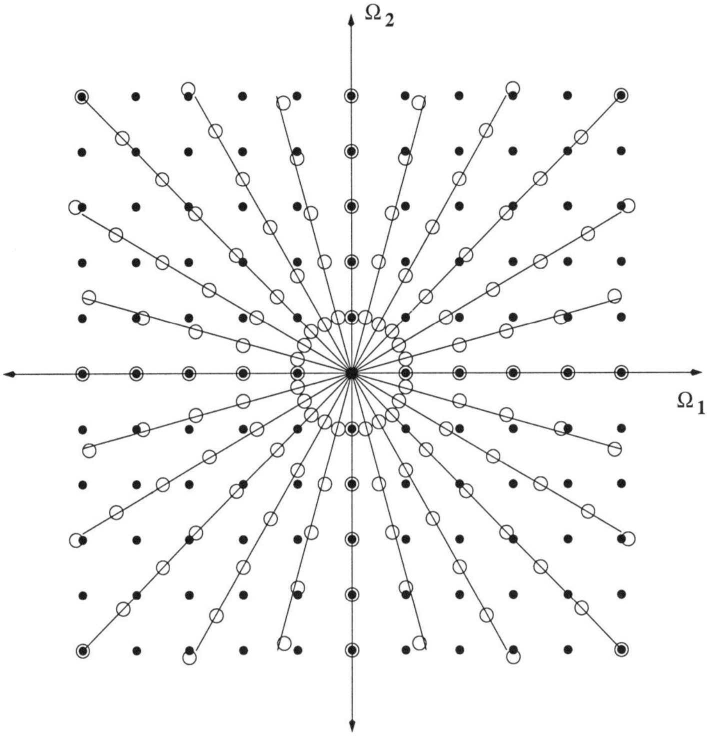

Figure 11.5 Graphical illustration of sample points for the projections (open circles) overlaid on the 2-D DFT lattice.

obtain a discrete representation of $x(t_1, t_2)$. Although this method is conceptually simple, most practical reconstruction implementations employ popular back projection techniques, which are alternate methods that achieve the same goal. Now that we have discussed the characteristics and composition of these image modalities, we consider next the issues concerning digital representation and telemedicine.

11.2 DICOM STANDARD AND TELEMEDICINE

Digital representations allow medical images to be stored, manipulated and displayed on computer systems. This leads naturally to the notion of telemedicine, where medical images of various types can be exchanged via the data communications networks connecting computer systems. Standards for image representation are of critical importance in telemedicine. They define the format and protocol so that systems compliant with the standard can communicate with each other regardless of which manufacturer supplied the equipment.

In the field of telecommunications, vendors look to the International Organization for Standardization (ISO) and the International Telecommunication Union (ITU) for image, audio, and video compression and transmission standards. In the medical area, the leaders playing this role are the American College of Radiology (ACR) and the National Electrical Manufacturers Association (NEMA). In 1983, both organizations collaborated to create a standard for medical imaging devices, known as ARC-NEMA Version 1.0. The standardization activities underwent a revision in 1988 to Version 2.0, before converging to the current standard, Version 3.0. The standard is known formally as DICOM Version 3.0, or DICOM for short, which stands for Digital Image Communication in Medicine. Major medical imaging equipment manufacturers have now adopted the DICOM standard [232]. DICOM is a set of rules specifying transmission and storage formats for image data and identifying information. It establishes a standard for data exchange in a networked environment, thereby facilitating interoperability among equipment from different vendors. In telemedicine, DICOM is extremely important because all electronic transfer of medical records in the future will be done digitally.

11.3 COMPRESSION FOR TELEMEDICINE

Medical image and data transmission is the essence of telemedicine. Given that the common transmission channels are bandlimited, data must be compressed prior to transmission so that real-time or near real-time transmission can be realized. In the medical community, there is great concern that any form of compression may affect the diagnostic quality of the reconstructed image. This is obviously an important issue, both from health care and legal perspectives.

Fortunately, there are options available that can address these concerns, either partially or completely. Compression can be performed in a way that guarantees that diagnostic quality is preserved. The most non-controversial method is via *lossless compression.* Lossless compression algorithms allow perfect reconstruction of the original image from the compressed bit stream. Lossless methods have been under study for many years and many fine tutorial sources are available on this subject. Although it would take us too far afield to discuss these algorithms, we will briefly outline the principles on a conceptual level in the next section.

In many telemedicine applications the compression produced by lossless coders is not sufficient for real-time transmission and economical storage. Higher compression ratios can be achieved by relaxing the requirements of perfect image reconstruction — in other words, *lossy compression.* In lossy coding, some information is discarded and the challenge is to preserve the essential information that allows for image reconstruction with only minimal distortion. Lossy coders typically achieve compression ratios of 10 to 50 and more. Currently, lossy compression has limited application in primary diagnosis, owing to concern for the legal consequences of an incorrect diagnosis. However, one can argue that it is possible to preserve diagnostic quality without having lossless performance. High quality coders that can serve this function are often called *visually lossless* and *diagnostically lossless* methods. *Visually lossless* compression results in no subjective visible loss under normal viewing conditions [32, 59]. Similarly, *diagnostically lossless* compression results in no loss of

Table 11.1 Example of two possible entropy table assignments.

Case A		Case B	
L_1	0	L_1	1
L_2	10	L_2	01
L_3	110	L_3	100
L_4	1110	L_4	1000

diagnostically significant information. In the next two sections, we will discuss lossless and lossy techniques that are suitable for telemedicine.

11.4 LOSSLESS METHODS FOR MEDICAL DATA

Lossless compression techniques exploit information redundancy within the input data. Entropy coders are an important subclass of lossless methods, based on the idea that data may be represented by binary codewords of different lengths. The assignment of codewords to symbols is made such that symbols that occur frequently are represented with short words, while symbols occurring infrequently are assigned codewords that are long. The optimal realization of this concept is what an entropy coder attempts to do.

We define the entropy of a memoryless N-symbol source as

$$H = -\sum_{i=1}^{N} P[x_i] \log_2 P[x_i],$$

where x_i is a symbol in an N-symbol alphabet and $P[x_i]$ is the symbol probability. The entropy represents the lower bound on bit rate that can be achieved for a memoryless source while preserving exact reconstruction. By examining the entropy equation, we can see that if we choose the binary codeword length for a symbol x_i to be $-\log_2 P[x_i]$, we can achieve the entropy of the discrete memoryless input. In order for these entropy codes to be useful in practice, they should be uniquely and instantaneously decodable. For example, assume that we have four symbols, L_1, L_2, L_3, L_4, and we wish to code the sequence

$$L_4, L_1, L_2, L_1, L_1, L_1, L_1, L_1.$$

We can consider the two codeword assignments shown in Table 11.1. Using the assignment in case A, we obtain

$$111001000000.$$

It is a simple matter to observe that this binary string can be decoded in only one way; hence it is uniquely decodable. However, if the sequence is coded using the assignment in case B, we obtain

$$100010111111.$$

Observe that using the assignment in case B, this string could also be decoded as

$$L_3, L_2, L_2, L_1, L_1, L_1, L_1, L_1.$$

Thus, the assignment in case B corresponds to a code that is not uniquely decodable. For a code to be uniquely and instantaneously decodable, it should satisfy the so-called prefix condition. That is, no codeword should be the prefix of any other codeword in the table. All entropy coding techniques that are currently employed for lossless compression are uniquely decodable. Perhaps the most popular of the entropy coders is the well-known Huffman coder. Huffman coding defines a relatively simple procedure for constructing a codeword table, like the one in Table 11.1, based on the symbol probabilities. Huffman coding and all of its variations always result in a uniquely decodable code.

Arithmetic coding is another type of entropy coder. It has emerged recently and has gained tremendous popularity and acceptance. It often performs better than Huffman coding in terms of reaching the entropy. The difference in performance arises because arithmetic coding assigns a variable-length codeword dynamically to an input stream, while Huffman coding is based on assigning a variable-length codeword to a set of symbols or fixed-length words. Algorithms for arithmetic coding exist in the public domain. Discussions on this topic may be found in [103].

Dictionary-based coders are another class of lossless algorithms. These coders are not truly entropy coders, since they do not use explicit knowledge of the input probability distribution. Rather, they are based on building a dictionary during the encoding and decoding process. The best known of the dictionary coders is the one due to Lempel and Ziv. The Lempel-Ziv coder is used in the "compress" command available in the UNIX operating system.

Run-length and differential encoders refer to two other classes of lossless coding methods. The basic idea underlying run-length encoding is to represent a sequence of numbers by a sequence of symbol pairs, $(\text{value}_i, \text{runlength}_i)$. The first symbol, denoted by "value," is the value of the given pixel. The second symbol, which is denoted by "run-length," is the number of times that symbol value is repeated. The precise format of run-length encoding may vary slightly from implementation to implementation. The DICOM standard supports a lossless compression algorithm called "packbits" which is based on run-length encoding (RLE) [103]. Packbits is especially effective in compressing the large blank marginal areas in ultrasound, CT, and MRIs and achieves compression ratios of 2:1 to 3:1.

Differential encoders generally involve taking some kind of difference between present and previous samples. If the samples are correlated, the resulting difference signal has reduced energy, making it more efficient to code. Since differential operators can be made invertible, reconstruction can be performed without loss. A good review of lossless and lossy medical image compression techniques is presented in [361]. In general, lossless compression methods are limited in terms of the amount of compression they can perform. The amount of compression depends on the input image. For x-ray images, compression ratios between

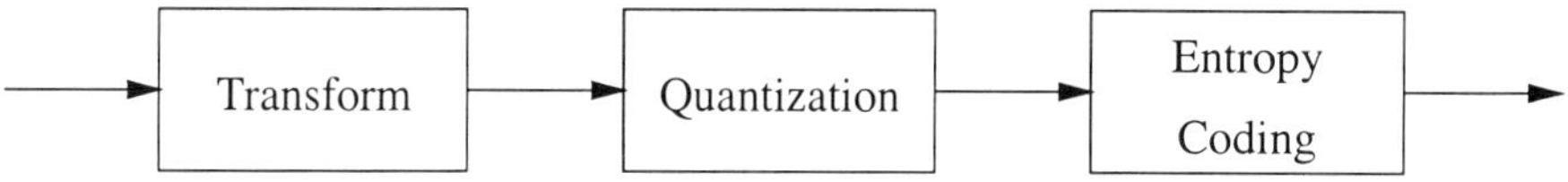

Figure 11.6 Typical functional blocks in an image coder.

3 and 4.8 have been reported [271, 185, 259]. For MRIs, the compression ratios are typically much smaller, such as 1.3 as reported in [271, 259]. MRIs tend to be noisy, which makes them difficult to compress. For ultrasound, images can often be compressed by a factor of 4 [259]. To achieve the compression ratios commonly needed for practical telemedicine, lossy compression algorithms must be considered.

11.5 LOSSY COMPRESSION METHODS FOR MEDICAL IMAGES

Generally speaking, in the medical community lossy compression is viewed in a negative light. Interestingly, however, many imaging systems and procedures which are employed in the medical profession involve inherent loss of information. Radiologists often read recorded ultrasound video clips from VHS or SVHS tapes, which introduce a significant amount of distortion or loss. In [172], it is shown that SVHS image quality is comparable to 26:1 JPEG[1] compressed image quality. In x-ray angiography, images are often acquired at 1024 by 1024 pixels and stored as downsampled 512 by 512 pixel images. This represents information loss in the form of aliasing. Given that these are standard practices, it is not hard to imagine that high quality lossy compression will gain acceptance in the near future. In fact, there are strong signs now indicating that this is happening.

Lossy compression is essential if higher compression ratios are to be achieved. For simplicity, we can picture an image coding system as containing the three components sketched in Figure 11.6. The transform block shown in the figure exploits the inter-pixel correlation. For classical transform coders, this is achieved by concentrating the image energy in a small number of transform coefficients. For predictive coders, inter-pixel correlation is exploited by reducing the pixel variance. The quantizer block shown in the figure maps the transformed pixel values into symbols of finite alphabet suitable for digital representation. Quantizers are often designed to exploit psychovisual redundancy by quantizing the transform coefficients according to visual importance. The entropy coder shown in the figure reduces the number of bits necessary to represent the image by encoding frequently occurring symbols using fewer bits and infrequent symbols using a larger number of bits. The result of the overall three component process is an output bit stream that is compressed, typically by a significant margin. The challenge is to preserve the essential information that allows for image reconstruction with only negligible distortions.

[1] JPEG is the ISO/CCITT international standard for image compression. It is discussed in the next subsection.

The most popular lossy compression coder is the baseline JPEG algorithm [250], which is based on the 2-D discrete cosine transform. Other block-based techniques include other block transforms [260] and various flavors of vector quantization [181, 269].

Block Transform Methods. Block transform compression methods fit most naturally into the above outlined format. The block transforms that generally receive the most attention in the context of image compression are the discrete cosine transform (DCT), the DFT, and the discrete sine transform (DST), with the overwhelming favorite being the DCT. The DCT is an N-point block transform with several variates. The most common for image compression is the 2-D 8×8 DCT-II, defined by

$$X[k_1, k_2] = \frac{1}{4}\sum_{n_1=0}^{7}\sum_{n_2=0}^{7} C_{k_1} C_{k_2} x[n_1, n_2] \cos\left(\frac{(2n_1+1)k_1\pi}{16}\right) \cos\left(\frac{(2n_2+1)k_2\pi}{16}\right)$$

where $X[k_1, k_2]$ is the 2-D 8×8 DCT output, $x[n_1, n_2]$ is an 8×8 input block, and

$$C_{k_1},\ C_{k_2} = \begin{cases} \frac{1}{\sqrt{2}} & \text{for } k_1, k_2 = 0 \\ 1 & \text{otherwise.} \end{cases}$$

The inverse transform is given by

$$x[n_1, n_2] = \frac{1}{4}\sum_{k_1=0}^{7}\sum_{k_2=0}^{7} C_{k_1} C_{k_2} X[k_1, k_2] \cos\left(\frac{(2n_1+1)k_1\pi}{16}\right) \cos\left(\frac{(2n_2+1)k_2\pi}{16}\right).$$

There are many fast algorithms for implementing the 2-D 8×8 DCT in fixed-precision integer arithmetic, which makes the DCT attractive. The primary attraction, however, is that for natural images the DCT tends to compact the energy of the input into a small number of transform coefficients. This property, called *compaction*, is directly related to the input having a predominantly low frequency characteristic. As it turns out, it can be shown that the DCT is the best of the data independent block transforms in terms of compaction for natural images. This is why the DCT is used in the JPEG standard for image compression.

The JPEG (Joint Photographics Experts Group) standard is a product of collaborative efforts by the ISO, the CCITT (International Telegraph and Telephone Consultative Committee)[2] and the IEC (International Electrotechnical Commission). Although JPEG development was driven primarily by applications involving natural images, it is often considered for coding medical images. In fact, DICOM supports JPEG compression.

[2]The CCITT is now the ITU (International Telecommunication Union).

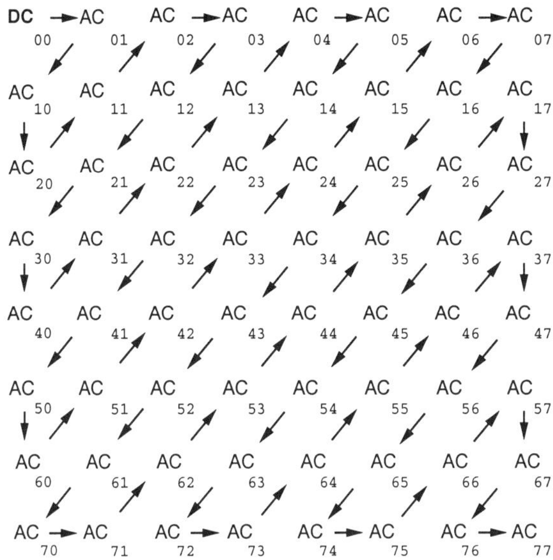

Figure 11.7 Zig-zag scan order used in JPEG.

On a conceptual level, the JPEG algorithm can be described in the following way. The input image is partitioned into contiguous blocks. The convention is to use square $N \times N$ blocks where $N = 8$. Next, the DCT is computed for each block. The transform coefficients are then quantized with a uniform quantizer. The process of taking the transform and performing quantization results in a sparse array of coefficients. That is, many of the coefficients assume a zero value. The block of quantized coefficients is then unwrapped using the zig-zag unwrapping scheme illustrated in Figure 11.7. This results in a 1-D string of numbers, typically with many zeros. In each block, the first transform coefficient (which is the DC coefficient) is treated separately. The DC coefficients are coded across blocks by computing the first backward difference, $d[n]$. That is, a sequence of values DC$[n]$ is formed by extracting the DC component from each block scanning from left to right. The first backward difference, defined by

$$d[n] = \mathrm{DC}[n] - \mathrm{DC}[n-1],$$

is computed and then coded by a DC Huffman table.

The AC coefficients (which are all the other coefficients except the DC) are coded on a block-by-block basis. Coding the AC coefficients is perhaps the most interesting part of the JPEG algorithm and the part responsible for the bulk of the compression. Run-length encoding is used to code the AC components, where the runs used are the number of zeros preceding each non-zero coefficient.

JPEG performs well when the bit rates are sufficiently high. However, at lower bit rates the distortions are prominent. To illustrate this point, Figure 11.8 shows an MRI image coded with the JPEG algorithm. The picture shown

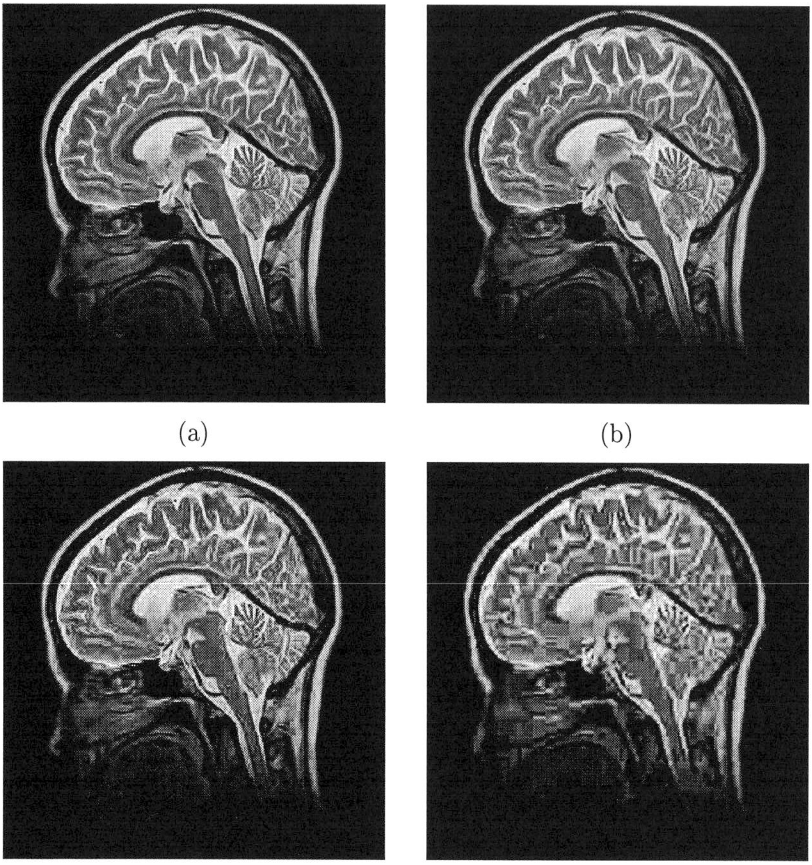

(a) (b)

(c) (d)

Figure 11.8 Example of JPEG compression on MRIs: (a) An original MRI test image; (b) MRI image coded at 1 bit/pixel; (c) MRI coded at 0.5 bits/pixel; (d) MRI coded at 0.25 bits/pixel.

in (a) is the original. The pictures in (b), (c), and (d) are reconstructed JPEG images coded at 1 bit/pixel, 0.5 bits/pixel, and 0.25 bits/pixel, respectively. It is immediately evident that JPEG performs well at the higher rate of 1 bit/pixel, but suffers noticeable quality degradation at rates below that. The nature of the distortion is commonly called blocking, which is most visible at the bottom of the coded images in Figure 11.8. Blocking artifacts are typical of images coded by block transform methods such as JPEG. To relieve the impact of blocking artifacts in medical image compression, some designers have considered larger block sizes, even full-frame DCT coding. Although moving to a transform coder where the DCT is the size of the full image does solve the blocking artifact problem, it results in a loss in compression efficiency. Full-frame DCT coding does not allow bits to be distributed locally according to the

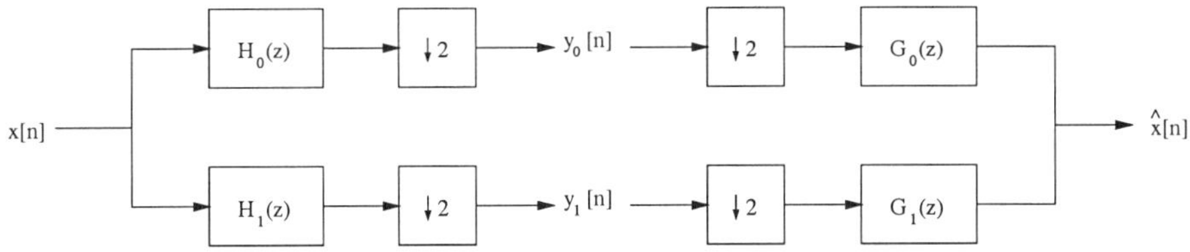

Figure 11.9 A two-band analysis-synthesis filter bank.

local information content. Hence, the preferred compromise is to have small transform blocks. Given this dilemma, the medical community is seeking better techniques for incorporation in the DICOM standard.

11.5.1 Subband/Wavelet Methods for Still Images

Subband image coding [363, 342, 357] eliminates the blocking artifacts by processing the image as a whole. Subband coding is presently the most promising image compression technique. The basic idea is to decompose the image into a number of frequency subbands and then code the subbands. Reference [208] presents a discussion of applications of subband coding to medical images. Typically, the baseband contains the low frequency intensity information and the upper bands contain edge and texture information.

The simplest and the most widely used subband image decomposition is the separable decomposition based on 1-D two-band filter banks, an example of which is shown in Figure 11.9. These filter banks decompose the N-sample signal $x[n]$ into an $N/2$-sample low frequency subband $y_0[n]$ and an $N/2$-sample upper subband $y_1[n]$. The filters $H_0(z)$, $H_1(z)$, $G_0(z)$, and $G_1(z)$, are designed to achieve perfect or near reconstruction of the input signal from the subband components. A number of solutions have been proposed, including perfect reconstruction conjugate quadrature filters (CQFs) [300] and quadrature mirror filters (QMFs) [160]. Subband filters for image decomposition should be designed to preserve localization of image features in subbands and to avoid annoying ringing effects around edges.

In separable subband image decompositions, 1-D filter banks are applied to the rows and columns of an image to obtain four subbands. This process can be applied successively to subband images to obtain a particular decomposition structure. Popular structures are the uniform decomposition and the octave-tree decomposition, sketched in Figure 11.10.

Filtering finite-support signals, such as image rows, produces an output that is longer than the input. This is counterproductive in subband coding, where the purpose of processing is data reduction. The most frequently adopted solutions to the data expansion problem are the circular and symmetric periodic extension methods [301, 302].

The early subband coders used differential pulse coded modulation (DPCM) to encode the baseband and pulse coded modulation (PCM) to encode the upper bands [357], followed by entropy coding. Subband coefficients have been

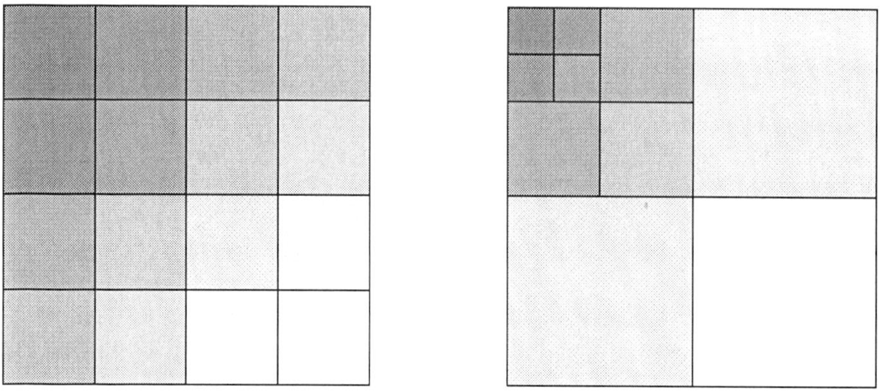

(a) (b)

Figure 11.10 (a) A two-level uniform decomposition; (b) A three-level octave tree decomposition.

modeled as Laplacian [58] and generalized Gaussian [357] random variables, for which optimal scalar quantizers have been designed. Later, vector quantization was used in different configurations [51]. An important issue in coding the subband coefficients is bit allocation among subbands. It has been shown that optimal allocations can be constructed by choosing subband rates with equal slopes on the individual rate-distortion ($R(D)$) curves. Many effective subband image coders have been built around this strategy. More recently, subband coders have been proposed that exploit both inter-band correlations as well as intra-band dependencies [180, 179, 182]. These coders involve training on a set of test images and become optimized for those images. Although they require the extra step of training, they can be tuned to specific classes of medical images, resulting in higher performance.

Zerotree coding [289, 276] is now one of the most popular subband coding techniques. It is being considered for coding medical images and has many attractive features. In particular, these coders are very fast and can support progressive transmission. Zerotree coders work by exploiting the dependencies among subbands by predicting the absence of significant information across subbands.

11.5.2 Motion Compensated Prediction and Transform Methods for Medical Video

Medical images such as ultrasound are often in the form of video sequences. When faced with sequences of correlated medical images, higher compression can be achieved by exploiting the redundancies from frame to frame in addition to those within the frame. A host of video algorithms are available for coding

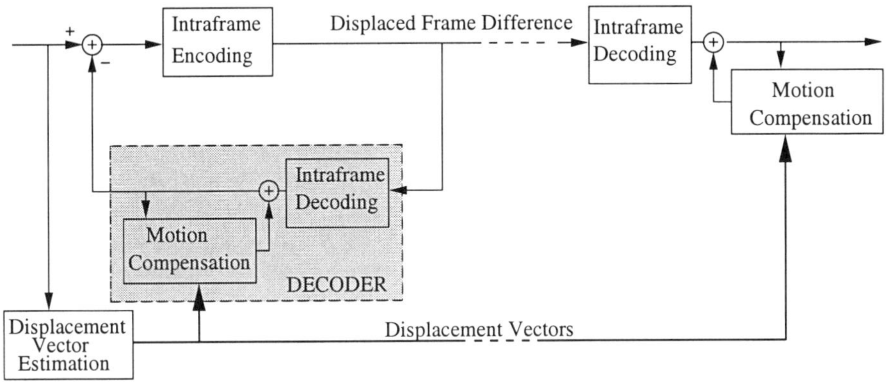

Figure 11.11 The generic motion compensation-based video codec.

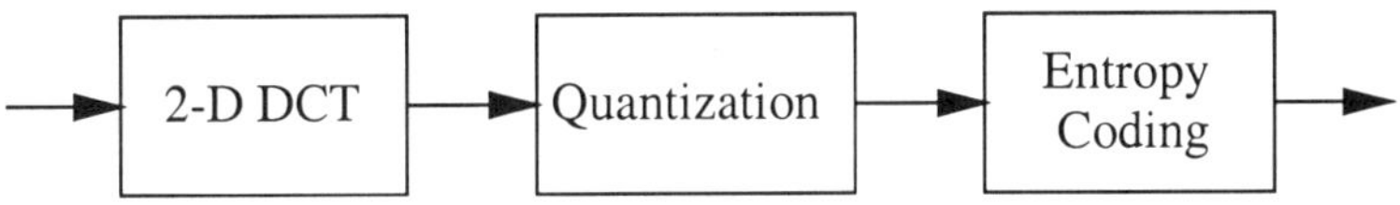

Figure 11.12 The intra-frame coding system.

image sequences. However, the majority of these coders are based on classical motion compensation (MC) as depicted in Figure 11.11. Motion compensation is used to take advantage of the dependencies between consecutive images in the sequence. For MC-based video coding, each frame is decomposed into square blocks. Then, for each block, a displacement vector (v_i, v_j) is generated that points to the best corresponding block in the previously encoded frame. This method is commonly known as block-matching motion estimation. By tiling together all the best corresponding blocks from the previous image, we obtain a prediction of the current image. The difference between the current image and its prediction is called the displaced-frame difference, and it is encoded and transmitted together with the displacement vectors. The displaced-frame difference is encoded using intra-frame coding (JPEG), as illustrated in Figure 11.11. The intra-frame coding system is depicted in Figure 11.12. The frame is decomposed into square blocks and the two-dimensional DCT of each block is taken. The DCT coefficients are quantized using a uniform quantizer with variable step size. The quantization step is controlled by a buffer-control algorithm that tries to keep constant the number of bits for each encoded frame. The quantization indices are then entropy coded using either Huffman or arithmetic coding.

Motion compensation-based techniques are the standard for coding any type of video signal. Thus, such techniques are the first to be considered for ultrasound sequences. However, these coding algorithms do not perform well on ultrasound video. We have evaluated the standard MC algorithms, MPEG-I, MPEG-II, and H.263 [144, 153] on ultrasound and have observed that the performance is extremely poor — an observation that is consistent with those reported in [178]. This is due in part to the high-frequency nature of ultrasound and the very rapid changes in high-frequency noise from frame to frame. These characteristic properties cause motion compensation methods to fall apart and often increase the entropy rather than decrease it. Another reason for the poor performance is the type of motion present in ultrasound. The three-dimensional motion of the organs does not fit well into the translational motion model implied by MC coders. In addition to this, and perhaps most significant is that many medical image sequences like ultrasound are very noisy. This causes MC-based methods to have difficulty. To help address this, Strintzis *et al.* [308] have used the noise models proposed in [183] to introduce a maximum-likelihood motion estimator. This motion estimator, tuned to the statistical properties of ultrasound images, surpasses MPEG-I by more than 2 dB in terms of reconstructed image quality. This result supports the notion that medical image modalities are sufficiently different from natural image sequences, and that gains can be achieved by exploiting the unique characteristics of these modalities. Interestingly, further improvement in performance can be achieved by using subband methods.

11.5.3 Subband Methods for Medical Video Sequences

Since video can be viewed as a 3-D extension of a 2-D image, one might consider the direct extension of 2-D transforms to 3-D. It is straightforward to see that video coders can be constructed in this way. A straight extension of this idea for DCT-based coding was suggested in [15], where an $8 \times 8 \times 8$-point 3-D DCT is used. This results in coefficients that decay in amplitude. Like all block-based techniques, however, the performance of this type of coder at low bit rates suffers from blocking artifacts. Three-dimensional subband coding overcomes the problem of block artifacts at low bit rates, because no block discontinuities are present in the structure. Three-dimensional subband coding was proposed for the first time by Karlsson and Vetterli in 1987 [169] in the context of video transmission over packet-switched networks. The 3-D decompositions can be obtained relatively efficiently by applying the 1-D two-band filter bank in Figure 11.9 along the rows, columns, and temporal direction of the 3-D input signal. Perhaps the simplest 3-D subband decomposition is the two-frame, 11-band configuration [169, 201, 36, 37, 196, 257, 253, 154, 31], shown in Figure 11.13(a). In the temporal domain, two-tap filters are used. Specifically,

$$H_0(z) = \frac{1}{2}(1 + z^{-1}), \qquad G_0(z) = (1 + z^{-1}),$$
$$H_1(z) = \frac{1}{2}(1 - z^{-1}), \qquad G_1(z) = (1 - z^{-1}).$$

The advantages of using the two-tap temporal filters are that the filters are simple, they introduce a very low delay, they require low storage space (the same number of pixels as in MC-based coding), and they have a crude but reasonable frequency response. Longer temporal filters, such as 10-tap QMFs, and block

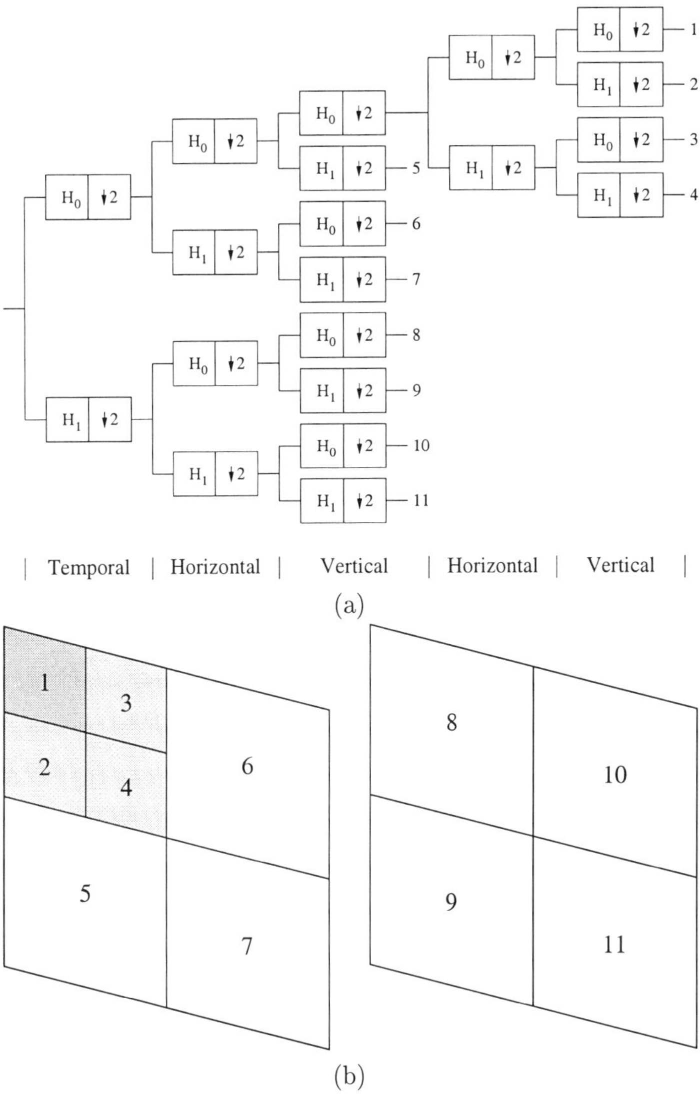

Figure 11.13 Two-frame 11-band 3-D subband decomposition: (a) filter bank configuration; (b) transform domain tiling.

transforms have been used [35, 233] to obtain finer grain decompositions in the temporal frequency dimension.

Other, more complex decompositions, such as the 3-D octave-tree decomposition shown in Figure 11.14 (b), have also been considered. A 3-D octave-tree decomposition can be obtained by applying the one-level subband split depicted in Figure 11.14 (a) to a 3-D block of pixels in a repeated fashion that continues to split the lower frequency subband cube successively. Different decompositions provide different mixes of spatio-temporal resolution. Some decompositions are better than others. Thus, it helps (to the extent possible) to match the decomposition to the particular class of input video.

Adaptive decompositions have been proposed for both the spatial and temporal directions. The idea here is that instead of having the decomposition fixed at the onset, we allow the decomposition to adapt to the input. Kim et al. [175] have proposed using 2, 4, or 8 temporal blocks adaptively. Smaller blocks are used when the amount of motion is larger. They also have proposed the use of 2-D wavelet packets to decompose the temporal subbands.

In addition to the decomposition, many choices are available for the 1-D two-band filter. For example, short-kernel subband filters [187, 169], the five- and nine-tap QMFs, 8-, 10-, 12- and 16-tap QMFs [36, 196, 257, 253, 233, 233, 37], and the 5/7 biorthogonal filters [201] are among some that have been considered previously. In general, computational efficiency is often the dominant concern with 3-D subband coders, which makes the simpler filters attractive.

Clearly, there are many viable possibilities for 3-D subband filters and decompositions. Similarly, there are many ways to code the subbands. To code the baseband, one might consider PCM [190, 154], 1-D DPCM [169], 2-D DPCM [257, 37], unbalanced tree-structured vector quantization (TSVQ) [253], and block DCT coding [196], all of which have been proposed previously. In addition, one can consider using one of the MPEG or H.261 standard video coders to code the moving subbands. More specifically, an off-the-shelf H.261 coder may be applied directly to the sequence of baseband subband images [35]. The advantage here is that the baseband images are small, hence the complexity of the H.261 coder is reduced dramatically. For the high frequency bands, PCM [37], DPCM, ADPCM [253], lattice VQ [241], switched-codebook VQ [35], or geometric VQ (GVQ) [253, 154] may be considered. Greater efficiency can sometimes be realized in coding the high frequency subbands by extracting a prediction from the baseband. Often, the significant values in the high frequency bands are clustered around edges. Thus, edges in the baseband can be used to predict edge energy in the higher frequency bands [196]. Rather than restricting the scheme to be static, one can further consider an adaptive version of this approach where the same structure is employed but with adaptive switching between different scalar quantizers based on the edge information extracted from the baseband [201].

The computational simplicity of the zerotree method can be used for video compression by extending it to three dimensions. Three dimensional versions of the zerotree method have been explored and found to work well [34, 174]. However, it is often desirable to incorporate visual perception into the algorithm, since the true measure of success is how the reconstructed sequence appears to the observer. This is not a trivial task in general. An approach in this direction is to use the just-noticeable distortion (JND) measure as suggested by Safranek and Johnston [274, 37]. Here, based on a local activity measure, a JND profile is first computed that reflects the visual effect of the quantization error corresponding to each subband coefficient. Adaptive-step quantization of the coefficients is then performed based on the computed profile. One can

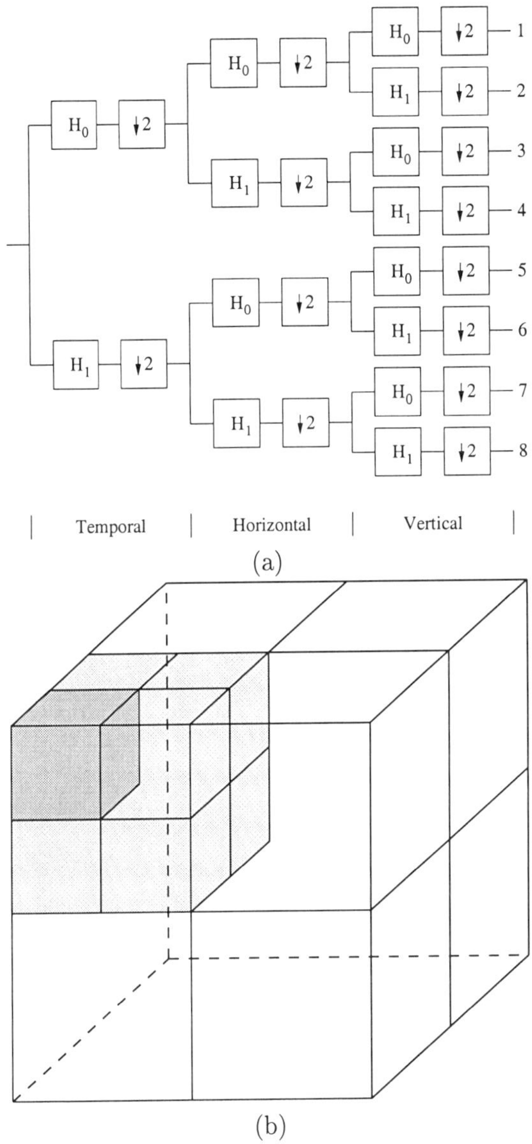

Figure 11.14 The 3-D octave-tree subband decompositions: (a) filter bank structure for a one-level decomposition; (b) transform domain tiling for a two-level decomposition.

also exploit the masking properties of the human visual system by properly weighting the spatial detail versus the temporal detail subbands, as suggested in [170]. In areas with substantial motion, we can reduce spatial resolution based on the fact that spatial contrast sensitivity of the human visual system decreases with motion speed. Much work remains to be done in the area of perceptual coding of video.

Whether one designs a 3-D subband coder to minimize an error distortion or to maximize perceptual performance, 3-D subband coding has some attractive advantages in terms of parallelism [117], performance, and compatibility. As a case in point, subband coding is suitable for adaptive multirate transmission, which is desired in variable-rate channels such as packet-switched networks [169]. Coding can be layered such that each layer is composed of partially coded subbands. The partial coding of the subbands can be performed by using multistage quantizers, tree structured quantizers, bit plane quantizers, as well as a variety of others. Highly-granular rate scalability may be obtained by using a variable number of subbands combined with multirate quantizers [31].

Three dimensional subband coding can be made to be resilient to channel errors. Error resilience is higher for 3-D subband coding than for MC-based coders, in general. Layered protection can further improve the performance by applying unequal error protection to bits with different perceptual significance [36]. This way a "graceful degradation" of performance in the presence of noise can be achieved.

11.6 SUBBAND CODING OF MEDICAL VOLUMETRIC DATA

Medical imaging modalities such as MRI, CT, and PET (Positron Emission Tomography) produce 3-D data by imaging cross-sectional "slices" of organs. These are generally acquired in sets, where slices may be taken every few millimeters apart. Efficient volumetric data coders take advantage of inter-slice correlation. An obvious way to code 3-D volumetric data is to use a standard motion compensation-based video coder [178], as we described in the previous section. Better results can be achieved by using affine transformations applied to triangular patches as part of the motion compensation model [238].

Like in the 2-D case, blocking artifacts can limit the performance of block-based coders. This has motivated the consideration of 3-D transform coders for volumetric data. Unlike video signals, volumetric image data is isotropic, and straightforward extensions of two-dimensional algorithms can be applied. For example, in [14], full-frame 3-D DCT coding of subbands is used, followed by scalar quantization.

A number of 3-D subband coders have also been proposed, building on the underlying theory of multidimensional subband decompositions developed by Vetterli [342]. The decomposition used by all these coders is similar to the octave-tree decomposition in Figure 11.14 (b), with two [14] and three [202] levels. The popular subband filters used are the four-tap CQF filter banks [14, 155], or the 9/7-tap QMFs [352, 202]. The authors in [352] use scalar quantization followed by run-length and Huffman encoding to code the subband coefficients. The 3-D zerotree coding algorithm mentioned earlier is used in [202]. An adaptive scheme is introduced in [155], where the authors propose a 3-D nonlinear DPCM coding of the baseband and an octave-tree coding of high frequency bands. The minimum size cubes are $2 \times 2 \times 2$ pixels and are vector quantized.

Using 3-D instead of 2-D decompositions lead to an increase in compression ratio by 70% for CT images and by 35% for MR images at a peak signal to noise ratio (PSNR) of 50 dB [14]. An increase of 4 to 6 dB in CT image reconstruction quality is reported for 3-D subband coding in [202] for bit rates in the range 0.25 to 1.25 bits/pixel. Thus, it is abundantly clear that inter-slice redundancies in medical volumetric data can be exploited to improve compression performance.

11.7 CLOSING REMARKS

The area of telemedicine is very much in its infant stages. It benefits from the vast amount of research that has been performed in the context of video teleconferencing, computer networks, multimedia applications, and the underlying microelectronics technology that facilitates fast computing. At the same time, telemedicine has not received the attention it is due as an area of research in its own right. As we have mentioned repeatedly throughout this chapter, medical images have unique characteristics and are not well matched to the types of natural images that motivate much of the research in teleconferencing and multimedia systems. The popular medical modalities often involve images of very large size, images with high amplitude resolutions (e.g. 12 bits/pixel versus 8 bits/pixel), and may contain a high concentration of high spatial frequencies. Moreover, telemedicine systems must deliver a high level of quality, in most cases diagnostic quality, and that quality must be preserved under the viewing conditions that are common in the medical community.

The display and quality assessment of 12 bit images are issues that never arise in compression scenarios involving natural images. Such issues are important for medical images. A typical approach to viewing 12 bit medical images is called *window and level*, and is typically applied in an interactive way. The window and level feature allows a physician to control the interval of gray levels to be displayed on the monitor. The center amplitude value is called the *level value* and the range is called the *window value*. This technique is especially useful for 12-bit images because the human visual system cannot distinguish more than 256 levels (8 bits) of gray in an image. As illustrated in Figure 11.15, the solution offered by the window and level approach is to map a particular window of gray levels in the input image to the 256 levels that can be displayed on a monitor and that can be distinguished by the eye.

In addition, stations for viewing medical images may include zooming and scrolling features. These features are typically offered for interactive use via a mouse-type device for better localizing the region of interest (ROI) for diagnoses. The ROI can be square, circular, or can have arbitrary shape. Both window/level and zoom/scroll operations can be taken into account when designing an efficient image compression algorithm by allocating a larger percentage of the bit rate to windows and regions of interest. Current off-the-shelf compression-transmission algorithms were not designed with these image characteristics and viewing specifications in mind. Hence, the telemedicine community must assume the role of technology providers rather than consumers if telemedicine is to reach its full potential.

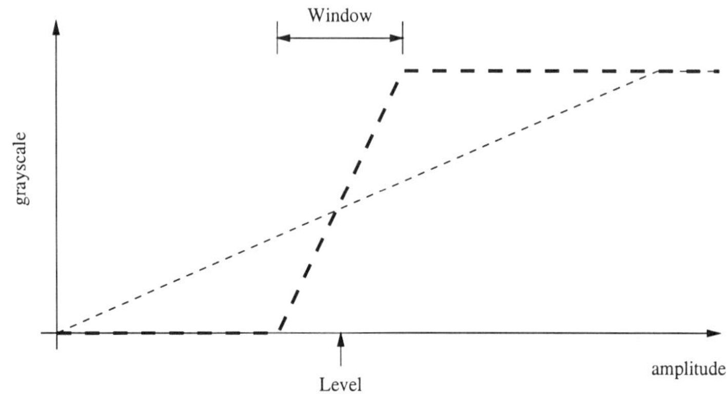

Figure 11.15 The window and level display technique.

In the quest for improved quality, speed, and functionality in the telemedicine environment, transforms are clearly important. Transforms, particularly linear transforms such as block DCTs and DFTs, and subband decompositions are important components of the compression-transmission and image formation phases of the various medical systems. They provide a means for efficient coefficient compactions, a means to enhance functionality via multiresolution and layered processing, and even perhaps the potential to combine aspects of image formation and enhancement with compression. Without doubt, there lie many challenges ahead.

References

[1] A. N. Akansu, P. Duhamel, X. Lin, and M. de Courville. Orthogonal transmultiplexers in communication: A review. *IEEE Transactions on Signal Processing*, 46(4):979–995, April 1998. Special Issue on Wavelets and Filter Banks.

[2] A. N. Akansu and R. A. Haddad. *Multiresolution Signal Decomposition: Transforms, Subbands, Wavelets.* Academic Press, 1992.

[3] A. N. Akansu, M. V. Tazebay, and R. A. Haddad. A new look at digital orthogonal transmultiplexers for CDMA communications. *IEEE Transactions on Signal Processing*, 45(1):263–267, January 1997.

[4] M. Alard. Transformée IOTA. In *Journées d'étude de la Société des Électriciens et des Électroniciens*, Paris, France, March 1996.

[5] J. B. Allen and S. T. Neely. Micromechanical models of the cochlea. *Physics Today*, pages 40–47, July 1992.

[6] ANSI, T1.413. Network and customer installation interfaces - asymmetric digital subscriber line (ADSL) metallic interface, 1995.

[7] ANSI, T1.601. Integrated services digital network (ISDN) basic access interface for use on metallic loops for application on the network side of the NT (Layer 1 specification), 1992.

[8] M. Antonini, M. Barlaud, and P. Mathieu. Image coding using wavelet transform. *IEEE Transactions on Image Processing*, 1(2):205–220, April 1992.

[9] A. V. Balakrishnan. A contribution to the sphere-packing problem of communication theory. *Journal of Mathematical Analysis and Application*, 3:485–506, December 1961.

[10] I. Balasingham and T. A. Ramstad. On the relevance of the regularity constraint in subband image coding. In *Asilomar Conference on Signals, Systems and Computers*, Pacific Grove, November 1997.

[11] F. Bao and N. Erdol. On the discrete wavelet transform and shiftability. In *Asilomar Conference on Signals, Systems and Computers*, pages 1442–1445, Pacific Grove, CA, November 1993.

[12] F. Bao and N. Erdol. Optimal initial phase wavelet transform. In *IEEE Digital Signal Processing Workshop*, pages 187–190, October 1994.

[13] F. Bao and N. Erdol. The optimal wavelet transform and translation invariance. In *IEEE International Conference on Acoustics, Speech and Signal Processing*, volume 3, pages 13–16, 1994.

[14] A. Baskurt et al. Coding of 3D medical images using 3D wavelet decompositions. In *IEEE International Conference on Acoustics, Speech and Signal Processing*, volume 5, page 562, April 1993.

[15] M. Bauer and K. Sayood. Video coding using 3-D DCT and dynamic code selection. In *IEEE Data Compression Conference*, page 451, 1995.

[16] T. C. Bell, J. G. Cleary, and I. H. Witten. *Text Compression*. Prentice Hall, Englewood Cliffs, New Jersey, 1990.

[17] M. G. Bellanger. On computational complexity in digital transmultiplexer filters. *IEEE Transactions on Communications*, 30:1461–1465, July 1982.

[18] M. G. Bellanger, G. Bonnerot, and M. Coudreuse. Digital filtering by polyphase network: Application to sample-rate alteration and filter banks. *IEEE Transactions on Acoustics, Speech and Signal Processing*, 24(2):109–114, April 1976.

[19] M. G. Bellanger and J. Daguet. TDM-FDM transmultiplexer: Digital polyphase and FFT. *IEEE Transactions on Communications*, 22(9), September 1974.

[20] W. Bender, D. Gruhl, and N. Morimoto. Techniques for data hiding. Technical Report, MIT Media Lab, 1994.

[21] S. A. Benno and J. M. Moura. Nearly shiftable scaling functions. In *IEEE International Conference on Acoustics, Speech and Signal Processing*, volume 2, pages 1097–1100, Detroit, 1995.

[22] T. Berger. *Rate Distortion Theory*. Prentice Hall, Englewood Cliffs, NJ, 1971.

[23] Blauert. *Spatial Hearing*. MIT Press, 1983.

[24] F. Boland, J. O. Ruanaidh, and C. Dautzenberg. Watermarking digital images for copyright protection. In *IEE International Conference on Image Processing and Its Applications*, pages 321–326, Edinburgh, Scotland, 1995.

[25] L. Boney, A. H. Tewfik, and K. N. Hamdy. Digital watermarks for audio signals. In *IEEE International Conference on Multimedia Computing and Systems*, pages 473–480, 1996.

[26] M. Bosi, K. Brandenburg, S. Quackenbush, L. Fielder, K. Akagiri, H. Fuchs, M. Dietz, J. Herre, G. Davidson, and Y. Oikawa. ISO/IEC MPEG-2 advanced audio coding. In *AES Convention Record*, Los Angeles, November 1996.

[27] K. Brandenburg and G. Stoll. ISO-MPEG-1 audio: A generic standard for coding of high quality digital audio. In N. Gilchrist and C. Grewin, editors, *Collected Papers on Digital Audio Bit-Rate Reduction*, pages 31–42. Audio Engineering Society, New York, 1996.

[28] K. Brandenburg, G. Stoll, Y. Dehery, J. D. Johnston, L. Van Der Kerkhof, and E. Schoeder. The ISO-MPEG-1 audio codec: A generic standard for coding of high quality digital audio. In *AES Convention Record*, Vienna, 1992. Preprint 3336.

[29] C. M. Brislawn. Classification of nonexpansive symmetric extension transforms for multirate filter banks. *Applied and Computational Harmonic Analysis*, 3:337–357, 1996.

[30] E. Chalom and V. M. Bove, Jr. Segmentation of an image sequence using multi-dimensional image attributes. In *International Conference on Image Processing*, volume 2, pages 525–528, Lausanne, Switzerland, 1996.

[31] E. Chang and A. Zakhor. Scalable video coding using 3-D subband velocity coding and multirate quantization. In *IEEE International Conference on Acoustics, Speech and Signal Processing*, volume 5, page 574, April 1993.

[32] K. Chen and T. Ramabadran. Near-lossless compression of medical images through entropy-coded DPCM. *IEEE Transactions on Image Processing*, page 538, September 1994.

[33] W. H. Chen and C. H. Smith. Adaptive coding of monochrome and color images. *IEEE Transactions on Communications*, 25(11):1285–1292, November 1977.

[34] Y. Chen and W. A. Pearlman. Three-dimensional subband coding of video using the zero-tree method. In *SPIE Proceedings on Visual Communications and Image Processing*, page 1302, 1996.

[35] W. L. Chooi and K. N. Ngan. 3-D subband coder for very low bit rates. In *IEEE International Conference on Acoustics, Speech and Signal Processing*, volume 5, page 405, 1994.

[36] C. Chou and C. Chen. A perceptually optimized 3-D subband codec for video communication over wireless channels. *IEEE Transactions on Circuits and Systems for Video Technology*, page 143, April 1996.

[37] C. H. Chou. Low bit-rate 3-D subband video coding based on the allocation of just noticeable distortion. In *International Conference on Image Processing*, volume 1, page 637, 1996.

[38] P. A. Chou, T. Lookabaugh, and R. M. Gray. Entropy-constrained vector quantization. *IEEE Transactions on Information Theory*, 37(1):31–42, January 1989.

[39] J. S. Chow, J. C. Tu, and J. M. Cioffi. A discrete multitone transceiver system for HDSL applications. *IEEE Journal on Selected Areas in Communications*, 9(6):895–908, August 1991.

[40] C. Chrysafis and A. Ortega. Efficient context based entropy coding for lossy wavelet image compression. In *IEEE Data Compression Conference*, pages 241–250, Snowbird, Utah, 1997.

[41] C. K. Chui. *Wavelets and Applications.* Academic Press, 1991.

[42] C. K. Chui. *Wavelets: A Tutorial in Theory and Applications.* Academic Press, 1992.

[43] J. Cioffi, G. P. Dudevoir, M. V. Eyuboglu, and G. D. Forney. MMSE decision-feedback equalizers and coding, Parts I and II. *IEEE Transactions on Communications*, 43(10), October 1995.

[44] D. Cochran and H. Sokbom. Orthonormal basis of bandlimited wavelets with frequency overlap. In *IEEE International Conference on Acoustics, Speech and Signal Processing*, pages 1541–1544, Detroit, 1995.

[45] D. Cochran and C. Wei. Scale based coding of digitial communication signals. In *IEEE-SP International Symposium on Time-Frequency and Time-Scale Analysis*, pages 455–458, October 1992.

[46] R. R. Coifman and Y. Meyer. Nouvelles bases orthonormées de $l^2(\mathbf{r})$ ayant la structure du système de Walsh. Technical Report Preprint, Department of Mathematics, Yale University, 1989.

[47] R. R. Coifman, Y. Meyer, S. Quake, and M. Wickerhauser. Signal processing with wavelet packets. Technical report, Numerical Algorithm Research Group, Yale University, 1990.

[48] R. R. Coifman and M. V. Wickerhauser. Entropy based algorithms for best basis selection. *IEEE Transactions on Information Theory*, 32:712–718, March 1992.

[49] ETSI Normalization Commitee. Radio broadcasting systems, digital audio broadcasting (DAB) to mobile, portable and fixed receivers. Norme

ETSI, document ETS 300 401, European Telecommunications Standards Institute, Sophia-Antipolis, Valbonne, France, 1995–1997.

[50] C. E. Cook and H. S. Marsh. An introduction to spread spectrum. *IEEE Communications Magazine*, pages 8–16, March 1983.

[51] P. C. Cosman, R. M. Gray, and M. Vetterli. Vector quantization of image subbands: A survey, June 1995.

[52] I. Cox, J. Kilian, T. Leighton, and T. Shamoon. Secure spread spectrum watermarking for multimedia. Technical Report 95-10, NEC Research Institute, 1995.

[53] S. Craver, N. Memon, B.-L. Yeo, and M. Yeung. Can invisible watermarks resolve rightful ownerships? IBM Research Technical Report RC 20509, *IBM CyberJournal*, July 1996.

[54] S. Craver, N. Memon, B.-L. Yeo, and M. Yeung. Resolving rightful ownerships with invisible watermarking techniques: Limitations, attacks, and implications. IBM Research Technical Report RC 20755, *IBM CyberJournal*, March 1997.

[55] A. Croisier, D. Esteban, and C. Galand. Perfect channel splitting by use of interpolation/decimation tree decomposition techniques. In *IEEE International Symposium on Circuits and Systems*, Patras, Greece, 1976.

[56] M. Crouse and K. Ramchandran. Joint thresholding and quantizer selection for decoder-compatible baseline JPEG. In *IEEE International Conference on Acoustics, Speech and Signal Processing*, May 1995.

[57] Zoran Cvetković and Martin Vetterli. Oversampled filter banks. *IEEE Transactions on Signal Processing*, 46(5):1245–1255, May 1998.

[58] J. C. Darragh. *Subband and Transform Coding of Images.* PhD thesis, University of California, Los Angeles, 1989.

[59] M. Das, D. L. Neuhoff, and C. L. Lin. Near-lossless compression of medical images. In *IEEE International Conference on Acoustics, Speech and Signal Processing*, page 2347, Detroit, 1995.

[60] I. Daubechies. Orthonormal basis of compactly supported wavelets. *Communications in Pure and Applied Mathematics*, XLI:909–996, 1988.

[61] I. Daubechies. *Ten Lectures on Wavelets.* Society for Industrial and Applied Mathematics, Philadelphia, 1992.

[62] S. Davidovici and E. G. Kanterakis. Narrow-band interference rejection using real-time Fourier transforms. *IEEE Transactions on Communications*, 37(7):713–722, July 1989.

[63] G. A. Davidson and M. Bosi. AC-2: High quality audio coding for broadcasting and storage. In *Annual Broadcast Engineering Conference Record*, pages 98–105, Las Vegas, March 1992.

[64] G. Davis and A. Nosratinia. Wavelet-based image coding: An overview. *Applied and Computational Control, Signals, and Circuits*, 1:205–269, 1998.

[65] G. M. Davis. The wavelet image compression construction kit. http://www.cs.dartmouth.edu/~gdavis/wavelet/wavelet.html.

[66] G. M. Davis and S. Chawla. Image coding using optimized significance tree quantization. In J. A. Storer and M. Cohn, editors, *IEEE Data Compression Conference*, pages 387–396, March 1997.

[67] M. Davis. The AC-3 multichannel coder. In *AES Convention Record*, New York, October 1993.

[68] M. de Courville. *Utilisation de Bases Orthogonales pour l'Algorithmique Adaptative et l'Égalisation des Systèmes Multiporteuses.* PhD thesis, École Nationale Supérieure des Télécommunications, October 1996.

[69] M. de Courville, P. Duhamel, P. Madec, and J. Palicot. Blind equalization of OFDM systems based on the minimization of a quadratic criterion. In *IEEE International Conference on Communications*, volume 3, pages 1318–1321, Dallas, June 1996.

[70] M. de Courville, P. Duhamel, P. Madec, and J. Palicot. A least mean squares blind equalization technique for OFDM systems. *Les Annales des Télécommunications*, 52(1–2):12–20, January 1997.

[71] Y. F. Dehery. MUSICAM source coding. In *AES Convention Record*, pages 71–80, 1991.

[72] S. Del Marco, P. Heller, and J. Weiss. An M-band, 2-dimensional translation invariant wavelet transform and applications. In *IEEE International Conference on Acoustics, Speech and Signal Processing*, volume 2, pages 1077–1080, Detroit, 1995.

[73] G. Deslauriers and S. Dubuc. Symmetric iterative interpolation processes. *Constructive Approximation*, 5(1):49–68, 1989.

[74] R. A. DeVore, B. Jawerth, and B. J. Lucier. Image compression through wavelet transform coding. *IEEE Transactions on Information Theory*, 38(2):719–746, March 1992.

[75] R. A. Dillard and G. M. Dillard. *Detectability of Spread Spectrum Signals.* Artech House, Norwood, MA, 1989.

[76] R. C. DiPietro. An FFT based technique for suppressing narrow-band interference in PN spread spectrum communications systems. In *IEEE International Conference on Acoustics, Speech and Signal Processing*, pages 1360–1363, 1989.

[77] R. Dixon. *Spread Spectrum Systems.* John Wiley and Sons, New York, 1976.

[78] Bellcore Document. Boc notes on the lec networks, April 1994. Bellcore Document SR-TSV-002275, Issue 2.

[79] D. L. Duttweiler and C. Chamzas. Probability estimation in arithmetic and adaptive-Huffman entropy coders. *IEEE Transactions on Image Processing*, 4(3):237–246, March 1995.

[80] B. Edler. Coding of audio signals with overlapping block transform and adaptive window functions. *Frequenz*, 43:252–256, 1989. (in German).

[81] B. Edler. Aliasing reduction in sub-bands of cascaded filter banks with decimation. *Electronics Letters*, 28:1104–1105, 1992.

[82] O. Egger, T. Ebrahumi, and M. Kunt. Arbitrarily-shaped wavelet packets for zerotree coding. In *IEEE International Conference on Acoustics, Speech and Signal Processing*, volume 4, page 2337, Atlanta, 1996.

[83] H. F. Engler and D. H. Howard. A compendium of analytic models for coherent and noncoherent receivers. Report Number: AFWAL-TR-85-1118, September 1985.

[84] N. Erdol, F. Bao, and F. Basbug. Optimal receiver design with wavelet bases. In *IEEE International Conference on Acoustics, Speech and Signal Processing*, volume 4, pages 121–124, 1994.

[85] D. Esteban and C. Galand. Application of quadrature mirror filters to split band voice coding schemes. In *IEEE International Conference on Acoustics, Speech and Signal Processing*, pages 191–195, May 1977.

[86] G. Evangelista. Orthogonal wavelet transforms and filter banks. In *Asilomar Conference on Signals, Systems and Computers*, 1989.

[87] M. Eyuboglu and G. D. Forney, Jr. Lattice and trellis quantization with lattice- and trellis-bounded codebooks – High-rate theory for memoryless sources. *IEEE Transactions on Information Theory*, 39:46–59, January 1993.

[88] T. C. Farrell and G. Prescott. A method of finding orthogonal wavelet filters with good energy tiling characteristics. *IEEE Transactions on Signal Processing*, 47(1), January 1999.

[89] T. C. Farrell, G. Prescott, and S. Chakrabarti. A potential use for artificial neural networks in the detection of frequency hopped low probability

of intercept signals. In *IEEE Wichita Conference on Communications, Networking and Signal Processing*, pages 25–30, Wichita, KA, April 1994.

[90] N. Farvardin and J. W. Modestino. Optimum quantizer performance for a class of non-Gaussian memoryless sources. *IEEE Transactions on Information Theory*, 30:485–497, May 1984.

[91] L. D. Fielder, M. Bosi, G. A. Davidson, M. F. Davis, C. Todd, and S. V. Vernon. AC-2 and AC-3: Low-complexity transform-based audio coding. In N. Gilchrist and C. Grewin, editors, *Collected Papers on Digital Audio Bit-Rate Reduction*, pages 54–72. Audio Engineering Society, New York, 1996.

[92] P. Flandrin. Some aspects of non-stationary signal processing with emphasis on time-frequency and time-scale methods. In M. Combes, A. Grossman, and P. Tchamitchian, editors, *Wavelets: Time Frequency Methods and Phase Space*, pages 68–98. Springer-Verlag, Berlin, 1987.

[93] P. Flandrin. Wavelet analysis and synthesis of fractional brownian motion. *IEEE Transactions on Information Theory*, 38(2):910–917, March 1992.

[94] H. Fletcher. *The ASA Edition of Speech and Hearing in Communication.* Acoustical Society of America, Woodbury, NY, 1995.

[95] N. J. Fliege. Orthogonal multiple carrier data transmission. *European Transactions on Telecommunications*, 3(3):225–253, May 1992.

[96] N. J. Fliege. *Multirate Digital Signal Processing: Multirate Systems, Filter Banks, Wavelets.* John Wiley and Sons, New York, 1994.

[97] B. Friedlander and B. Porat. Detection of transient signals by the Gabor representation. *IEEE Transactions on Acoustics, Speech and Signal Processing*, 37:169–180, February 1989.

[98] M. Frisch and H. Messer. The use of the wavelet transform in the detection of an unknown transient signal. *IEEE Transactions on Information Theory*, 38:892–897, March 1992.

[99] H. Fuchs. Improving MPEG audio coding by backward adaptive linear stereo prediction. In *AES Convention Record*, New York, October 1995.

[100] J. Garcia-Frias and J. D. Villasenor. An analytical treatment of channel-induced distortion in run length coded image subbands. In *IEEE Data Compression Conference*, pages 52–61, Snowbird, UT, March 1997.

[101] W. A. Gardner. *Statistical Spectral Analysis.* Prentice-Hall, Englewood Cliffs, NJ, 1988.

[102] A. Gersho. Asymptotically optimal block quantization. *IEEE Transactions on Information Theory*, 25(4):373–380, July 1979.

[103] A. Gersho and R. Gray. *Vector Quantization and Signal Compression*. Kluwer, Boston, MA, 1992.

[104] J. Gevargiz, P. Das, and L. B. Milstein. Performance of a transform domain processing DS intercept receiver in the presence of finite bandwidth interference. In *IEEE Global Telecommunications Conference*, pages 21.5.1–21.5.5, December 1986.

[105] J. Gevargiz, M. Rosenmann, P. Das, and L. B. Milstein. A comparison of weighted and non-weighted transform domain processing systems for narrowband interference excision. In *IEEE Military Communications Conference*, pages 32.3.1–32.3.4, October 1984.

[106] G. B. Giannakis. Filterbanks for blind channel identification and equalization. *IEEE Signal Processing Letters*, 4(6):184–187, June 1997.

[107] N. Gilchrist and C. Grewin, editors. *Collected Papers on Digital Audio Bit-Rate Reduction*. Audio Engineering Society, New York, 1996.

[108] B. Girod. The information theoretical significance of spatial and temporal masking in video signals. In *SPIE Proceedings on Human Vision, Visual Processing, and Digital Display*, volume 1077, pages 178–187, 1989.

[109] H. Gish and J. N. Pierce. Asymptotically efficient quantizing. *IEEE Transactions on Information Theory*, 14(5):676–683, September 1968.

[110] R. Gold. Optimal binary sequences for spread spectrum multiplexing. *IEEE Transactions on Information Theory*, pages 619–621, October 1967.

[111] S. Goldwasser and M. Bellare. Lecture notes on cryptography. Preprint, July 1996.

[112] R. M. Gray. On the asymptotic eigenvalue distribution of Toeplitz matrices. *IEEE Transactions on Information Theory*, 18(6):725–230, November 1972.

[113] A. Grossman, J. Morlet, and T. Paul. Transforms associated to square integrable group representation, I. General results. *Journal of Mathematical Physics*, 27:2437–2479, 1985.

[114] R. A. Haddad, A. N. Akansu, and A. Benyassine. Time-frequency localization in M-band filter banks and wavelets: A critical review. *Journal of Optical Engineering*, 32(7):1411–1429, July 1993.

[115] F. J. Harris. On the use of windows for harmonic analysis of the discrete Fourier transform. *Proceedings of the IEEE*, 66:51–83, January 1974.

[116] F. Hartung and B. Girod. Digital watermarking of raw and compressed video. In *SPIE Proceedings on Digital Computing Techniques and Systems for Video Communications*, volume 2952, pages 205–213, October 1996.

[117] J. Hartung. Architecture for the real-time implementation of three-dimensional subband video coding. In *IEEE International Conference on Acoustics, Speech and Signal Processing*, volume 3, page 225, March 1992.

[118] S. Haykin. *Digital Communications*. John Wiley and Sons, New York, 1988.

[119] C. Heil and D. Walnut. Continuous and discrete wavelet transforms. *SIAM Review*, 31:628–666.

[120] P. N. Heller, H. L. Resnikoff, and R. O. Wells. Wavelet matrices and the representation of discrete functions. In C. K. Chui, editor, *Wavelets: A Tutorial in Theory and Applications*. Academic Press, Boston, 1992.

[121] R. P. Hellman. Asymmetry of masking between noise and tone. *Perception and Psychophysics*, 2:241–246, 1972.

[122] C. W. Helstrom. *Elements of Detection and Modulation Theory*. Prentice-Hall, 1995.

[123] J. L. Hennessy and D. A. Patterson. *Computer Architecture – A Quantative Approach*. Morgan Kaufman, San Mateo, CA, 1990.

[124] C. Herley. Boundary filters for finite-length signals and time-varying filter banks. *IEEE Transactions on Circuits and Systems*, 42(2):102–114, February 1995.

[125] C. Herley, J. Kovačević, K. Ramchandran, and M. Vetterli. Tilings of the time-frequency plane: Construction of arbitrary orthogonal bases and fast tiling algorithms. *IEEE Transactions on Signal Processing*, 41(12):3341–3359, December 1993.

[126] C. Herley and M. Vetterli. Orthogonal time-varying filter banks and wavelets. In *IEEE International Symposium on Circuits and Systems*, volume 1, pages 391–394, May 1993.

[127] C. Herley, Z. Xiong, K. Ramchandran, and M. T. Orchard. Joint space-frequency segmentation using balanced wavelet packet trees for least-cost image representation. *IEEE Transactions on Image Processing*, September 1997.

[128] J. Herre and J. D. Johnston. Enhancing the performance of perceptual audio coders by using temporal noise shaping (TNS). In *AES Convention Record*, Los Angeles, November 1996.

[129] K. Hetling, G. Saulnier, and P. Das. Optimized filter design for PR-QMF based spread spectrum communications. In *Proceedings of the IEEE International Conference on Communications*, pages 1350–1354, 1995.

[130] K. Hetling, G. Saulnier, and P. Das. Optimized PR-QMF based codes for multiuser communications. In H. Szu, editor, *SPIE Proceedings on Wavelet Applications for Dual Use*, volume 2491, pages 248–259, Orlando, FL, April 1995.

[131] K. Hetling, G. Saulnier, and P. Das. PR-QMF based codes for multipath/multiuser communications. In *IEEE Global Telecommunications Conference*, 1995.

[132] K. Hetling, G. Saulnier, and P. Das. Performance of filter bank-based spreading codes for multipath/multiuser interference. In H. Szu, editor, *SPIE Proceedings on Wavelet Applications*, 1996.

[133] K. J. Hetling. *Multirate Filter Banks for Spread Spectrum Waveform Design*. PhD thesis, Rensselaer Polytechnic Institute, 1996.

[134] K. J. Hetling, G. J. Saulnier, and P. K. Das. Performance of filter bank-based spreading codes for cellular and micro-cellular channels. In H. Szu, editor, *SPIE Proceedings on Wavelet Applications*, 1997.

[135] F. Hlawatsch and G. F. Boudreaux-Bartels. Linear and quadratic time-frequency signal representations. *IEEE Signal Processing Magazine*, pages 21–67, April 1992.

[136] R. Hleiss, P. Duhamel, and M. Charbit. Oversampled OFDM systems. In *International Conference on Digital Signal Processing*, volume 1, pages 329–332, Santorini, Hellas, Greece, July 1997.

[137] S. J. Howard. Narrowband interference rejection using small FFT block sizes. In *IEEE Military Communications Conference*, pages 26.3.1–26.3.5, 1992.

[138] F. M. Hsu and A. A. Giordano. Digital whitening techniques for improving spread spectrum communications performance in the presence of narrowband jamming and interference. *IEEE Transactions on Communications*, 26:209–216, February 1978.

[139] M. Hussey. *Basic Physics and Technology of Medical Diagnostic Ultrasound*. Elsevier, 1985.

[140] R. A. Iltis and L. B. Milstein. Performance analysis of narrow-band interference rejection techniques in DS spread-spectrum systems. *IEEE Transactions on Communications*, 32(11):1169–1177, November 1984.

[141] European Telecommunications Standards Institute. Integrated services digital network (ISDN) basic rate access digital transmission system on metallic local lines. ETSI ETR-080, European Telecommunications Standards Institute, July 1993.

[142] European Telecommunications Standards Institute. Transmission and multiplexing (TM); High bit rate digital subscriber line (HDSL) transmission system on metallic local lines; HDSL core specification and applications for 2048 kb/s based access digital sections. ETSI ETR-152, European Telecommunications Standards Institute, November 1997.

[143] ISO/IEC. JTC1/SC29 WG11, Coding of moving pictures and associated audio for digital storage media at up to 1.5 Mbit/s, Part 3: Audio, 1992. International Standard IS 11172-3.

[144] ISO/IEC. ISO/IEC 13818-2, Generic coding of moving pictures and associated audio, recommendation H.262 (MPEG-2), March 1994. Draft International Standard.

[145] ISO/IEC. JTC1/SC29 WG11, Information technology - generic coding of moving pictures and associated audio, Part 3: Audio, 1994. International Standard IS 13818-3.

[146] ISO/IEC. ISO/IEC JTC1/SC29/WG11 N1419, Report on the formal subjective listening tests of MPEG-2 NBC multichannel audio coding, November 1996.

[147] ISO/IEC. ISO/IEC JTC1/SC29/WG1 N715, Report on the requirements and profiles of JPEG-2000, November 1997.

[148] ISO/IEC. ISO/IEC JTC1/SC29/WG11 N1650, IS 13818-7 (MPEG-2 Advanced audio coding, AAC), April 1997.

[149] ISO/IEC. ISO/IEC JTC1/SC29/WG11 W1886, MPEG-4 Requirements document, October 1997.

[150] ISO/IEC. JTC1/SC29 WG11 MPEG, coding of moving pictures and audio, Part 7: Advanced audio coding, 1997. International Standard IS 13818-7.

[151] ISO/IEC. ISO/IEC JTC1/SC29/WG11 MPEG98/W02502, Text of ISO/IEC FDIS 14496-2, October 1998.

[152] ITU-5/SG15/LBC. Draft recommendation h.26p video coding for narrow telecommunication channels at below 64 kbps, March 1995.

[153] ITU Telecom. Standardization Sector Study Group 15. Document LBC-95 Working Party 15/1 Question 2/15, 1995. Draft Recommendation H.263, Expert's Group on Very Low Bitrate Video Telephony, Leidschendam, 7 April 1995.

[154] A. Jacquin and C. Podilchuk. Very low bit rate 3D subband-based video coding with a dynamic bit allocation. *SPIE Proceedings*, 1977:156, 1993.

[155] E. S. Jang and M. Venkatraman. Subband coding of 3D MRI images using octtrees. In *IEEE International Conference on Acoustics, Speech and Signal Processing*, page 2217, Detroit, 1995.

[156] N. Jayant, J. Johnston, and R. Safranek. Signal compression based on models of human perception. *Proceedings of the IEEE*, 81(10):1385–1421, October 1993.

[157] N. S. Jayant and P. Noll. *Digital Coding of Waveforms.* Prentice-Hall, Englewood Cliffs, NJ, 1984.

[158] J. Johnston and A. J. Ferreira. Sum-difference stereo transform coding. *IEEE International Conference on Acoustics, Speech and Signal Processing*, pages 569–571, March 1992.

[159] J. Johnston, C. Podilchuk, and R. Chen. Digital coding (data reduction) methods. In R. K. Jurgen, editor, *Digital Consumer Electronics Handbook*, chapter 3. McGraw-Hill, New York, 1987.

[160] J. D. Johnston. A filter family designed for use in quadrature mirror filter banks. In *IEEE International Conference on Acoustics, Speech and Signal Processing*, pages 291–294, April 1980.

[161] J. D. Johnston. Filter banks in audio coding. In A. N. Akansu and M. J. T. Smith, editors, *Subband and Wavelet Transforms: Design and Applications*, chapter 9. Kluwer, Norwell, MA, 1996.

[162] J. D. Johnston and K. Brandenburg. Wideband coding - Perceptual considerations for speech and music. In S. Furui and M. Sondhi, editors, *Advances in Speech Signal Processing.* Marcel Dekker, New York, 1992.

[163] J. D. Johnston, J. Herre, M. Davis, and U. Gbur. MPEG-2 NBC audio - stereo and multichannel coding methods, November 1996.

[164] J. D. Johnston, D. Sinha, S. Dorward, and S. R. Quackenbush. AT&T perceptual audio coder. In N. Gilchrist and C. Grewin, editors, *Collected Papers on Digital Audio Bit-Rate Reduction.* Audio Engineering Society, New York, 1996.

[165] W. W. Jones and K. R. Jones. Narrowband interference suppression using filter-bank analysis/synthesis techniques. In *IEEE Military Communications Conference*, pages 898–902, October 1992.

[166] R. L. Joshi, H. Jafarkhani, J. H. Kasner, T. R. Fisher, N. Farvardin, M. W. Marcellin, and R. H. Bamberger. Comparison of different methods of classification in subband coding of images. *IEEE Transactions on Image Processing*, 6:1473–1486, November 1997.

[167] I. Kalet. The multitone channel. *IEEE Transactions on Communications*, 37(2):119–124, February 1989.

[168] I. Kalet. Multitone modulation. In A. N. Akansu and M. J. T. Smith, editors, *Subband and Wavelet Transforms: Design and Applications*, chapter 13. Kluwer, Norwell, MA, 1996.

[169] G. Karlsson and M. Vetterli. Sub-band coding of video signals for packet-switched networks. In *SPIE Proceedings on Visual Communications and Image Processing*, volume 845, page 446, 1987.

[170] G. Karlsson and M. Vetterli. Three-dimensional subband coding of video. In *IEEE International Conference on Acoustics, Speech and Signal Processing*, pages 1100–1103, April 1988.

[171] T. Karp and N. J. Fliege. MDFT filter banks with perfect reconstruction. In *IEEE International Symposium on Circuits and Systems*, Seattle, May 1995.

[172] T. Karson. Digital storage of echocardiograms offers superior image quality to analog storage, even with 20:1 digital compression. In *Digital Cardiac Imaging: A Primer*. Digital Imaging and Communications in Medicine, 1997.

[173] J. W. Ketchum and J. G. Proakis. Adaptive algorithms for estimating and suppressing narrow-band interference in PN spread-spectrum systems. *IEEE Transactions on Communications*, 30:913–924, May 1982.

[174] B.-J. Kim and W. A. Pearlman. An embedded wavelet video coder using three-dimensional set partitioning in hierarchical trees (SPIHT). In *IEEE Data Compression Conference*, page 251, 1997.

[175] Y. K. Kim, R. C. Kim, and S. U. Lee. On the adaptive 3D subband video coding. In *SPIE Proceedings on Visual Communications and Image Processing*, page 123, 1996.

[176] E. Koch and J. Zhao. Towards robust and hidden image copyright labeling. In *IEEE Nonlinear Signal Processing Workshop*, pages 452–455, Thessaloniki, Greece, 1995.

[177] R. D. Koilpillai and P. P. Vaidyanathan. Cosine-modulated FIR filter banks satisfying perfect reconstruction. *IEEE Transactions on Signal Processing*, 40:770–783, April 1992.

[178] J. I. Koo, H. S. Lee, and Y. Kim. Application of 2-D and 3-D compression algorithms to ultrasound images. *SPIE Proceedings on Image Capture, Formatting and Display*, pages 434–439, 1992.

[179] F. Kossentini, W. Chung, and M. J. T. Smith. Subband image coding with intra- and inter-band subband quantization. In *Asilomar Conference on Signals, Systems and Computers*, Pacific Grove, CA, November 1993.

[180] F. Kossentini, W. Chung, and M. J. T. Smith. A jointly optimized subband coder. *IEEE Transactions on Image Processing*, August 1996.

[181] F. Kossentini, M. J. T. Smith, and C. Barnes. Entropy-constrained residual vector quantization. In *IEEE International Conference on Acoustics, Speech and Signal Processing*, volume 5, pages 598–601, April 1993.

[182] F. Kossentini, M. J. T. Smith, A. Scales, and D. Tucker. Medical image compression using a new subband coding method. In *SPIE Proceedings on Medical Imaging*, volume 2431, pages 550–560, 1995.

[183] C. Kotropoulos, X. Magnisalis, I. Pitas, and M. Strintzis. Nonlinear ultrasonic image processing based on signal adaptive filters and self-organizing neural networks. *IEEE Transactions on Image Processing*, 3(1):65–78, January 1994.

[184] J. Kovačević and W. Sweldens. Interpolating filter banks and wavelets in arbitrary dimensions. Technical report, Lucent Technologies, Murray Hill, NJ, 1997.

[185] G. R. Kuduvalli and R. M. Rangayyan. Performance analysis of reversible image compression techniques for high-resolution digital teleradiology. *IEEE Transactions on Medical Imaging*, (2):430, 1992.

[186] B. Le Floch, M. Alard, and C. Berrou. Coded orthogonal frequency division multiplex (TV broadcasting). *Proceedings of the IEEE*, 83(6):982–996, June 1995.

[187] D. Le Gall and A. Tabatabai. Subband coding of digital images using short kernel filters and arithmetic coding techniques. In *IEEE International Conference on Acoustics, Speech and Signal Processing*, pages 761–764, April 1988.

[188] D. J. Le Gall. MPEG: A video compression standard for multimedia applications. *Communications of the ACM*, 34(4):47–58, April 1991.

[189] G. E. Legge and J. M. Foley. Contrast masking in human vision. *Journal of the Optical Society of America*, 70(12):1458–1471, 1980.

[190] A. S. Lewis and G. Knowles. Video compression using 3-D wavelet transforms. *Electronic Letters*, page 396, March 1990.

[191] A. S. Lewis and G. Knowles. Image compression using the 2-D wavelet transform. *IEEE Transactions on Image Processing*, 1(2):244–250, April 1992.

[192] X. Lin and A. N. Akansu. A distortion analysis and optimal design of orthogonal basis for DMT transceivers. In *IEEE International Conference on Acoustics, Speech and Signal Processing*, volume 3, pages 1475–1478, Atlanta, May 1996.

[193] Y-P. Lin and P. P. Vaidyanathan. Linear phase cosine modulated maximally decimated filter banks with perfect reconstruction. In *IEEE International Symposium on Circuits and Systems*, London, May 1994.

[194] Y. Linde, A. Buzo, and R. Gray. An algorithm for vector quantizer design. *IEEE Transactions on Communications*, 26(1):84–95, January 1980.

[195] A. R. Lindsey. *Generalized Orthogonally Multiplexed Communications.* PhD thesis, Ohio University, June 1995.

[196] R. Llados-Bernaus and R. L. Stevenson. Robust low-bit-rate 3D subband codec. In *SPIE Proceedings on Visual Communications and Image Processing*, page 610, 1997.

[197] S. P. Lloyd. Least squares quantization in PCM. *IEEE Transactions on Information Theory*, 28:127–135, March 1982.

[198] S.P. Lloyd. Least squares quantization in PCM. Unpublished Bell Labs Technical Note, September 1957.

[199] S. M. LoPresto, K. Ramchandran, and M. T. Orchard. Image coding based on mixture modeling of wavelet coefficients and a fast estimation-quantization framework. In *IEEE Data Compression Conference*, pages 221–230, Snowbird, Utah, 1997.

[200] J. Lu, A. Nosratinia, and B. Aazhang. Progressive source-channel coding of images over bursty error channels. In *International Conference on Image Processing*, Chicago, October 1998.

[201] J. Luo, C. Chen, K. Parker, and T. S. Huang. Three dimensional subband video analysis and synthesis with adaptive clustering in high frequency subbands. In *International Conference on Image Processing*, volume 3, page 255, 1994.

[202] L. Luo et al. Volumetric medical image compression with three-dimensional wavelet transform and octave zerotree coding. In *SPIE Proceedings on Visual Communications and Image Processing*, page 579, 1996.

[203] S. Mallat. A theory for multiresolution signal decomposition: The wavelet representation. *IEEE Transactions on Pattern Analysis and Machine Intelligence*, 11(7):674–693, July 1989.

[204] S. Mallat and F. Falzon. Analysis of low bit image transform coding. *IEEE Transactions on Signal Processing*, April 1998.

[205] S. G. Mallat. Multiresolution approximations and wavelet orthonormal bases of $l^2(\mathrm{R})$. *Transactions of the American Mathematical Society*, 315:69–87, 1989.

[206] H. S. Malvar. Modulated QMF filter banks with perfect reconstruction. *Electronics Letters*, 26:906–907, June 1990.

[207] H. S. Malvar. *Signal Processing with Lapped Transforms.* Artech House, Norwood, MA, 1992.

[208] A. Manduca. Compressing images with wavelet/subband coding. *IEEE Engineering in Medicine and Biology*, (5):639, 1995.

[209] M. W. Marcellin. On entropy-constrained trellis coded quantization. *IEEE Transactions on Communications*, 42(1):14–16, January 1994.

[210] M. W. Marcellin and T. R. Fischer. Trellis coded quantization of memoryless and Gauss-Markov sources. *IEEE Transactions on Communications*, 38(1):82–93, January 1990.

[211] S. A. Martucci and I. Sodagar. Zerotree entropy coding of wavelet coefficients for very low bit rate video. *International Conference on Image Processing*, 2, September 1996.

[212] S. A. Martucci, I. Sodagar, T. Chiang, and Y.-Q. Zhang. A zerotree wavelet video coder. *IEEE Transactions on Circuits and Systems for Video Technology*, 7, February 1997. Special issue on MPEG-4.

[213] J. Max. Quantizaing for minimum distortion. *IEEE Transactions on Information Theory*, pages 7–12, March 1960.

[214] D. Meares, K. Watanabe, and E. Scheirer. Report on the MPEG-2 AAC stereo verification tests, 1998. ISO/IEC MPEG-2 document N2006.

[215] M. J. Medley. *Adaptive Narrow-Band Interference Suppression Using Linear Transforms and Multirate Filter Banks.* PhD thesis, Rensselaer Polytechnic Institute, December 1995.

[216] M. J. Medley, G. J. Saulnier, and P. K. Das. The application of wavelet-domain adaptive filtering to spread spectrum communications. In H. Szu, editor, *SPIE Proceedings on Wavelet Applications for Dual-Use*, volume 2491, pages 233–247, Orlando, FL, April 1995.

[217] M. J. Medley, G. J. Saulnier, and P. K. Das. Narrow-band interference excision in spread spectrum systems using lapped transforms. *IEEE Transactions on Communications*, 45(11):1444–1455, November 1997.

[218] M. Mettke, M. J. Medley, G. J. Saulnier, and P. K. Das. Wavelet transform excision using IIR filters in spread spectrum communication systems. In *IEEE Global Telecommunications Conference*, pages 1627–1631, November 1994.

[219] Y. Meyer. Ondelettes et functions splines, 1986. Seminaire EDP.

[220] L. B. Milstein. Interference rejection techniques in spread spectrum communications. *Proceedings of the IEEE*, 76(6):657–671, June 1988.

[221] L. B. Milstein and P. K. Das. An analysis of a real-time transform domain filtering digital communication system - Part I: Narrowband interference rejection. *IEEE Transactions on Communications*, 28(6):816–824, June 1980.

[222] L. B. Milstein and P. K. Das. An analysis of a real-time transform domain filtering digital communications system - Part II: Wideband interference rejection. *IEEE Transactions on Communications*, 31:21–27, June 1983.

[223] L. B. Milstein, S. Davidovici, and D. L. Schilling. The effect of multiple-tone interfering signals on direct sequence spread spectrum communication system. *IEEE Transactions on Communications*, 30(3):436–446, March 1982.

[224] R. A. Monzingo and T. W. Miller. *Introduction to Adaptive Arrays.* John Wiley and Sons, 1980.

[225] B. C. J. Moore. *An Introduction to the Psychology of Hearing.* Academic Press, New York, third edition, 1989.

[226] P. Moulin. A multiscale relaxation algorithm for SNR maximization in nonorthogonal subband coding. *IEEE Transactions on Image Processing*, 4(9):1269–1281, September 1995.

[227] T. Nakaya and N. Harashima. Motion compensation based on spatial transformations. *IEEE Transactions on Circuits and Systems for Video Technology*, June 1994.

[228] J. Nam and A. H. Tewfik. Combined audio and visual streams analysis for video sequence segmentation. In *IEEE International Conference on Acoustics, Speech and Signal Processing*, pages 2665–2668, April 1997.

[229] M. J. Narasimha and A. M. Peterson. Design of a 24-channel transmultiplexer. *IEEE Transactions on Acoustics, Speech and Signal Processing*, 27, December 1979.

[230] K. Nayebi, I. Sodagar, and T. P. Barnwell. The wavelet transform and time-varying tiling of the time-frequency plane. *IEEE-SP International Symposium on Time-Frequency and Time-Scale Analysis*, October 1992.

[231] G. A. Nelson, L. L. Pfeifer, and R. C. Wood. High speed octave band digital filtering. *IEEE Transactions on Audio and Electroacoustics*, pages 8–65, March 1972.

[232] National Electrical Manufacturer Association (NEMA). Digital imaging and communications in medicine (DICOM) standard, 1994.

[233] K. Ngan and W. Chooi. Very low bit rate video coding using 3D subband approach. *IEEE Transactions on Circuits and Systems for Video Technology*, page 309, June 1994.

[234] T. Q. Nguyen. A quadratic-constrained least-squares approach to the design of digital filter banks. In *IEEE International Symposium on Circuits and Systems*, pages 1344–1347, San Diego, May 1992.

[235] T. Q. Nguyen. The design of arbitrary FIR digital filters using the eigenfilter method. *IEEE Transactions on Signal Processing*, 41(3):1128–1139, March 1993.

[236] T. Q. Nguyen et al. ftp://eceserv0.ece.wisc.edu/pub/nguyen/SOFTWARE/UIFBD.

[237] D. L. Nicholson. *Spread Spectrum Signal Design.* Computer Science Press, 1988.

[238] Nosratinia, Mohsenian, Orchard, and Liu. Interframe coding of magnetic resonance images. *IEEE Transactions on Medical Imaging*, (5):639, 1996.

[239] A. Nosratinia and M. T. Orchard. A multi-resolution framework for backward motion compensation. In *SPIE Proceedings on Electronic Imaging*, San Jose, CA, February 1995.

[240] P. O. Oakley et al. A Fourier-domain formula for the least-squares projection of a function onto a repetitive basis in N-dimensional space. *IEEE Transactions on Acoustics, Speech and Signal Processing*, 38(1):114–120, January 1990.

[241] J. R. Ohm. Three-dimensional motion-compensated subband coding. In *SPIE Proceedings*, volume 1977, page 188, 1993.

[242] M. Ohta and S. Nogaki. Hybrid picture coding with wavelet transform and overlapped motion-compensated interframe prediction coding. *IEEE Transactions on Signal Processing*, 41(12):3416–3424, December 1993.

[243] A. V. Oppenheim and R. W. Schafer. *Digital Signal Processing.* Prentice-Hall Signal Processing Series. Prentice-Hall, Englewood Cliffs, NJ, 1975.

[244] R. S. Orr, C. Pike, and M. J. Lyall. Wavelet transform domain communication systems. In H. Szu, editor, *SPIE Proceedings on Wavelet Applications for Dual Use*, volume 2491, pages 217–282, Orlando, FL, April 1995.

[245] R. S. Orr, C. Pike, M. Tzannes, S. Sandberg, and M. Bates. Covert communications employing wavelet technology. In *Asilomar Conference on Signals, Systems and Computers*, pages 523–527, Pacific Grove, CA, November 1993.

[246] A. Papoulis. *Probability, Random Variables and Stochastic Processes.* McGraw-Hill, New York, 1991.

[247] E. Parzen. *Stochastic Processes.* Holden-Day, San Francisco, 1962.

[248] J. Patti, S. Roberts, and M. Amin. Adaptive and block excisions in spread spectrum communication systems using the wavelet transform. In *Asilomar Conference on Signals, Systems and Computers*, pages 293–297, Pacific Grove, CA, November 1994.

[249] A. Peled and A. Ruiz. Frequency domain transmission using reduced computational complexity algorithms. *IEEE International Conference on Acoustics, Speech and Signal Processing*, pages 964–967, 1980.

[250] W. B. Pennebaker and J. L. Mitchell. *JPEG Still Image Data Compression Standard.* Van Nostrand Reinhold, New York, 1993.

[251] R. L. Pickholtz, D. L. Schilling, and L. B. Milstein. Theory of spread-spectrum communications – A tutorial. *IEEE Transactions on Communications*, 30(5):855–884, May 1982.

[252] I. Pitas and T. Kaskalis. Applying signatures on digital images. In *IEEE Nonlinear Signal Processing Workshop Proceedings*, pages 460–463, Thessaloniki, Greece, 1995.

[253] C. Podilchuk and N. S. Jayant annd N. Farvardin. Three-dimensional subband coding of video. *IEEE Transactions on Image Processing*, page 125, February 1995.

[254] J. Princen and J. D. Johnston. Audio coding with signal adaptive filterbanks. In *IEEE International Conference on Acoustics, Speech and Signal Processing*, pages 3071–3074, Detroit, 1995.

[255] J. Princen, J. D. Johnston, and A. Bradley. Subband/transform coding using filter bank designs based on time-domain aliasing cancellation. In *IEEE International Conference on Acoustics, Speech and Signal Processing*, pages 2161–2164, Dallas, April 1987.

[256] J. G. Proakis. *Digital Communications*. McGraw-Hill, New York, 2nd edition, 1989.

[257] M. P. Queluz. A 3-dimensional subband coding scheme with motion-adaptive subband selection. In *EUSIPCO*, page 1263, September 1992.

[258] L. R. Rabiner and R. W. Schafer. *Digital Processing of Speech Signals*. Prentice-Hall Signal Processing Series. Prentice-Hall, Englewood Cliffs, NJ, 1978.

[259] T. V. Ramabadran and K. Chen. The use of contextual information in the reversible compression of medical images. *IEEE Transactions on Medical Imaging*, (2):185, 1992.

[260] A. Ramaswami and W. B. Mikhael. A mixed transform approach for efficient compression of medical images. *IEEE Transactions on Medical Imaging*, (3):343, 1996.

[261] K. Ramchandran and M. Vetterli. Best wavelet packet bases in a rate-distortion sense. *IEEE Transactions on Image Processing*, 2(2):160–175, 1992.

[262] T. A. Ramstad. Cosine modulated analysis-synthesis filter bank with critical sampling and perfect reconstruction. In *IEEE International Conference on Acoustics, Speech and Signal Processing*, pages 1789–1792, Toronto, May 1991.

[263] D. Reed and M. Wickert. Minimization of detection of symbol-rate spectral lines by delay and multiply receivers. *IEEE Transactions on Communications*, 36(1), January 1988.

[264] S. L. Regunathan, K. Rose, and S. Gadkari. Multimode image coding for noisy channels. In *IEEE Data Compression Conference*, pages 82–90, Snowbird, UT, March 1997.

[265] R. Rifkin. Comments on "Narrow-band interference rejection using real-time Fourier transforms". *IEEE Transactions on Communications*, 39(9):1292–1294, September 1991.

[266] O. Rioul. Regular wavelets: A discrete-time approach. *IEEE Transactions on Signal Processing*, 41(12):3572–3579, December 1993.

[267] O. Rioul and P. Duhamel. Fast algorithms for discrete and continuous wavelet transforms. *IEEE Transactions on Information Theory*, 38(2):569–586, March 1992.

[268] O. Rioul and M. Vetterli. Wavelets and signal processing. *IEEE Signal Processing Magazine*, 8(4):14–38, October 1991.

[269] E. A. Riskin et al. Variable rate VQ for medical image compression. *IEEE Transactions on Medical Imaging*, (9):290, 1990.

[270] R. Rivest. Cryptography. In J. Van Leeuwen, editor, *Handbook of Theoretical Computer Science*, volume 1, chapter 13, pages 717–755. MIT Press, Cambridge, MA, 1990.

[271] P. Roos et al. Reversible intraframe compression of medical images. *IEEE Transactions on Medical Imaging*, (4):328, 1988.

[272] M. Rosenmann, J. Gevargiz, P. Das, and L. B. Milstein. Probability of error measurement for an interference resistant transform domain processing receiver. In *IEEE Military Communications Conference*, pages 638–640, October 1983.

[273] J. H. Rothweiler. Polyphase quadrature filters – A new subband coding technique. In *IEEE International Conference on Acoustics, Speech and Signal Processing*, pages 1280–1283, 1983.

[274] R. J. Safranek and J. D. Johnston. A perceptually tuned sub-band image coder with image dependent quantization and post-quantization data compression. In *IEEE International Conference on Acoustics, Speech and Signal Processing*, pages 1945–1948, 1989.

[275] A. Said and W. A. Pearlman. An image multiresolution represenation for lossless and lossy compression. *IEEE Transactions on Image Processing*, 5(9):1303–1310, September 1996.

[276] A. Said and W. A. Pearlman. A new fast and efficient image codec based on set partitioning in hierarchical trees. *IEEE Transactions on Circuits and Systems for Video Technology*, 6(2):243–250, June 1996.

[277] S. Sandberg, M. Tzannes, P. Heller, R. S. Orr, C. Pike, and M. Bates. A family of wavelet-related sequences as a basis for an LPI/D communication prototype. *IEEE Military Communications Conference*, 2:537–542, October 1993.

[278] S. D. Sandberg. Adapted demodulation for spread-spectrum receivers which employ transform-domain interference excision. *IEEE Transactions on Communications*, 43(9):2502–2510, September 1995.

[279] S. D. Sandberg, S. Del Marco, K. Jagler, and M. A. Tzannes. Some alternatives in transform-domain suppression of narrow-band interference for signal detection and demodulation. *IEEE Transactions on Communications*, 43(12):3025–3036, December 1995.

[280] H. Sari, G. Karam, and I. Jeanclaude. Transmission techniques for digital terrestrial TV broadcasting. *IEEE Communications Magazine*, pages 100–109, February 1995.

[281] G. J. Saulnier, P. Das, and L.B. Milstein. An adaptive digital suppression filter for direct sequence spread spectrum communications. *IEEE Journal on Selected Areas in Communications*, 3(5):676–686, September 1985.

[282] R. W. Schafer, L. R. Rabiner, and O. Herrmann. FIR digital filter banks for speech analysis. *Bell System Technical Journal*, pages 531–544, March 1975.

[283] B. Scharf. Critical bands. In J. Tobias, editor, *Foundations of Modery Auditory Theory*, pages 159–202. Academic Press, New York, 1970.

[284] M. R. Schroeder et al. JASA, December 1979.

[285] S. Servetto, K. Ramchandran, and M. Orchard. Wavelet based image coding via morphological prediction of significance. In *International Conference on Image Processing*, Washington, DC, 1995.

[286] Criminal Justice Information Services. *WSQ Gray-Scale Fingerprint Image Compression Specification (version 2.0)*. Federal Bureau of Investigation, February 1993.

[287] D. Sevic and M. Popovic. A new efficient implementation of the oddly-stacked Princen-Bradley filter bank. *IEEE Signal Processing Letters*, 1:166–168, November 1994.

[288] C. E. Shannon. Coding theorems for a discrete source with a fidelity criterion. In *IRE National Convention Record*, volume 4, pages 142–163, March 1959.

[289] J. M. Shapiro. An embedded wavelet hierarchical image coder. In *IEEE International Conference on Acoustics, Speech and Signal Processing*, pages 657–660, March 1992.

[290] J. M. Shapiro. Embedded image coding using zerotrees of wavelet coefficients. *IEEE Transactions on Signal Processing*, 41:3445–3462, December 1993.

[291] K. Shenoi. *Digital Signal Processing in Telecommunications.* Prentice-Hall, 1995.

[292] P. G. Sherwood and K. Zeger. Progressive image coding for noisy channels. *IEEE Signal Processing Letters*, 4:189–191, July 1997.

[293] Y. Shoham and A. Gersho. Efficient bit allocation for an arbitrary set of quantizers. *IEEE Transactions on Acoustics, Speech and Signal Processing*, 36(9):1445–1453, September 1988.

[294] E. P. Simoncelli, W. T. Freeman, F. H. Adelson, and D. J. Heeger. Shiftable multiscale transforms. *ieeetit*, 38:587–607, March 1992.

[295] B. Sklar. *Digital Communications: Fundamentals and Applications.* Prentice-Hall, Englewood Cliffs, NJ, 1988.

[296] B. Sklar. A primer on turbo codes. *IEEE Communications Magazine*, 35(12):94–102, December 1997.

[297] D. Slepian and H. J. Landau. A note on the eigenvalues of hermitian matrices. *SIAM Journal of Mathematical Analysis*, 9(2), April 1978.

[298] J. R. Smith and S. F. Chang. Frequency and spatially adaptive wavelet packets. In *IEEE International Conference on Acoustics, Speech and Signal Processing*, May 1995.

[299] M. J. T. Smith and T. P. Barnwell, III. A procedure for desiging exact reconstruction filter banks for tree structured subband coders. In *IEEE International Conference on Acoustics, Speech and Signal Processing*, pages 27.1.1–27.1.4, San Diego, CA, March 1984.

[300] M. J. T. Smith and T.P. Barnwell, III. Exact reconstruction techniques for tree-structured subband coders. *IEEE Transactions on Acoustics, Speech and Signal Processing*, 34:434–441, June 1986.

[301] M. J. T. Smith and S. Eddins. Subband coding of images with octave band tree structures. In *IEEE International Conference on Acoustics, Speech and Signal Processing*, pages 1382–1385, Dallas, April 1987.

[302] M. J. T. Smith and S. L. Eddins. Analysis/synthesis techniques for subband image coding. *IEEE Transactions on Acoustics, Speech and Signal Processing*, 38(8):1446–1456, August 1990.

[303] I. Sodagar, K. Nayebi, and T. P. Barnwell. Time-varying analysis-synthesis systems based on filter banks and post filtering. *IEEE Transactions on Signal Processing*, November 1995.

[304] I. Sodagar, K. Nayebi, T. P. Barnwell, and M. J. T. Smith. Time-varying filter banks and wavelets. *IEEE Transactions on Signal Processing*, November 1994.

[305] A. K. Soman and P. P. Vaidyanathan. Paraunitary filter banks and wavelet packets. *IEEE International Conference on Acoustics, Speech and Signal Processing*, March 1992.

[306] G. Stoll, G. Thiele, S. Nielsen, A. Silzle, M. Link, R. Sedlmayer, and A. Breford. Extension of ISO/MPEG-Audio layer II to multi-channel coding: The future standard for broadcasting, telecommunication, and multimedia applications. In *AES Convention Record*, Berlin, 1994.

[307] G. Strang. Wavelets and dilation equations: A brief introduction. *SIAM Review*, pages 614–626, December 1989.

[308] M. Strintzis and I. Kokkinidis. Maximum likelihood motion estimation in ultrasound image sequences. *Signal Processing Letters*, 1997.

[309] M. D. Swanson, B. Zhu, and A. Tewfik. Object-based transparent video watermarking. In *IEEE Multimedia Signal Processing Workshop*, pages 369–374, 1997.

[310] M. D. Swanson, B. Zhu, and A. Tewfik. Multiresolution video watermarking using perceptual models and scene segmentation. to appear *IEEE Journal on Selected Areas in Communications*, 1998.

[311] M. D. Swanson, B. Zhu, and A. H. Tewfik. Transparent robust image watermarking. In *International Conference on Image Processing*, volume 3, pages 211–214, 1996.

[312] W. Sweldens. The lifting scheme: A new philosophy in biorthogonal wavelet constructions. In A. F. Laine and M. Unser, editors, *SPIE Proceedings on Wavelet Applications in Signal and Image Processing III*, volume 2569, pages 68–79, 1995.

[313] W. Sweldens and P. Schröder. Building your own wavelets at home. Technical Report 1995:5, Industrial Mathematics Initiative, Mathematics Department, University of South Carolina, 1995.

[314] Committee T1. A technical report on high-bit-rate digital subscriber lines (HDSL), February 1994. Committee T1-Telecommunications Technical Report No. 28.

[315] Committee T1. A technical report on high-bit-rate digital subscriber lines (HDSL), April 1996. Committee T1-Telecommunications Technical Report, Issue 2, No. T1E1.4/96-006R2.

[316] Committee T1. Draft T1.413, Issue 2, September 1997. Committee T1-Telecommunications Letter Ballot, LB 652.

[317] Committee T1. Draft T1 technical report on spectral compatibility, November 1998. T1E1.4 Contribution, T1E1.4/98-002R1.

[318] Committee T1. Draft technical report for single carrier rate adaptive digital subscriber line (RADSL), September 1998. Committee T1-Telecommunications Letter Ballot, LB 715.

[319] K. Tanaka, Y. Nakamura, and K. Matsui. Embedding secret information into a dithered multi-level image. In *IEEE Military Communications Conference*, pages 216–220, 1990.

[320] D. Taubman and A. Zakhor. Multirate 3-D subband coding of video. *IEEE Transactions on Image Processing*, 3(5):572–588, September 1994.

[321] M. Tazebay and A. Akansu. Progressive optimality in hierarchical filter banks. In *International Conference on Image Processing*, pages 825–829, 1994.

[322] M. V. Tazebay and A. N. Akansu. Adaptive subband transforms in time-frequency excisers for DSSS communications systems. *IEEE Transactions on Signal Processing*, 43(11):2776–2782, November 1995.

[323] M. V. Tazebay and A. N. Akansu. Performance analysis of direct sequence spread spectrum communications system employing interference excision. In *IEEE Digital Signal Processing Workshop*, pages 125–128, September 1996.

[324] A. H. Tewfik and M. Kim. Correlation structure of the discrete wavelet coefficients of fractional brownian motions. *IEEE Transactions on Information Theory*, 38(2):904–909, March 1992.

[325] A. H. Tewfik, D. Sinha, and P. E. Jorgensen. On the optimal choice of a wavelet for signal representation. *IEEE Transactions on Information Theory*, pages 747–765, March 1992.

[326] J. F. Tilki and A. A. Beex. Encoding a hidden digital signature onto an audio signal using psychoacoustic masking. In *International Conference on Signal Processing Applications and Technology*, pages 476–480, Boston, 1996.

[327] P. Topiwala. *Wavelet image and video compression*. Kluwer, Boston, MA, 1998.

[328] D. J. Torrieri. *Principles of Secure Communications Systems*. Artech House, Norwood, MA, 1985.

[329] M. Tsai, J. Villasenor, and F. Chen. Stack-run image coding. *IEEE Transactions on Circuits and Systems for Video Technology*, 6:519–521, October 1996.

[330] M. K. Tsatsanis and G. B. Giannakis. Transmitter induced cyclostationarity for blind channel equalization. *IEEE Transactions on Signal Processing*, 45(7):1785–1794, March 1997.

[331] F. B. Tuteur. Wavelet transformations in signal detection. In *IEEE International Conference on Acoustics, Speech and Signal Processing*, pages 1435–1438, April 1988.

[332] G. Ungerboeck. Channel coding with multilevel/phase signals. *IEEE Transactions on Information Theory*, 28(1):55–67, January 1982.

[333] M. Unser. An extension of the Karhunen-Loève transform for wavelets and perfect reconstruction filterbanks. In *SPIE Proceedings on Mathematical Imaging*, volume 2034, pages 45–56, 1993.

[334] M. Unser. Approximation power of biorthogonal wavelet expansions. *IEEE Transactions on Signal Processing*, 44(3):519–527, March 1996.

[335] P. P. Vaidyanathan. Theory and design of M-channel maximally decimated quadrature mirror filters with arbitrary M, having the perfect-reconstruction property. *IEEE Transactions on Acoustics, Speech and Signal Processing*, 35(4):476–492, April 1987.

[336] P. P. Vaidyanathan. Lossless systems in wavelet transforms. *IEEE International Symposium on Circuits and Systems*, pages 116–119, June 1991.

[337] P. P. Vaidyanathan. *Multirate Systems and Filter Banks.* Prentice-Hall Signal Processing Series. Prentice-Hall, Englewood Cliffs, NJ, 1993.

[338] P. P. Vaidyanathan and T. Q. Nguyen. Eigenfilters: A new approach to least-squares FIR filter design and applications including Nyquist filters. *IEEE Transactions on Circuits and Systems*, 34(1):11–23, January 1987.

[339] R. G. Van Schyndel, A. Z. Tirkel, and C. F. Osborne. A digital watermark. In *International Conference on Image Processing*, volume 2, pages 86–90, 1994.

[340] H. L. Van Trees. *Detection, Estimation, and Modulation Theory - Part I*, volume 1. John Wiley and Sons, New York, 1968.

[341] P. Vary. On the design of digital filter banks based on a modified principle of polyphase. *AEU*, 33:293–300, 1979.

[342] M. Vetterli. Multi-dimensional subband coding: Some theory and algorithms. *Signal Processing*, pages 97–112, February 1984.

[343] M. Vetterli. Splitting a signal into subband channels allowing perfect reconstruction. In *IASTED Conference on Applied Signal Processing*, Paris, France, June 1985.

[344] M. Vetterli. Filter banks allowing perfect reconstruction. *Signal Processing*, 10:219–244, April 1986.

[345] M. Vetterli. Perfect transmultiplexers. In *IEEE International Conference on Acoustics, Speech and Signal Processing*, pages 2567–2570, April 1986.

[346] M. Vetterli and C. Herley. Wavelets and filter banks: Relationships and new results. *IEEE International Conference on Acoustics, Speech and Signal Processing*, pages 1723–1726, April 1990.

[347] M. Vetterli and C. Herley. Wavelets and filter banks: Theory and design. *IEEE Transactions on Signal Processing*, 40(9):2207–2232, September 1992.

[348] M. Vetterli and J. Kovačević. *Wavelets and Subband Coding.* Prentice-Hall, Englewood Cliffs, NJ, 1995.

[349] J. Villasenor, B. Belzer, and J. Liao. Wavelet filter evaluation for image compression. *IEEE Transactions on Image Processing*, 2:1053–1060, August 1995.

[350] Waal and R.N.J. Veldhuis. Subband coding of stereophonic digital audio signals. In *IEEE International Conference on Acoustics, Speech and Signal Processing*, pages 3601–3604, Toronto, May 1991.

[351] B. A. Wandell. *Foundations of Vision.* Sinauer Associates, Sunderland, MA, 1995.

[352] J. Wang and H. K. Huang. Medical image compression using 3-D wavelet transformations. *IEEE Transactions on Medical Imaging*, (4):547, 1996.

[353] A. B. Watson, G. Y. Yang, J. A. Soloman, and J. Villasenor. Visual thresholds for wavelet quantization error. In *SPIE Proceedings*, volume 2657, pages 382–392, 1996.

[354] S. B. Weinstein and P. M. Ebert. Data transmisison by frequency-division multiplexing using the discrete fourier transform. *IEEE Transactions on Communications*, 19(5):628–634, October 1971.

[355] J. J. Werner. The HDSL environment. *IEEE Journal on Selected Areas in Communications*, 9(6):785–800, August 1991.

[356] J. J. Werner. Tutorial on carrierless AM/PM - Part II: Performance of bandwidth-efficient line codes. Technical report, AT&T Network Systems Contribution T1E1.4/93-058, March 1993.

[357] P. H. Westerink. *Subband Coding of Images.* PhD thesis, T. U. Delft, October 1989.

[358] M. V. Wickerhauser. *Adapted Wavelet Analysis from Theory to Software.* A. K. Peters, Wellesley, MA, 1994.

[359] I. H. Witten, R. M. Neal, and J. G. Cleary. Arithmetic coding for data compression. *Communications of the ACM*, 30:520–540, June 1987.

[360] R. Wolfgang and E. Delp. A watermark for digital images. In *International Conference on Image Processing*, volume 3, pages 219–222, Lausanne, Switzerland, 1996.

[361] S. Wong et al. Radiologic image compression - A review. *Proceedings of the IEEE*, 83(2):194, February 1995.

[362] D. Woodring and J. D. Edell. *Detectability Calculation Techniques.* United States Naval Research Laboratory, Washington, DC, September 1977.

[363] J. W. Woods, editor. *Subband Image Coding.* Kluwer, Boston, 1991.

[364] J. W. Woods and T. Naveen. A filter based allocation scheme for subband compression of HDTV. *IEEE Transactions on Image Processing*, 1:436–440, July 1992.

[365] J. W. Woods and S. O'Neal. Subband coding of images. *IEEE Transactions on Acoustics, Speech and Signal Processing*, 34(10):1278–1288, October 1986.

[366] G. W. Wornell. A Karhunen-Loève-like expansion for $1/f$ processes via wavelets. *IEEE Transactions on Information Theory*, 36:859–861, July 1990.

[367] G. W. Wornell. Emerging applications of multirate signal processing and wavelets in digital communications. *Proceedings of the IEEE*, 84(4):586–603, April 1996.

[368] G. W. Wornell and A. V. Oppenheim. Wavelet-based representations for a class of self-similar signals with application to fractal modulation. *IEEE Transactions on Information Theory*, 38(2):785–800, March 1992.

[369] G. W. Wornell and A. V. Oppenheim. Wavelet-based representations for the $1/f$ family of fractal processes. *Proceedings of the IEEE*, 81(10):1428–1450, October 1993.

[370] G. A. Wright. Magnetic resonance imaging. *IEEE Signal Processing Magazine*, page 56, January 1997.

[371] X. Wu. High-order context modeling and embedded conditional entropy coding of wavelet coefficients for image compression. In *Asilomar Conference on Signals, Systems and Computers*, Pacific Grove, CA, November 1997.

[372] X. Wu. Lossless compression of continuous-tone images via context selection, quantization, and modeling. *IEEE Transactions on Image Processing*, 6:656–664, May 1997.

[373] Z. Xiong, K. Ramchandran, and M. Orchard. Wavelet packets image coding using space-frequency quantization. *IEEE Transactions on Image Processing*, 7:892–898, June 1998.

[374] Z. Xiong, K. Ramchandran, and M. T. Orchard. Joint optimization of scalar and tree-structured quantization of wavelet image decomposition. In *Asilomar Conference on Signals, Systems and Computers*, pages 891–895, Pacific Grove, CA, November 1993.

[375] Z. Xiong, K. Ramchandran, and M. T. Orchard. Space-frequency quantization for wavelet image coding. *IEEE Transactions on Image Processing*, 6(5):677–693, May 1997.

[376] Z. Xiong and X. Wu. Wavelet image coding using trellis coded space-frequency quantization. In *IEEE Multimedia Signal Processing Workshop*, Los Angeles, CA, December 1998.

[377] W. Yost. *Fundamentals of Hearing, An Introduction.* Academic Press, New York, third edition, 1994.

[378] P. L. Zador. Asymptotic quantization error of continuous signals and the quantization dimension. *IEEE Transactions on Information Theory*, 28(2):139–149, March 1982.

[379] A. Zandi, J. D. Allen, E. L. Schwartz, and M. Boliek. CREW: Compression with reversible embedded wavelets. In *IEEE Data Compression Conference*, pages 212–221, Snowbird, UT, March 1995.

[380] B. Zhu, A. Tewfik, and O. Gerek. Low bit rate near-transparent image coding. In H. Szu, editor, *SPIE Proceedings on Wavelet Applications for Dual Use*, volume 2491, pages 173–184, Orlando, FL, April 1995.

[381] R. E. Ziemer and R. L. Peterson. *Digital Communications and Spread Spectrum Systems.* Macmillan, New York, 1985.

[382] H. Zou and A. H. Tewfik. Parametrization of compactly supported orthonormal wavelets. *IEEE Transactions on Signal Processing*, 41(3):1428–1431, March 1993.

[383] Zwicker and Fastl. *Psychoacoustics, Facts and Models.* Springer, 1990.

Index